HIGH PERFORMANCE CONCRETE

HIGH

PERFORMANCE

CONCRETE

From material to structure

EDITED BY

Yves Malier

Professor of Civil Engineering
ENPC Paris and ENS Cachan
France

CRC Press
Taylor & Francis Group
Boca Raton London New York

CRC Press is an imprint of the
Taylor & Francis Group, an **informa** business
A TAYLOR & FRANCIS BOOK

CRC Press
Taylor & Francis Group
6000 Broken Sound Parkway NW, Suite 300
Boca Raton, FL 33487-2742

First issued in paperback 2019

Original French language edition – *Les Bétons à Hautes Performances* –
© 1990, Presses de l'Ecole National des Ports et Chaussées, Départment
Edition de l'Association Amicale des Ingénieurs Anciens Elèves de l'Ecole
Nationale des Ponts et Chaussées, 28 rue des Saint-Peres, 75007 Paris,
France.

ISBN-13: 978-0-419-17600-8 (hbk)
ISBN-13: 978-0-367-86505-4 (pbk)

Visit the Taylor & Francis Web site at
http://www.taylorandfrancis.com

and the CRC Press Web site at
http://www.crcpress.com

A catalogue record for this book is available from the British Library

Library of Congress Cataloging-in-Publication data available

Contents

CONTRIBUTORS

P. Acker, LCPC, Paris, France

P.-C. Aitcin, Sherbrooke University, Quebec, Canada

G. Bernier, ENS de Cachan, France

D. Brazillier, Direction Departementale de l'Equipement de Saône et
    Loire, France

J.-M. Brocherieux, SPIE Batignolles, France

G. Cadoret, Technodes SA Groupe Ciments Français, France

A. Carles-Gibergues, INSA, Toulouse, France

G. Causse, Scetauroute, France

J.-F. de Champs, Campenon Bernard, Paris, France

G. Chardin, PPB Saret, France

H. Charif, IBAP, EPFL, Lausanne, Switzerland

J.-L. Costaz, Electricité de France, Septen, France

C. Courtel, Groupe Bouyges, Scientific Division, Paris, France

A. Criaud, Technodes SA Groupe Ciments Français, France

J.-C. Faure, Sogea, Paris, France

R. Favre, IBAP, EPFL, Lausanne, Switzerland

R. Gagne, Sherbrooke University, Quebec, Canada

J. Grandet, INSA, Toulouse, France

F. Hanus, Sogea, Paris, France

G. Ithurralde, Electricité de France, Septen, France

J.-P. Jaccoud, IBAP, EPFL, Lausanne, Switzerland

P. Laplante, Sherbrooke University, Quebec, Canada

F. de Larrald, LCPC, Paris, France

J.-P. Le Boulicaut, Bétons Granulats Lafarge, France

R. Le Roy, LCPC, Paris, France

M. Lessard, Sherbrooke University, Quebec, Canada

C. Levy, Lafarge Coppée Recherche, France

M. Lorrain, INSA, Toulouse, France

Y. Malier, ENS Cachan, France

P. Monachon, Campenon Bernard, Paris, France

S. Montens, Scetauroute, France

M. Moranville-Regourd, ENS de Cachan, France

G. Peiffer, Sogea, Paris, France

D. Perraton, Sherbrooke University, Quebec, Canada

M. Pigeon, Laval University, Quebec, Canada

R. Pleau, Laval University, Quebec, Canada

L. Pliskin, L.P. Paris, France

P. Richard, Groupe Bouygues, Scientific Division, Paris, France

S. Roi, Enterprise Dalla Vera, France

I. Schaller, LCPC, Paris, France

J.-P. Sudret, LCPC, Paris, France

C. Valenchon, Bouygues Offshore, Paris, France

C. Vernet, Technodes SA Groupe Ciments Français, France

FOREWORD

By bringing together, in 1986, thirty partners from industry and research, all highly qualified, particularly responsible and justifiably ambitious, the Project 'New Ways for Concrete' has, from the outset, displayed four clear ambitions.

1. To decompartmentalize certain innovative processes by combining, in a single scientific policy, finalized research, experimentation on actual structures, and improvement of regulations and codes.
2. To treat our new materials at the same time microscopically, macroscopically, and according to how they are used in the structures, with the aim that each one of us should move beyond his specialization in one of these three areas, becoming involved in, participating in, and ultimately taking charge of work in the other two sectors, which formerly did not appear to concern him.
3. To bring about, among the partners, a fruitful pooling of a large number of research policies, studies and results.
4. To make available (in due course, and respecting everybody's interests) to all members of the profession (owners, architects, builders, inspectors, researchers, teachers, etc.) the greater part of the results of our work.

This collective publication presents contributions from the most highly qualified representatives of the thirty partners in the Project and by international experts with whom we regularly co-operate. It is to a large extent based on papers presented at seminars organized by the Project. It deals in turn with:

- the formulation and structural placing and use of different high performance concretes;
- the characterization and modelling of their mechanical behaviour, their durability with respect to internal changes and external aggressive agents;
- their application in very different structures (bridges, tunnels, offshore installations, industrial buildings, prefabricated components, etc.).

The publication also constitutes proof that, in the scientific field of high performance concretes there now exists a common language, a basis for dialogue, mutual recognition . . . in short, the foundations of common understanding between engineers and researchers.

We hardly need a crystal ball to predict that we have not yet completely evaluated all the consequences of such a frame of mind, including those for the economy and quality of our constructions.

I wish to express my sincere gratitude to all those who, over the past four years, have maintained their enthusiasm for what seemed, to so many others, merely a pleasing Utopian.

My gratitude is also especially due to those of our partners and overseas colleagues who have contributed to the realization of this first publication on high performance concretes. It is also due to Lysiane Alasluquetas, without whom this work could certainly never have been achieved.

Yves Malier
Cachan

INTRODUCTION

Y. MALIER
ENS de Cachan, France

1 Ways to obtain high performance

Smeaton (1756), Vicat (1818) and Apsdin (1825) all contributed to inventing modern concrete. Monier and Lambot (1848), Coignet (1852) and Hennebique (1880) put it to use in the first reinforced concrete buildings.

Then, for a century, concrete remained a mixture of aggregates, cement and water. This third ingredient played two essential roles: ensuring hydration of cement and participating actively in the workability of fresh concrete by giving the material satisfactory rheological properties.

During the last ten years, numerous scientific investigations have shown the detrimental effects of excess non-hydrated water on the **strength** and **durability** of concrete. Nevertheless, water is essential to obtain effective rheological properties for placing. This requirement therefore points to the need to explore ways of **reducing the water content** so as to improve the engineering properties of concrete.

At the same time, other research scientists have been focussing on reconstituting a monolithic or solid rock-like material from a **very compact mix**, placing emphasis on mix design.

So, very quickly, two approaches stood out as ways to obtain high performance. They differ in their physical and chemical nature:

(a) Deflocculation of cement grains
Deflocculation is achieved by using organic products (condensates of formaldehyde and sulphonate melamine or of formaldehyde and naphthalene sulphonate). This is the process by which the cement grains in suspension in water can recover their initial grain size (between 5 and 50 um for the most part). This first approach leads to an appreciable reduction in the quantity of water necessary, since quite a lot of this water is no longer trapped in the cement grain flakes (as it would be in traditional concrete where its contribution to workability is then negligible).

High Performance Concrete: From material to structure. Edited by Yves Malier. © 1992 Taylor & Francis.
Published by Taylor & Francis, 2 Park Square, Milton Park, Abingdon, Oxon, OX14 4RN. ISBN 0 419 17600 4.

(b) Widening the range of grain size

This extension is achieved by using extremely fine chemically reactive materials (silica fume, calcareous fillers even black carbon, etc), so that they fill the microvoids in grain packing, thus improving the compactness of the material and at the same time improving the rheological properties of the fresh mix. It follows that the quantity of water necessary for placing the concrete can be further reduced.

Table 1   The two ways to obtain H.P.

REDUCE THE FLOCCULATION OF CEMENT GRAINS	WIDEN THE RANGE OF GRAIN SIZE
PLASTICIZERS : - FORMALDEHYPE AND SULPHONATE MELAMINE - FORMALDEHYDE AND SULPHONATE NAPHTHALENE	CEMENT ADDITIVES : - SILICA FUME - CALCAREOUS FILLERS - ETC

The first approach can be used alone and leads to interesting gains in engineering properties, workability and durability. Obviously, the second approach implies simultaneous recourse to the first, since it is naturally useless to complete the grain size range of the granular material towards very fine elements if priority has not been given to reducing flocculation.

As parts of large-scale projects, different experimental programmes have confirmed that, where materials available locally are used, respecting these simple principles offers the possibility of obtaining high performance concrete, measured in terms of characteristic compressive strength, with values between 60 and 80 MPa (as used in the Joigny bridge). And these values can be obtained without any real increase in the basic cost of the concrete.

Furthermore, a more precise approach, a stricter choice of basic ingredients, acceptance of a more noticeable increase in cost, absolute obligation to use the two approaches already described, can even now make it possible, using industrial production methods, to obtain strengths between 90 and 140 MPa (Seattle 118 MPa) which the designer may consider essential for a project.

Finally, a different type of approach calling upon carefully selected ingredients (cements and aggregates of exceptional quality, inclusion of polymers, etc), new production processes (compaction, autoclaving, etc), new structural design (constraint, etc) can ensure mechanical strengths of several hundred MPa for new applications for projects where the designer can exceed the usual costs.

This is the way to open the field to new applications of these hyper-performance concretes, especially in other industrial sectors where their relatively low cost will often be very competitive with that of the more noble materials usually chosen.

2 High performance (H.P.C.) rather than high strength (H.S.C.)

From the very beginning, traditional concrete was characterized essentially by its compressive strength. But this must now change with new concretes since many other properties are improved and can therefore become decisive in the choice of solutions made in engineering projects.

Let us briefly analyse the progress accomplished over the last few years in our knowledge of high strength concrete.

(a) Microstructure

The research undertaken within the framework of the French National Project has clearly defined the links between the improvement of concrete performance and the densification of the matrix and the cement paste-aggregate interface.

Observation of microstructure has confirmed two aspects:

- in a 65 MPa HPC with no silica fume (Joigny bridge) the capillary porosity is lower than that of an ordinary concrete. The texture of the hydrates however remains the same and the cement paste-aggregate interface is still somewhat porous and crystallized,

- in a 105 MPa VHPC containing silica fume, the matrix is perfectly homogeneous, apparently amorphous. The particles of silica fume, evenly distributed between the cement grains, become the loci of hydrate nucleation. The capillary porosity is diminished and discontinuous, contrary to that of other concretes where it is interconnected. The silica fume particles play the role of filler or exert pozzolanic reaction densifying the cement paste-aggregate interface. Failure then occurs across the grains and not between or around them as in other concretes. Moreover, silica fume adsorbs the excess water molecules which no longer migrate towards the aggregate. There is no bleeding, therefore no transition zone at the cement paste-aggregate interface.

(b) Placing

Eliminating the shear threshold in the fresh cement paste by adding a plasticizer leads to a concrete which flows easily, although it appears viscous and "sticky". Placing and pumping operations are made much easier.

Furthermore, precise investigations have been carried out on creep of HPC loaded at early age. These demonstrate that the high strength obtained during the first hours and first days leads to a very different approach to the scheduling of site work. Formwork removal and prestressing can be undertaken very rapidly, implying important saving and simplification.

In certain specific cases, it can be interesting to retard setting for several hours. This can be done whithout detrimental effect thanks to the very high thixotropy of the paste which also avoids any segregation (cf. Joigny bridge).

On the other hand, the fact that there is no bleeding water leads to early and intense surface desiccation. Careful curing is therefore essential since it is the only way to avoid surface cracking due to plastic shrinkage.

(c) Mechanical behaviour

Apart from the gain in compressive strength, emphasis must be given to the advantage of the increase in tensile and shear strength in all the situations enhancing these properties (resistance of beams to lateral shear, crosswalls, problems of point loading and impact, etc).

The increase in the elastic modulus must be analysed by the designer in the light of maintaining, even increasing fracture toughness and with extensive improvement in the quality of bond between the steel and the concrete. Provided the quantities of passive reinforcement remain more or less the same (i.e. greater in percentage or volume), this will result in improved resistance to cracking and better ductility of the "reinforced concrete composite" in normal service conditions in the structure.

Finally, the creep HPCs obtained with the second approach (widening the grain size range)is appreciably reduced.

(d) Durability

Concrete is a porous material: it is characterized by the range of pore sizes and their type of connections, by the discontinuities in the microtexture such as joints in grains, and by the crystalline nature of hydrates. This porosity implies permeability which allows movement of fluid liable to cause expansion, cracking and corrosion of reinforcement.

HPCs and VHPCs show better resistance to chemical attacks than traditional concretes. They are also recommended in the case of a potential reaction between the alkalis of the pore solution and reactive aggregate (alkali-silica reaction for example). The most usual silica fume content is around 10 %. Scandinavians and Icelanders have near 20 years experience of using such concretes. Now carbonation can destroy the passive film on steels: the rate of corrosion of reinforcement then depends on the electrical resistivity of the concrete. Carbonation was studied closely during testing for the Joigny bridge, but its influence is negligible compared with reactions on ordinary concrete, VHPC containing silica fume shows very good performance as regards accelerated carbonation. Norwegian research parallel to work in our teams confirms that silica fume increases the electrical resistivity of concrete, which restricts the galvanic current and therefore steel corrosion.

Our research team has also obtained confirmation of satisfactory behaviour of HPC and VHPC subjected to freeze-thaw cycles. The pore structure of these concretes is so fine that ice cannot form during freezing even if the concrete is saturated with water. These results confirm those obtained by Canadian scientists - HPC and VHPC containing 5 - 10 % silica fume and a plasticizer have a network of air bubbles which remain stable under vibration. They resist scaling when subjected to deicing salts.

3 A new material calls for new structural design

It is quite common to dismiss the advantage of increased strength of concrete claiming that the cross sections used in structural design for normal concretes lead to dimensions quite compatible with the space needed to accommodate reinforcing steels, cables, vibrating pokers, etc.

In the past, similar negative remarks about using a new material in traditional design raised difficulties for Freyssinet and Magnel when they introduced prestressed concrete or even earlier for Hennebique and Coignet for reinforced concrete.

In fact, there is a need for a completely new and resolute approach to structural design. It is our opinion that if this is to be effective, the approach must be global. It must especially integrate data concerning:

- the materials (possible use of HPC, fibre-reinforced HPC, light-weight aggregate HPC, reinforcement with improved elastic properties, improved cables etc);

- the technology (enhancement of external prestressing to obtain more slender sections of higher strength and lighter in weight, development of composite construction where the problems of connections are different with HPC, revival of precast structures with new devices for assembly and connection, use of constraint multiplying the strength of certain members, etc);

- the construction processes (proper use of the outstanding workability of this concrete to fully develop pumping technology, use of short-term strength for a new approach, from the economic point of view, to formwork and precasting, use of the possibilities of partial prestressing at very short term, etc);

- the induced effects: to take two examples of different types - care in the quality of curing as soon as the HPC is placed, or emphasis on the specific characteristics of problems of prestressing diffusion resulting from the association of HPC and external prestressing, it being established that in certain conditions they lead to specific technological solutions, transverse prestressing for example);

- the shape of structures (a revival of funicular arch loading, research on lowering weight using steel construction models such as trusses, advances in light-weight bolted structures, etc);

- the specific conditions (a chemically aggressive environment, a location with total impermeability to air, resistance to friction and impact).

- stages in maintenance: the high durability of HPC will allow for the integration right from the start of the replacement of prestressing cables who technology will certainly evolue in the next two decades. Such a concept will make it possible to reconcile in a structure the lifetime of concrete and the permanence of the characteristics of cables.

Before choosing the appropriate concrete, a designer must endeavour to distinguish all the interactions between these different groups of parameters, rather than concentrating exclusively on just some of them.

4 Lessons from the recent past and future prospects

Examining about one hundred HPC structures built throughout the world will confirm the reality of this analysis. However, observing them through the sole criterion of high compressive strength alone would certainly only justifiy the choice of HPC for about 15 - 25 % of them from the economic point of view (examples in Table 2).

In other words, and to conclude, it must never be forgotten that the cost of a structure certainly includes the cost of the basic constituent materials, but also the amortization of the equipment essential for the building and the sum of the maintenance and the adaptations needed during service life.

The true advantage of HPC can only be judged in the light of the sum of these different costs.

Table 2

TYPES OF STRUCTURE	PROPERTIES IMPROVED	PRACTICAL EXAMPLES
BRIDGES	Short term strength, workability, durability, deferred deformation, strength	Joigny(F), Rance (F) Perthuiset (F), Louhans (F) Champs du Comte (F) Sylans (F), Ré (F), Auzon (F)
OFFSHORE STRUCTURES	Durability, compression and shear, workability, abrasion and impact	Gullfaks B,C (N) Tere Neuve (CAN) Terre Adélie (F)
HIGH-RISE BUILDINGS	Compression and shear, workability, short term strength, constraint	Water T. PL Chicago (USA) Nova Scotia Toronto (CAN) 2 Union Sq. Seattle (USA) 1 Wacker Chicago (USA) 225 Wacker Chigaco (USA) 181 Wacker Chicago (USA) NW Hospital Chicago (USA) Arche Paris (F) Chibune R.S. Osaka (JAP.)
TUNNELS	Durability, compression, short term strength	Villejust (F) Manche (F and G.B.) La Baume (F)
HIGHWAYS	Abrasion, impact, frost-thaw, shear, durability, workability	Valerenga Oslo (N) Highway E18-E6 (N) Ranasfoss BR. (N) Shestad TU. (N) Highway 86 Paris (F) Paris Airport (F)
PRECASTING OF STRUCTURAL MEMBERS	Short term compression, shear, workability	Precast joists (F) Precast floor slabs (F)
STEEL-CONCRETE COMPOSITE CONST.	Shear, compression, workability, constraint	La Roize (F) 2U Sq. Seattle (USA)
DRAINAGE	Durability, abrasion, compression, workability	Paris (F)
SPECIAL FOUNDATIONS UNDERPINNING	Compression, workability, short term strength, deferred deformation	Hassan Mosque (MAR.)
NUCLEAR STRUCTURES	Durability, strength, water tightness	Civeaux (research) (F)

Photo 1 : Bridge of the "Ile de Ré" (F)
(Short term strength)

Photo 2 : Bridge of "Joigny" (F)
(workability, durability)

Photo 3 : Precast plankings (F)
            (lightness)

Photo 4 : "Grande Arche", Paris (F)
            (workability, surface quality)

Photo 5 : Hassan Mosque, Casablanca (MAR)
  (short term strength, diferred deformation, compression)

Photo 6 : Bridge of "Perthuiset" (F)
  (durability, short term strength)

Photo 7 : Bridge of the "Rance" (F)
(compression, short term strength , workability)

Photo 8 : Bridge of "Sylans" (F)
(high strength, short term strength, diferred deformation,
workability)

Photo 9 : Roof of the "Grande Arche", Paris (F)
(lightness, high strength, workability)

Photo 10 : Bridge of "Champs du Comte" (F)
(durability, abrasion, frost-thaw)

PART ONE

KNOWLEDGE OF THE MATERIAL

1 MICROSTRUCTURE OF HIGH PERFORMANCE CONCRETE

M. MORANVILLE-REGOURD
ENS de Cachan, France

1 Introduction

Hardened concretes have mechanical characteristics lower than those of steel or alumina. The highest differences appear in the flexure strength and toughness (1). A Portland cement concrete is a porous and heterogeneous material. The matrix which embeds sand grains and aggregates is constituted by different hydrates. The most important of them are hydrated silicates C - S - H which can appear as fibers and Ca (OH)$_2$ which crystallises in massive superimposed hexagonal plates (Fig. 1). The total porosity of a Portland cement paste is between 25 and 30 % by volume for a cement / water ratio of 0.5. This porosity is decomposed into two types of cavities (i) C - S - H pores of several nanometer size (ii) capillary pores between hydrates, air bulles, cracks : their size is between 100 nm and several mm.

The low mechanical performances of concrete have been attributed to the capillary porosity and excess of water needed for the workability of fresh concrete. An improvement has been obtained by several processes which reduce the porosity (impregnation, pressure) and the water / cement ratio (use of superplasticizers). New products also appeared. They were MDF, Macro Defect Free cement and DSP, Densified System containing homogeneously arranged ultrafine Particules (2). The first one contains a polymer, the second silica fume.

2 High performances cement pastes

Filling in capillary pores or extracting the excess of water by pressure or reducing the water / cement ratio with superplasticizers are processes which densify the cement paste which therefore appears more homogeneous and more amorphous than the normal Portland cement paste.

High Performance Concrete: From material to structure. Edited by Yves Malier. © 1992 Taylor & Francis.
Published by Taylor & Francis, 2 Park Square, Milton Park, Abingdon, Oxon, OX14 4RN. ISBN 0 419 17600 4.

Figure 1: Portland cement paste, w/c = 0.5, (1) Fibrous C-S-H,
(2) Ca(OH)$_2$, (3) capillary pore.

2.1. Cement paste with a low water / cement ratio

In 1897, Féret (3) gave an expression of the compressive strength as follows :

$$Rc = A \left(\frac{c}{c + e + a} \right)^2$$

(1)

with c, w, a respectively the volume of cement, water and air. After this formula reducing the water/cement ratio leads to an increase in strength. However there is a limit of the water/cement ratio related to the workability of the fresh concrete.

2.1.1 Superplasticizers

Superplasticizers like naphtalene sulfonate, melamine, lignosufonate used for dispersing solid particles also involve a reduction of the water/cement ratio down to w/c = 0.16 (4). Studies with the proton nuclear magnetic resonance showed that the superplasticizer was adsorbed on the solid particles and formed a pellicle in which the water molecules were still mobile (4,6). To the pellicular effect is added the dispersion of solid particles. Both improve the rheology of the suspension. Compressive strengths as high as 200 MPa were achieved in that way. Porosity was about 5 % by volume and the matrix occurred homogeneous and like amorphous.

2.1.2 Pressure and vibration

Compressive strengths of 644 MPa were measured under pressure at high temperature (1020 MPa, 150°C). The total porosity was therefore 2 % in volume (7). Hydrates were identified as gels. The degree of cement hydration was 30 %. Silicates C - S - H embedding anhydrous cement grains behaved like a glue between dense particles. Both hydrates and clinker simultaneously contributed to the high strength of the hardened cement paste. The vibration eliminated air bubbles created during mixing which could behave as Griffith defects in the flexure tests.

2.2 Ultrafine particles DSP

DSP (4) contain Portland cement, silica fume and superplasticizer. Silica fume occurs as microspheres of 0.5 um average size which fill in interstitial spaces between cement grains of 30 - 100 um size. First silica fume plays a physical role of filler. All solid particles are well dispersed by the superplasticizer and there is no bleeding. During the cement hydration, silica fume spheres are sites of nucleation (8) for cement hydrates and then react as a pozzolanic material giving an homogeneous and like amorphous C - S - H (Fig. 2).

Figure 2: DSP with silica fume, (1) $Ca(OH)_2$, (2) amorphous C-S-H.

2.3 Cement MDF

The densification of the cement paste increases compressive strength. However the material still exhibits a low flexure strength.

A new approach of the porosity was done by Kendall, Howard and Birchall (9) who considered two pore families :

Pores of volume p : capillary pores
Pores of length 2c : cracks.

Applying the Griffith theory of the fracture mechanics, the criterion for crack extension becomes :

$$\sigma = \left(\frac{Eo\ Ro\ (1 - p)\ exp\ (-kp)}{\pi\ c} \right)^{0.5}$$

(2)

with Eo = Young modulus
Ro = fracture energy for p = 0
2c = crack length.

So there appears two possibilities for increasing strengths: either reduce capillary porosity or reduce crack length. This second way could increase the flexure strength when 2c < 1 mm. This approach leads to the MDF, Macro Defect Free cement (1).

In MDF a water soluble polymer (hydroxypropylmethyl cellulose or hydrolised polyvinylacetate) disperses and lubricates cement grains in the cement paste suspension. The polymer being able to form a rigid setting gel, a strong mixing is necessary. During setting and hardening, the polymer dihydrates while cement hydrates. In the hardened material the polymer is highly bound to cement grains and the final porostiy is about 1 % by volume.

A macro defect free cement is composed of 100 parts of cement (in weight), 7 parts of polymer and 10 parts of water.

The microstructure is close to that of cements with a low water / cement ratio. The main characteristic is a dense and amorphous matrix around clinker grains. Ca (OH)$_2$ crystals are in thin lamellae distributed in the cement paste contrary to the large and thick plates in a Portland cement paste (Fig. 3).

The limited space for the formation of large crystals avoids the appearance of cracks along the cleavage planes of superimposed plates. Flexure strengths as high as 150 MPa have been related to the absence of capillary pores and cracks.

MDF pastes can be moulded, extruded and rolled as plastics. They can be used in composite materials containing sand, metallic powder, fibers which increase their toughness and resistance to abrasion.

Figure 3: MDF cement paste Thin Ca(OH)$_2$ crystals in an amorphous and homogeneous matrix (1).

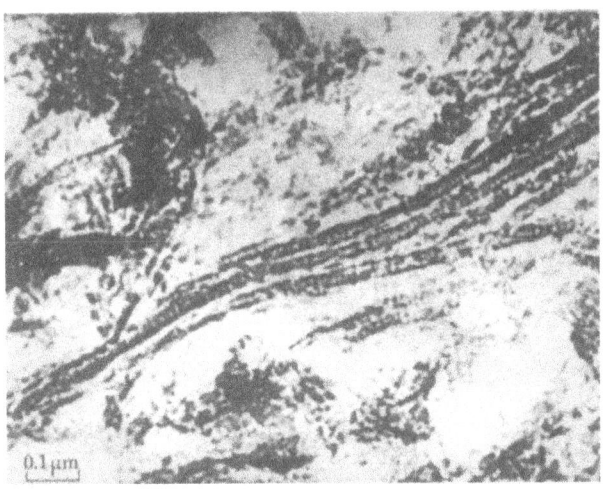

3 Autoclaved mortars

Autoclaving a cement and sand mixture results in a material with 200 MPa in compressive strength. This high strength has been related to a strong bond between matrix and aggregates. Inert sand grains at room temperature react during the heat treatment under pressure. Cement hydrates are different from C - S - H. They occur as tobermonite platelets or xonotlite fibers.

4 High performance concretes

Both characteristics involved in the mechanical strengths of concrete are the microstructure of the cement paste and the nature of the cement paste - aggregate interface.

4.1 Cement paste - aggregate interface

In a normal concrete the cement paste - aggregate called "auréole de transition" (10) is better crystallised, more porous and less resistant than the matrix because of an excess of water. Characteristics of this interface correspond to the fracture surface, cracking, composition and texture of hydrates. As an example, cracks progress first around siliceous aggregates and then through the cement matrix (Fig. 4). On a smooth aggregate surface, hydrates deposit is composed of Ca (OH)$_2$ amorphous film and C - S - H fibers. This deposit is weakly bound to the aggregate and can be torn away

Figure 4: Cement paste - aggregate interface in Portland cement concrete : porous interface and crack along aggregate, G = aggregates, C = cement.

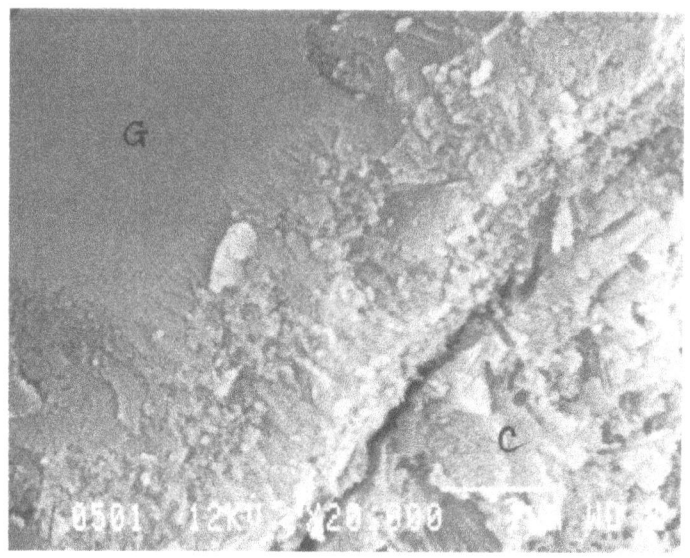

Figure 5: Cement paste - aggregate interface in Portland cement concrete, P = oriented crystals of portlandite $Ca(OH)_2$, G = aggregate.

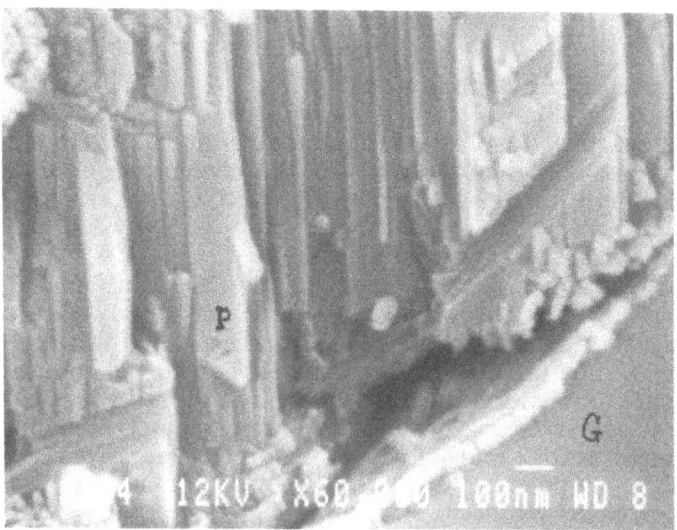

easily. An oriented Ca (OH)$_2$ crystallisation is also observed on siliceous aggregates (Fig. 5). On the contrary with limestone aggregates which are considered as reactive, cracks progress along aggregates but at a distance corresponding to the thickness of the reacted zone. The interfacial zone well known for his high porosity between 1 500 and 3 000 nn has been clearly improved by using silica fume.

4.2 DSP concretes
A DSP concrete (4) which developed 110 - 160 MPa in compression was constituted of :

    400 kg   Portland cement + 80 kg silica fume
    80 - 90 liters of water : w/c = 0.16 - 0.18.

Fracture surfaces of DSP concretes are transgranular proving a densification of the cement paste compared to normal with intergranular rupture. As shown in the Figure 6, there is no crack around aggregate and no Ca (OH)$_2$ oriented crystals at the interface. Strengths as high as 270 MPa were obtained with fused alumina aggregates (4).
Another characteristic of the microstructure is the homogeneous matrix due to the pozzolanic ractivity of the silica fume. The optimum content of silica fume has been found between 7 and 15 % by weight of cement. For high amounts like 40 %, silica fume is not totally hydrated and concrete can get brittle.

Figure 6: Transgranular fracture in DSP concrete, C = homogeneous matrix, no crack and no oriented crystals along aggregate (G).

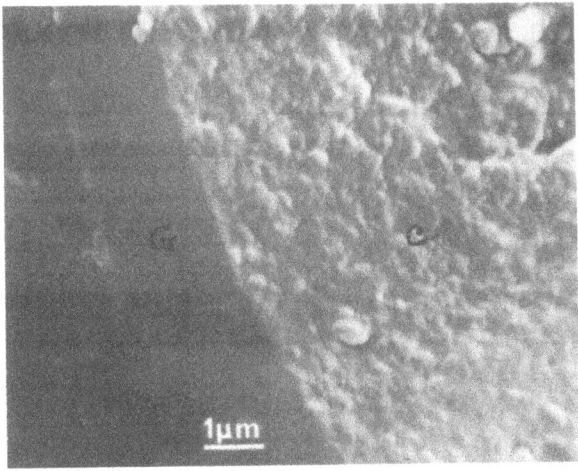

4.3 Microcracking

Microcracking of concrete can be estimated by observing under microscope polished sections impregnated with a colour dye (11). Microcracks are considered as surfaces or discontinuity zones in the cement paste and at the matrix - aggregate interface. The method of aleatory secants is able to quantify the map cracking (Table 1).

Table 1 - Density of microcracks and microdefects on concrete polished sections cut in two perpendicular directions and impregnated with a red dye (12).

Concrete	Thin cracks $n\ cm^{-1}$	Microporous interfaces $n\ cm^{-1}$	Mircroporous matrix $n\ cm^{-1}$
76 MPa	0 0.02	1.45 1.50	1.35 0.40
110 MPa	0.04 0.07	0.20 1.00	0.06 0.25

High strength (76 MPa) and very high strength (110 MPa) are little cracked if they are compared to normal concretes (Table 2). In silica fume concrete microcracks appear at short term and their density increases slowly. In normal concretes with no silica fume, the map cracking spreads out gradually but it remains less dense than that of silica fume concrete (13).

Table 2 - Distribution of microcracks and porous zones in three concretes of voussoirs in a tunnel (400 kg cement /m^3, w/c = 0.36, σ_c = MPa at 28 days) in three different environments

Micro defects	Microcracks		Microporous interfaces		Microporous matrix	
Density	$n\ cm^{-1}$		$n\ cm^{-1}$		$n\ cm^{-1}$	
Concrete	External	Internal	External	Internal	External	Internal
1	0.13	0.02	2.15	1.80	0.85	0.60
2	0.20	0.09	1.55	2.05	0.45	0.25
3	0.06	0.00	3.25	1.55	1.35	0.30

The largest difference between normal, high strength and very high strength concretes occur in the microporosity. The cement paste is less porous and interfaces between cement paste and aggregates are significantly improved.

4.4 Porosity

The total porosity of silica fume concrete has been measured by a mercury porosimeter and compared to that of a normal concrete (Fig. 7). The total mercury volume intruded under 150 MPa pressure is in normal concrete twice as large as that of high strength concrete (14). Two factors are here responsible for the reduction in porosity :

The presence of silica fume acting as a pozzolanic filler.
The use of a superplasticizer reducing the water/cement ratio from 0.56 to 0.21.

Figure 7: Total porosity and pore size distribution given by mercury porosimetry for three concretes with decreasing water/cement ratios (14).

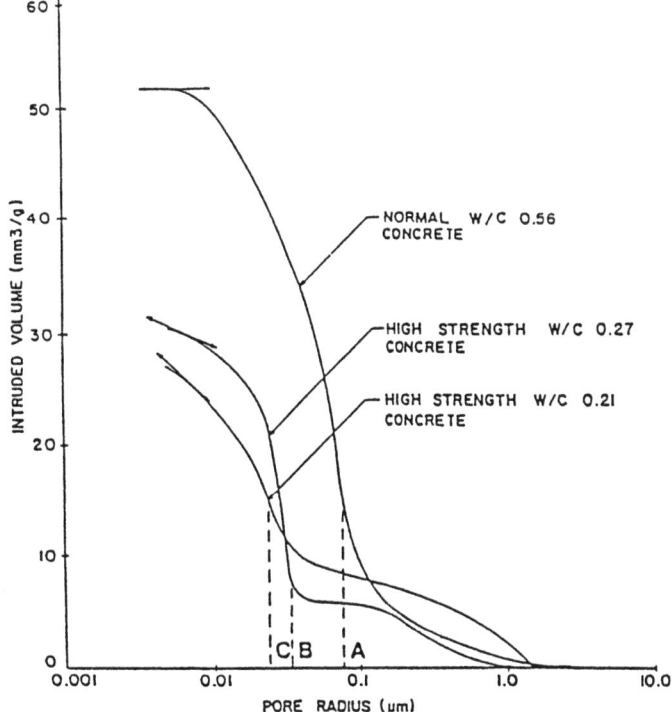

5 Conclusion

A high compressive strength and a better flexure strength/ compressive strength ratio can be reached by different ways such as:

- impregnation,
- lower water/cement ratio using superplasticizer,
- use of silica fume,
- reducing porosity and cracking as in Macrodefect Free Cement,
- thermal treatment : autoclave or hot pressure,
- adequate choice of aggregates (shape, roughness, hardness).

The improvement in mechanical performances can be achieved in modifying the microstructure through two ways :

1 Formation of a continuous lattice of dense particles : tobermonite and xonotlite in autoclaved mortars, clinker grains in cement pastes of low water cement/ratio and low degree of hydration. These particules participate in mechanical performances by their intrinsic strength and their intercrystalline bonds.

2 Formation of an amorphous and homogeneous matrix embedding clinker grains and aggregates. In DSP concretes fracture surfaces are transgranular. There is neither "auréole de transition" nor oriented and massive Ca $(OH)_2$ crystals. Mechanical performance is related to the low degree of crystallinity and to the disordered orientation of solids in an homogeneous matrix.

6 References

1. J.D. Birchall. Cement in the context of new materials for an energy - expressive future. Phil. Trans. R. Soc. Lond. A 310, 31-42, 1983
2. M. Regourd. Microstructure of high strength cement paste systems. Materials Research Society. Vol 42, Ed. J.F. Young, 3-17, 1985
3. R. Féret. Bull. Soc. Encour. Ind. Nat. Paris II, 1604, 1897
4. H.H. Bache. Densified cement/ultrafine particle based materials. Second Int. Conf. Superplasticizers in concrete. Ottawa, 1-39, 1981
5. M. Regourd. Evaluation of superplasticizers using proton magnetic resonance. A review. Materials Research Society, Vol 85, 245-254, 1987
6. J.C. Mac Tavish, L. Miljkovic, M.M. Pintar, R. Blinc and G. Lahajnar. Hydration of white cement by spin grouping NMR. Cem. Concr. Res. 15, 2, 367-377, 1985
7. D.M. Roy and G.R. Gouda. Optimization of strength in cement pastes. Cem. Concr. Res, 5, 153-162, 1975

8. P.K. Mehta. Pozzolanic and cementitious by-products in concrete. Another look. 3rd CANMET/ACI. Int. Conf. on Fly Ash, Silica Fume, Slag and Natural Pozzolanic in concrete. SP, 114-1, Vol 1, 1-44, 1989

9. K. Kendall, A.J. Howard and J.D. Birchall. The relation between porosity, microstructure and strength and the approach of advanced cement - based materials. Phil. Trans. R. Soc. Lond. A 310, 139-153, 1983

10. J. Farran, R. Javelas, J.C. Maso et B. Perrin. Existence d'une auréole de transition entre les granulats d'un mortier ou d'un béton et la masse de la pâte de ciment hydraté. Conséquences sur les propriétés mécaniques. C.R. Acad. Sci. Paris T 275, Série D, 1467-1468 (1972)

11. M. Moranville-Regourd. Détection de la microfissuration des bétons par une méthode d'imprégnation. Conf. européenne sur la fissuration des bétons et la durabilité des constructions. AFREM - CEE, St. Rémy-Les-Chevreuse, 37-44, 1988

12. M. Moranville-Regourd. Microfissuration des bétons de hautes performances. Projet National Voies Nouvelles du Béton - ACI Québec. Montréal, 1er mars 1990

13. S. Chatterji and A.D. Jensen. Investigation of old concrete containing microsilica. Ibid Ref. 8. Supplementary Papers, 419-430, 1989

14. S. Sarkar, E. Kassab and P.C. Aitcin. La microstructure des bétons à hautes performances. Séminaire Univ. Sherbrooke. Canada, Dec. 1988.

2 THE USE OF SUPERPLASTICIZERS IN HIGH PERFORMANCE CONCRETE

P-C. AITCIN
University of Sherbrooke, Canada

1 Introduction

In a previous series of papers [1, 2, 3], it was shown that in order to increase concrete compressive strength, it is necessary:

1 to decrease as much as possible concrete water/cement ratio, by lowering the amount of mixing water with a high dosage of superplasticizer;
2 to select, in a given place, the most efficient cement/superplasticizer combination in terms of rheology (slump loss) and strength;
3 to use the best aggregates from the strength and bond points of view in order to retard as much as possible the propagation of critical fissures when concrete is submitted to increased loads.

In this paper, we will concentrate only on the second point of view: how to select, in a given location, the best cement/superplasticizer combination in terms of rheology and strength.

In order to ensure a rapid success for high performance concrete, both technologically and economically the concrete industry must be able to produce and place economical concretes using present concrete technology. In fact, what concrete producer is interested in outlaying a considerable capital investment to produce and place high performance concrete for what is presently a quite small market?

What designer is interested in specifying a high performance concrete for a building if he knows that the concrete producer will have to face serious field problems because he has not mastered producing and placing such concrete?

What contractor will not charge more for placing high performance concrete, when he knows that he will face such placing problems from a material that cannot be trusted?

High Performance Concrete: From material to structure. Edited by Yves Malier. © 1992 Taylor & Francis.
Published by Taylor & Francis, 2 Park Square, Milton Park, Abingdon, Oxon, OX14 4RN. ISBN 0 419 17600 4.

What owner is ready to take a chance on an attractive but not finely tuned material?

It is essential in the 1990s to design high performance concrete that can be fabricated in standard concrete plants and placed as easily as an ordinary concrete with the transportation and placing equipment regularly used for normal-strength concrete.

It is in that sense that the selection of the right superplasticizer/cement combination is one of the most important things when a concrete producer wants to make a high performance concrete.

2 Portland cement and water

Water is an essential ingredient in concrete ranking alongside cement. In fact, water that is introduced into concrete during mixing has two functions: a physical function, to give the concrete the right rheological properties, and a chemical function, to contribute to the development of the hydration reaction. Concrete should ideally have only enough water to develop the final and full hydraulic potential of the cement while providing the rheology needed for easy placement.

Unfortunately, available Portland cements preclude attaining concrete with this characteristic. On one hand, cement particles, characterized by many unsaturated superficial electrical charges, have a strong tendancy to flocculate when in contact with a liquid as polar as water as can be seen in Figure 1 [4]. On the other hand, the hydration reaction does not wait until the concrete is in the forms before starting. Hydration begins as soon as Portland cement touches water due to the fact that certain cement compounds are very reactive and that Portland cement has a lot of very fine particles, some micrometers in diameter, that are of course very reactive.

This explains why when only cement and water are used, it is necessary to use more water than necessary to fully hydrate cement particles, given that concrete needs good workability.

3 Water reducers

In order to alleviate the flocculation tendancy and to reduce the amount of mixing water, certain organic molecules, well-known for their dispersing properties, can be used [5]. The first dispersing molecules, more commonly water reducers, were produced from papermill waste and were called lignosulfonates (Fig.2). This by-product was not very expensive and required only simple additional processing to be used successfully in concrete.

It was rapidly discovered, however, that the dosage could not be increased at will without secondary drawbacks [6]:

Figure 1: Deflocculation of cement grains by a superplasticizer (after Uchikawa [10])

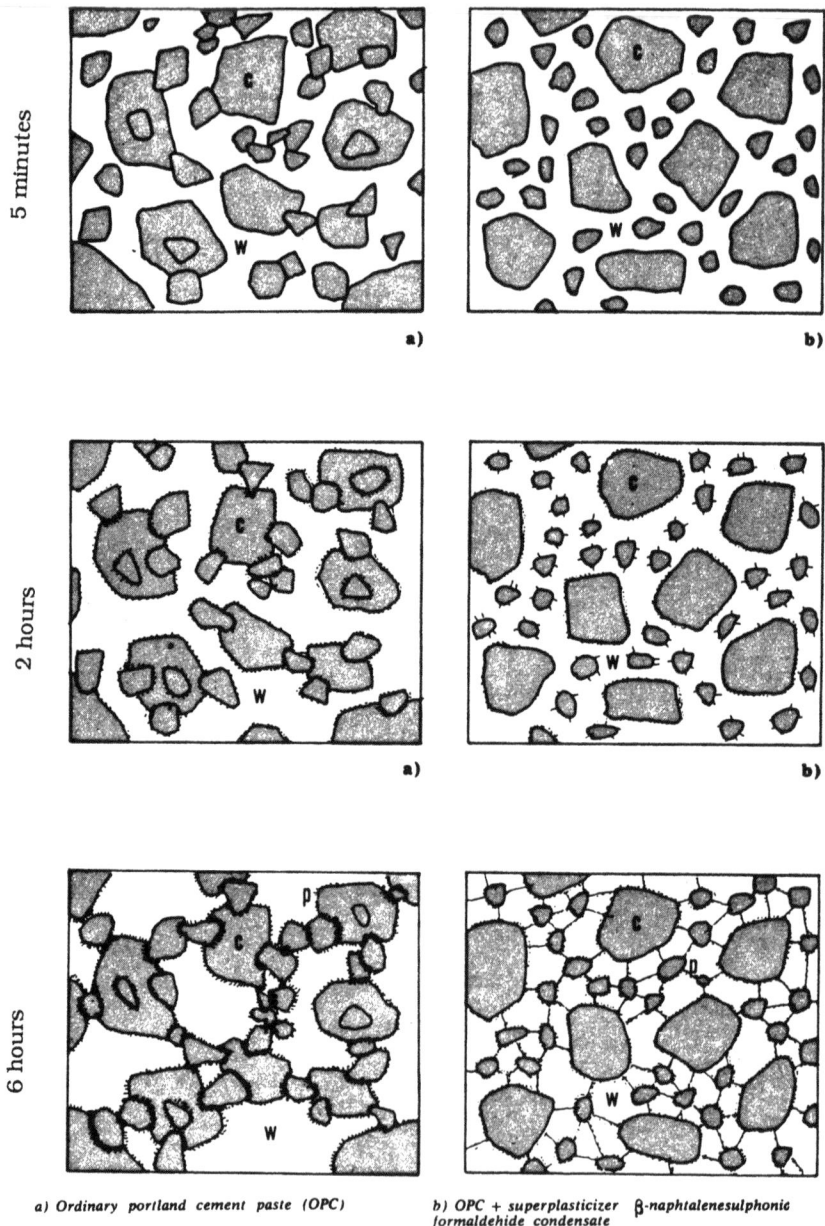

5 minutes

a)

b)

2 hours

a)

b)

6 hours

a) Ordinary portland cement paste (OPC)

b) OPC + superplasticizer β-naphtalenesulphonic formaldehide condensate

Figure 2. Schematic representation of a molecule of lignosulfonate. After Rixom (6).

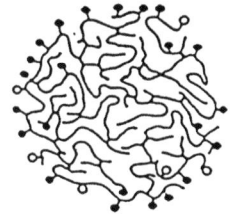

Molecule of
lignosulfonate

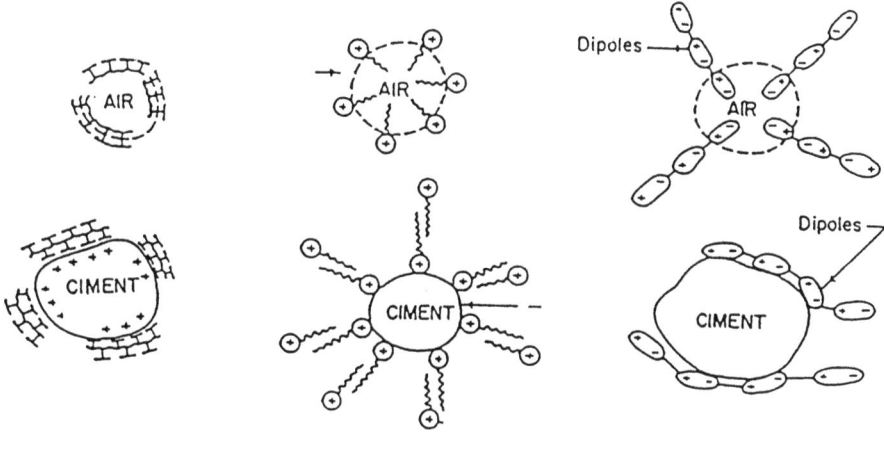

O   COOH Group

●   SO₃H Group

Ж   R-O-R Linkages

Figure 3. Mode of action of water reducers. After Joisel (5).

Dipoles

AIR

AIR

AIR

CIMENT

CIMENT

CIMENT

Dipoles

Dipoles

(a) anionic   (b) cationic   (c) non-ionic

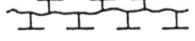

17

- excessive retardation due to the presence of sugars in the wood;
- entrapment of big air bubbles caused by surfactants.

Other types of molecules were unsuccessfully tried to overcome these drawbacks.

Chemically, water reducers can be anionic, cationic, or even nonionic. They are composed of molecules having a highly charged end that neutralizes one electrical site on a cement particle with an opposite charge. In the case of nonionic water reducers, these molecules act like dipoles that are glued to the cement grains. Figure 3 illustrates the mode of action of the three types of water reducer molecule.

4 Superplasticizers

For 30 years, the concrete industry worldwide was very happy with these first-generation water reducers, until almost simultaneously in Japan and in Germany [7], synthetic molecules with highly dispersive properties appeared on the market. These new products were salts of sulfonated naphthalene formaldehyde condensate (Japanese) or of sulfonated melamine formaldehyde condensate (German). These products are now known variously as superplasticizers, high-range water reducers, and fluidifiers (Fig. 4). At the beginning, these products were almost always used to fluidify concrete in the field [8]. Lignosulfonates were still used during the mixing at the plant. One of the major drawbacks of these first superplasticizers was their very brief action: they were only efficient for 15 to 30 minutes in the better cases [9].

However, superplasticizer fabrication technology has improved since then. Most of superplasticizers now can keep a high slump for 30 to 45 minutes, but there are still some Portland cements that don't behave so well with superplasticizers; the basis for this becomes clear when the mode of action of the superplasticizer is studied.

5 Portland cement reactivity and cement/superplasticizers compatibility

Even though it is not yet possible to explain in minute detail how superplasticizers work, the main lines governing their action on Portland cement are more or less well understood, especially the slump-loss problem that is fairly obvious when making high performance concrete at a very low water/cement ratio [10, 11]. It must be pointed out that high performance concretes are generally made with only 120 to 135 L/m^3 of mixing water compared to 160 to 180 L/m^3 in an ordinary concrete, depending on the entrained air content and maximum size of the coarse aggregate and its nature, and other factors.

Figure 4. Schematic representation of the 2 principal molecules of current commercial plasticizers.

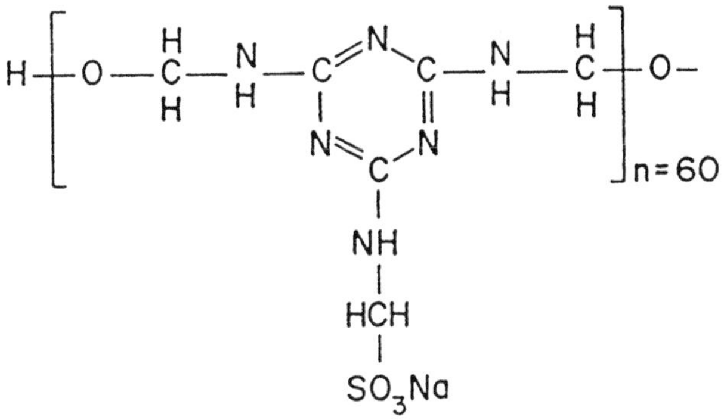

(a) Formaldehyde and melamine sulfonate condensate

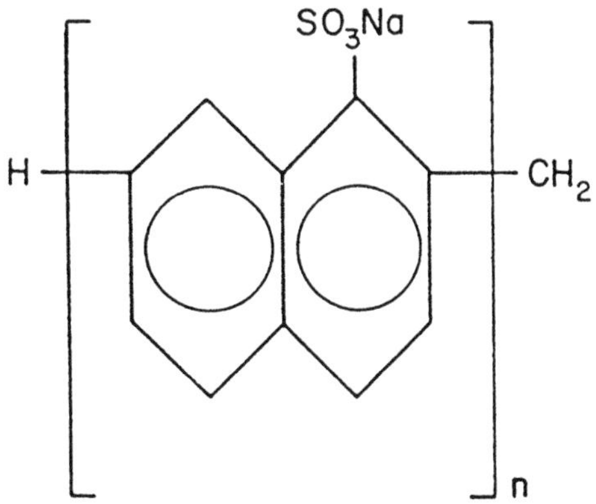

(b) Formaldehyde and naphthalene sulfonate condensate

High performance concrete rheology is mostly influenced, on the one hand, by the rate at which the different mineral phases that constitute Portland cement are able to react with water molecules, diverting them for their initial role as fluidifiing agent, and, on the other hand, by the rate at which certain superplasticizer molecules can be trapped by the new compound that forms immediately after the contact of water and Portland cement.

The two phases that present a rapid hydration and consume a substantial amount of water are calcium hemihydrate (plaster of Paris, $CaSO_4. 1/2 H_2O$) that could form during the dehydration of a certain amount of gypsum ($CaSO_4. 2H_2O$) during the grinding and the interstitial phase in which C3A and C4AF react with gypsum to give ettringite ($CaO. 3Al_2O_3. 3 CaSO_4. 32H_2O$) [10].

Most researchers believe that superplasticizer molecules are preferably and solidly absorbed on the di and tricalcium silicates, so that they can control very efficiently their hydration, and in some cases, retard it substantially [10, 12].

The rate at which cement consumes water molecules during the first moments after mixing corresponds to what is usually referred as the cement's rheological reactivity.

It appears that a certain amount of superplasticizer molecules are consumed during hydration.

The amount of superplasticizer molecules consumed during the first moments after mixing corresponds to what has been qualified as cement/superplasticizer compatibility.

Experience shows that, with some very reactive cements, it is difficult to make very low water/cement ratio high performance concrete which still has a 50 mm slump 30 minutes after mixing, even though the initial slump was 200 mm. The surface of these concretes becomes progressively shiny with such concretes it is not appropriate to try to increase their slump by adding more superplasticizer because it is clear such concretes are lacking in water. In fact, in such cases, it is better to add a small amount of water to recover a certain slump practically instantaneously, but this recovered slump can also be lost quite rapidly when the new addition of water has reacted.

So, when the composition of a high performance concrete has to be optimized in a given location, it is very important to find, firstly, the cement that has the lowest rheological reactivity, that is, the cement that will fix the least amount of water immediately following mixing, and secondly, the superplasticizer that will not compete with ettringite crystals, formed when the cement enters in contact with water, to neutralize the C3A. Unfortunately, there are no theoretical means to predict either of these behaviors. Concrete producers usually have to proceed by trial and error, monitoring the slump loss of their concrete, which can be a quite long and fastidious procedure.

At the University of Sherbrooke, a very simple empirical test has been developed which has proven very efficient and extremely reliable. It has been used with a great variety of Portland cements varying in their contents of C3A and C4AF alkalis, limestone filler, C3S and C2S, and so on as well as with different finenesses [13].

The test consists of using a small modified pump to recirculate a cement grout having water/cement ratio of 0.35 and containing a reference naphthalene superplasticizer dosed at a given solid content (depending on the type of the cement) for four minutes. The flow time of 1 L of such a grout through a Marsh cone used by the petroleum industry to check the fluidity of bentonite grout is measured (Fig. 5).

When the initial time has been measured, the grout is placed in a plastic container that is continuously agitated until the next measurement is taken 40 minutes later. It is possible to characterize the initial reactivity of cement and its compatibility with a particular superplasticizer in less than one hour by using less than one bag of cement. Rheopump have never failed whenever used in conjunction with slump-loss measurements on concrete [14].

Figure 5. Schematic of the Rheopump

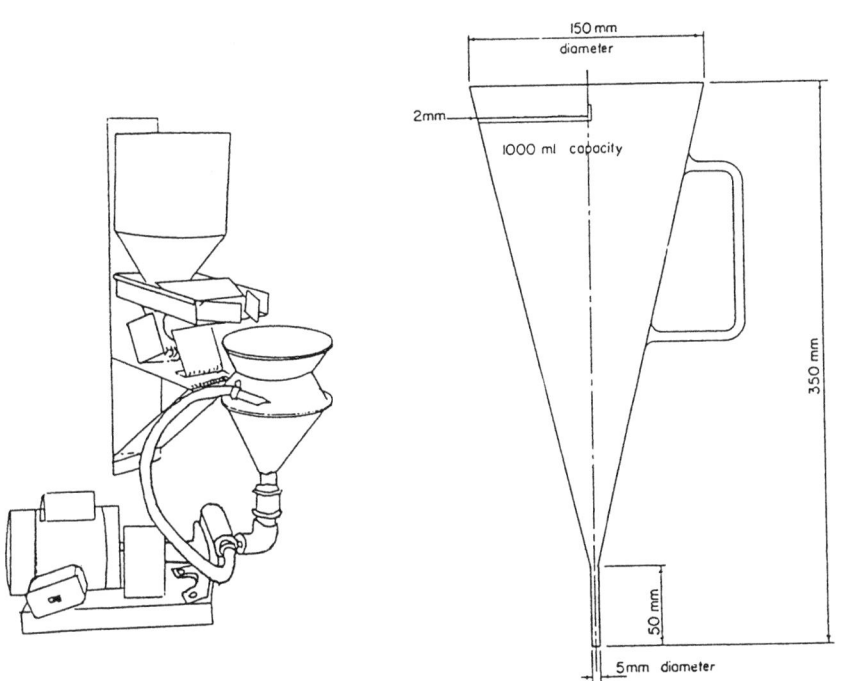

Experience has also shown that the rheopump could be used to find the appropriate retarder dosage when necessary in order to slow down the slump loss or to study the compatibility of different superplasticizers with a given cement.

6 Practical use of superplasticizers in high performance concrete

When a high performance concrete has to be made, it is very important to carefully select the different constituents. Unfortunately, concrete producers don't always have complete latitude in this area, because quite often, they have to face different situations of :

- a physical nature : facilities may allow little latitude when selecting materials because of limited number of bins and maintaining service to regular customers;.
- an economical nature : because of geographic location, local resources offer very few choices for selecting between economically available materials; or
- a societal nature : some concrete producers are tied to a cement group or quarry, which limits the type of cement that can be used. Since most major cement companies have their own affiliated admixture company, this also limits superplasticizer selection.

Our intent here is to discuss the simple general rules that each concrete producer should follow and adapt to his particular situation.

6.1 Selecting a type of superplasticizer

To begin with, should a naphthalene or a melamine superplasticizer be used?

Based solely on solid content, naphthalene superplasticizers should be preferred over melamine. The former is generally marketed with a solid content between 40 and 42 percent, while the latter has a solid content between 22 and 30 percent. Some new melamine superplasticizers are said to have a 40% solid content.

In practical terms, the selection of a particular type of superplasticizer should take into account other factors. Experience shows that superplasticizer efficiency depends, not only on the amount of solids, but also on the quality of the solid (the length of the molecular chains, the amount of impurities, the amount of residual sulphates, etc.) This explains why the final choice of a superplasticizer must not be based on solid content or price per liter, but on a comparison of its economic efficiency, that is, the amount of dollars that has to be spent in order to reach the given level of workability.

22

Finally, considerations other than technical or economical ones can also influence the final choice. For example, the quality of the service offered by the marketing team of the admixture company, the regularity of deliveries, the consistency in the quality of the superplasticizer, and the confidence in the admixture company, have to be considered before the final choice is made. There are few cases when the superplasticizer is selected solely on purely technical criteria: some precasters prefer melamine superplasticizers because they provide surfaces with fewer bubbles.

For strictly commercial reasons, the development of high performance concrete in North America was based on a melamine superplasticizer in the Chicago area. At that time, the marketing of melamine superplasticizer was well done. The contrary is now true, with an increasing number of high performance concretes across the continent using naphthalene superplasticizer.

Finally, it should be mentioned that, because the fabrication of naphthalene and melamine superplasticizers require complex processing the real number of superplasticizers on the market is quite limited. Admixture companies prefer to buy the basic molecules, possibly adding different ingredients that supposedly enhance the superplasticizer. It can be also mentioned that superplasticizer producers can produce both naphthalene- and melamine-based products by simply changing their raw materials and their processing: basic production equipment remains the same.

6.2 Selection of superplasticizer brand

As mentioned above, the number of naphthalene/melamine superplasticizer manufacturers is quite limited, so that a great number of commercial superplasticizers come from the same reactor and have quite similar properties. It is possible to recognize the origin of certain commercial superplasticizers by analysing their infrared spectrum and by subjecting their solids to differential thermal analysis (Fig. 6) [15]. The determination of the molecular chain length can be also very interesting, however, it is quite difficult to master liquid phase chromatography which is a technique with which most laboratories are not familiar[16].

This does not mean that all naphthalene superplasticizers, in particular, are as efficient when they have the same amount of solids. In some cases, superplasticizer producers exercise relative control over the processing and the use of raw materials of varying quality.

6.3 Selection of the type of superplasticizers in a given brand

When superplasticizers were first commercialized, concrete producers did not have much choice, because most admixture companies offered just one type of superplasticizer. Unfortunately, but sometimes conveniently, this is no longer the case. Choosing products of the same brand often isn't easy, especially when making

Figure 6.
(a) Thermogravimetric analysis of different plasticizers
(b) Infrared spectrum of the same plasticizers

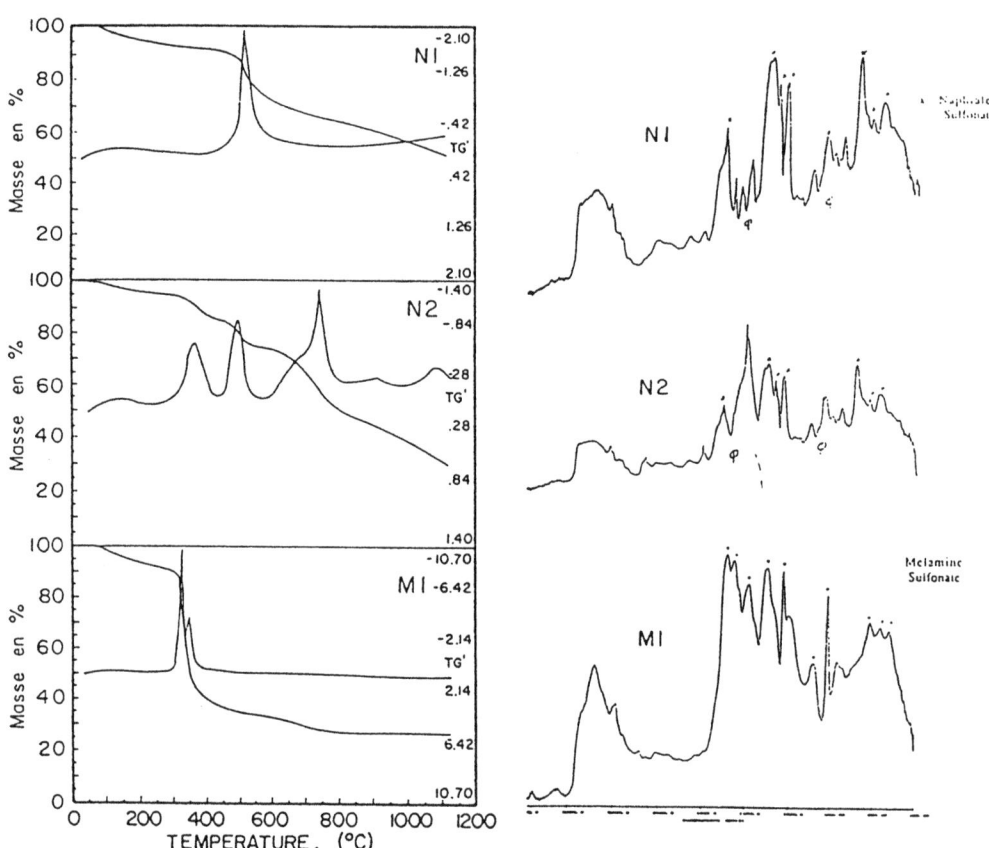

N_1 et N_2 Naphthalene

M_1 Melamine

high performance concrete, because it is well-known presently that a universal superplasticizer does not exist. Portland cements can be classified according to types, but within each type, their final character is influenced by a great number of factors. Superplasticizer variability seems to be less than that of Portland cement because they are fabricated from a limited number of fairly pure raw materials in a straightforward process. The main complexity of the process lies with ensuring that each step is closely adhered to and controlled very strictly.

In order to facilitate superplasticizer use admixture companies and superplasticizer producers have developed different formulations. In most cases, these formulations can be distinguished, particularly in the case of naphthalene superplasticizer, by the nature of the base that is used to neutralize the sulfonic acid (soda or lime), the length of the molecular chains, the amount of residual sulphate and the nature of some other admixtures that can be incorporated in the superplasticizer: lignosulfonate, gluconate, triethanolamine, etc. As superplasticizers lose their original simplicity, their use becomes more complicated.

The author thinks that, when possible, it is preferable to use only pure products and to add other admixtures, when necessary, in order to obtain the specific effect needed in the particular case, rather than using typical formulations proposed by admixture companies.

This attitude implies hiring a specialist who has a good knowledge of superplasticizer use and giving him a certain latitude of action. Of course, his employment can be justified only if the volume of concrete annually sold is sufficient to warrant his salary. When this is out of reach, the concrete producer should listen to the advice of the technical/sales agent from the admixture company selected.

6.4 Selection of the formulation : liquid or solid

Presently, most superplasticizers are offered in a liquid or solid form. Due to their ease of use and the very short mixing time in most concrete plants, it is normally preferable to use a liquid superplasticizer. Nevertheless, one must realize that superplasticizers are very sensitive to ambient temperature. Naphthalene sulphonates freeze at - 4° C; their viscosity increases very rapidly when the ambient temperature is below + 5° C, greatly affecting their efficiency. Frozen superplasticizers must be stored for 24 hours at a maximum temperature of 35°C.

Similarly, some superplasticizers do not perform very well at high temperature developing very bad odors and bacteria or fungi Normally, in order to take full advantage of the efficiency of the superplasticizer, it should be used and stored at temperatures between 10 and 30° C.

Of course, such storage problems do not occur with the solid form. The dissolution rate of solids is not instantaneous, so generally it is necessary to increase the solid content dosage when superplasticizer is used in this form during mixing.

6.5 Dosage
In concrete plants, the superplasticizer dosage has to be known in liters of commercial solution per cubic meter of concrete, in contrast to the scientific practice of expressing superplasticizer dosage in terms of solid content.

As it is very important in high performance concrete to control precisely the water/cement ratio, it is absolutely necessary to know how to take into account the amount of water that is contributed by the superplasticizer

6.5.1 Amount of water and solid contents contained in a given
     volume of superplasticizer
A commercial liquid superplasticizer has a specific gravity d_l and a solid content equal to X L/m^3 percent. What is the water correction required when X L/m^3 of the superplasticizer are used in concrete containing Y kg of cement per m^3. What is the solid content of superplasticizer used (compared to the mass of cement)?

The mass of superplasticizer M_f corresponding to X liters is

$$M_f = X \cdot d_l \tag{1}$$

The volume of water V_W contained in X liters of superplasticizer is

$$V_W = M_f \; \frac{100-s}{100} \tag{2}$$

$$= \frac{100-s}{100} \cdot X \cdot d_l \tag{3}$$

The amount of solids contained in X liters of superplasticizer is equal to

$$F = M_f \cdot \frac{s}{100} \tag{4}$$

$$= \frac{s}{100} \cdot X \cdot dl \tag{5}$$

The superplasticizer dosage S, expressed as a percentage of solid to the mass of cement, is

$$S = \frac{F}{Y} \cdot 100 \tag{6}$$

$$= \frac{s}{100} \cdot \frac{X}{Y} \cdot d_l \tag{7}$$

Sample calculation : 8.25 liters of naphthalene superplasticizer with a specific gravity of 1.21 and a solid content of 40% have been used in a concrete containing 400 kg of cement per cubic meter in order to get the desired workability.

What is the amount of water that has been added to the concrete when using the solution of commercial superplasticizer? What is the superplasticizer dosage in terms of solid content?

According to Equation (2), the amount of water is:

$$V_W = \frac{100\text{-}40}{100} \cdot 8.25 \cdot 1.21 = 6.0 \text{ L/m}^3 \tag{8}$$

According to Equation (7), the superplasticizer dosage is:

$$S = \frac{40}{100} \cdot \frac{8.25}{400} \cdot 1.21 = 1.0\% \tag{9}$$

6.5.2 Estimating the dosage of superplasticizer when making a high performance concrete

Selecting superplasticizer dosage is not easy the first time a concrete producer decides to make a high performance concrete. One has to wonder where to start.

As the rheological behavior of commercial Portland cements presently available on the market is quite different when they are used in a very low water/cement ratio mixture, and due to the great variability of the compatibility between superplasticizers and cements, there is no single formula that gives the right dosage. The following procedure could be used to determine the right dosage after two or three iterations.

Experience shows that a high-performance concrete with a 200 mm slump can be obtained from a Type I cement with a solid content dosage between 0.75 to 1.50%. When the dosage of solids is above 1.50% a delay in setting is usually observed. This delay usually varies from 12 to 24 hours, although greater values are possible.

It is suggested to start calculating the amount of superplasticizer taking into account a dosage of 1% and using the following formula:

$$X = \frac{100 \cdot S \cdot Y}{s \cdot d_l} \tag{10}$$

that can be deduced quite easily from Equation (7).

Sample calculation. Suppose that a concrete having a water/cement ratio of 0.27 has to be made using 135 liters of water per cubic meter. The superplasticizer has a specific gravity of 1.21 and a solid content of 40%. The cement dosage will be equal to:

$$\frac{135}{0.27} = 500 \text{ kg/m}^3 \tag{11}$$

Since the solid content is equal to 1%, 5 kg of solids are needed, which corresponds to a superplasticizer volume of:

$$X = 100 \cdot \frac{1}{100} \cdot 500 \cdot \frac{1}{40} \cdot \frac{1}{1.21} = 10.33 \text{ L/m}3 \qquad (12)$$

It should be taken into account that the superplasticizer will bring

$$V_W = \frac{100-40}{100} \cdot 10,33 \cdot 1.21 = 7.5 \text{ liters of water per m}3 \text{ to the}$$
$$\text{concrete} \qquad (13)$$

The calculated composition is made and its slump followed for one hour, if possible. The superplasticizer dosage is corrected by increasing or decreasing according to the results obtained in the first trial batch. If the amount of superplasticizer was not correct, it is good to make drastic corrections to be sure to exceed the optimum dosage. If the corrections are too small, finding the optimum dosage can take a long time.

For example, if the concrete made with 10.33 L/m3 of superplasticizer is too fluid and prone to segregation, the next step will be to use 7.5 L/m3. If the concrete lacks workability, a dosage of 13 liters should be used. With some cements, even these drastic modifications could be insufficient to exceed the optimum dosage.

NOTE : When calculating the composition of high performance concrete, the water/cement ratio needed to guarantee a certain level of compressive strength and the dosage of water required to guarantee a certain level of workability one hour after mixing have to be selected very carefully.

Some concrete with a low rheological activity (consuming little water during the first hour after mixing), require as little as possible as 120 L/m3 of water, while some very reactive cements resist yielding as 200 mm slump with less than 150 L of water per m3 , regardless of the amount of superplasticizer used. In such cases, too little water results in a very rapid slump loss and by a shiny appearance on the surface of the concrete.

If the reactivity of such a cement cannot be controlled by the addition of a retarding agent, as will be seen later, the only thing to be done is to increase the water/cement ratio, knowing that this increase will reduce compressive strength. Such a cement is not very good for high performance concrete. Whenever possible, use another cement.

6.6 Simultaneous utilization with a water reducer
In order to save money, some concrete producers add lignosulphonate-based water reducers dosed at their usual rate of 0.5 to 1.5 L/m3. This practice seems to be the rule in Norway and with certain North American producers that pretend to save the same amount of superplasticizer [17].

6.7 Simultaneous utilization with a retarder

Experience shows that, with some rheologically reactive cements, the incorporation of a certain amount of retarding agent early on in mixing, (about 5 to 10% of the total volume of superplasticizer that is supposed to be used), can solve the slump-loss problem without unduly retarding the setting of the concrete. It seems better to use a retarding agent based on sodium gluconate rather than a lignosulphonate retarding agent, because the former entraps fewer air bubbles in the concrete and seems to be more effective. Moreover, sodium gluconate is a chemical product more controlled than lignosulphonate when used as a retarding agent. When the simultaneous use of a retarder and superplasticizer is required, the optimum dosage of both admixtures has to be found in terms of cost and short-term compressive strength. Too low a compressive strength at 1 day can delay removing the forms.

The simultaneous use of a retarder and superplasticizer can be quite tricky. Climatic factors also have to be considered because cement reactivity is greatly influenced by the concrete temperature.

6.8 Double or single introduction

6.8.1 Precast concrete

If high performance concrete has to be used in precast plants, the superplasticizer is introduced in a single shot at the beginning of the mixing because, in most such facilities, concrete is placed within 15 to 20 minutes of mixing, which should allow slump-loss control. It is important in such cases to be sure that the selected dosage will not excessively retard the setting of the cement.

Short-molecule superplasticizers could be used because they will be adsorbed more rapidly on C_3A and on the first hydrates formed C_3S. Moreover, it seems that melamine-based superplasticizers are less retarding than the naphthalene variety.

6.8.2 Ready-mix concrete plant

If the high performance concrete is made in a ready-mix plant, it must be delivered in the field with sufficient workability to ensure that placing is as easy as that of a regular concrete. This is why, from a practical point of view, the most frequent major problem to solve when making high performance concrete is controlling the slump loss during the period corresponding to the delivery and placing of the concrete.

This objective can be reached in different ways: a certain amount of cement can be replaced by a less reactive supplementary cementitious material (fly ash [18], silica fume [19], slag [20], limestone filler, etc.) or by using a retarding agent (see paragraph 6.7). In this paper, only the second alternative and double introduction with or without a retarding agent will be discussed.

29

6.8.3 Principle of double introduction without retarder

In this case, mixing requires only enough superplasticizer to ensure a good dispersion and sufficient deflocculation of cement particles during the mixing so that the original slump at the arrival at the field is just between 50 to 100 mm. In fact, it is quite difficult to fluidify a concrete having a slump less than 50 mm with a second dosage of superplasticizer. Experience shows that when using this double dosage technique, a certain amount of superplasticizer is saved, because while the first deflocculation is incomplete, certain cement particles stay unhydrated during mixing. According to cement reactivity and the superplasticizer efficiency, the initial superplasticizer dosage during mixing can represent 60 to 75% of the total dosage. In general, such a dosage should give the concrete an initial slump of 100 to 150 mm.

It is difficult to be more precise because cement reactivity, superplasticizer efficiency, as well as the temperature of concrete during mixing and mixing duration.

6.8.4 Principle of double introduction with a retarder

In this case, a certain amount of retarder representing about 5 to 10% of the total volume of the superplasticizer is added during mixing to try to better control the slump during transportation to the site. In this case, too, the concrete should arrive in the field with a 50 to 100 mm slump which can be increased to 200 mm. Experience shows that, with some cements, substantial savings in the superplasticizer dosage can be made. In some cases, the use of 0.5 to 1 L of retarder during the inital mixing can result in a savings of 2 to 3 L of superplasticizer. It should be pointed out that the amount of retarder has to be adjusted carefully so that it does not excessively delay early strength development and lower excessively early compressive strength (12 to 24 hours).

7 Conclusion

There is no high performance concrete without superplasticizers. In fact, it is due to the efficiency of the dispersing action of superplasticizers we can now make high performance concrete. The facility with which high performance concrete can be made depends essentially on the behavior of the binder in the presence of the superplasticizer.

This behavior depends on the reactivity of the cement, that is, the rate at which it consumes water during the first moment following the mixing, and on the compatibility of the superplasticizer with the cement, that is, the rate at which some of its molecules are trapped within the new hydrates formed.

Some very reactive cements and inefficient superplasticizers require the use of a retarder. Under these conditions, concrete adjustments and delivery in the field must be finely tuned and under full control, neither of which is easy.

In order to facilitate concrete delivery and to save money, it is judicious to use the double introduction technique, in which a certain amount of superplasticizer is added at the plant in order to obtain at the end of mixing a slump of 100 to 150 mm. The remaining amount of superplasticizer can be added at the job site. To facilitate increasing the slump at the job site, the concrete should arrive with a slump between 50 to 100 mm making it fairly easy to plasticise a high performance concrete. It also seems to enhance superplasticizer performance.

The ideas and results presented here are probably not definitive, but they represent the present state of our knowledge. It should be mentioned that this knowledge is rapidly evolving. Researchers and concrete producers devote a lot of effort in this field because they are committed to delivering high-performance concrete with ever lower water/cement ratios.

It seems presently that it is not so difficult to deliver high performance concretes having water/cement ratio equal to 0.3 with practically any of the commercial cements produced in North America and to obtain, if quite good aggregates are available, compressive strengths of 70 and 100 MPa when using around 10 L of superplasticizer per m^3.

With some very efficient cements, it is possible to lower somewhat more the water/cement ratio (down to 0.25) with or without a retarding agent and to reach compressive strengths around 100 to 120 MPa when using 10 to 15 L /m^3.

It is only with Portland cements that present a very low rheological reactivity combined with the use of silica fume and some very efficient superplasticizers that it has been possible to lower the water/cement ratio below 0.25 and make high performance concrete having a 200 mm slump when they are delivered on site. The compressive strength of such concretes can exceed 150 MPa.

Trying to lower the water/cement ratio further is not necessarily advantageous since aggregates then usually represent the weakest link in the concrete as the site where microcracks start to develop.

The formulation, fabrication, and delivery of high performance concrete having a very low water/cement ratio represent a challenge for concrete technologists. Of course, the high performance concrete market is not now very large, but it is a market for the future, a market that will differentiate the good from the bad. It is a challenge that all concrete producers should dare meet It won't be easy since there are no two cements and not two superplasticizers that react the same way. But that's no reason to turn away.

It can be hoped that with the development of high performance concrete, cement producers will start developing new cements or new combinations of cementitious materials particularly well-suited for the fabrication of high performance concrete and that superplasticizer producers will do the same. So, within a few years, the production and delivery of high performance concrete having a water/cement ratio of 0.25 will be as easy as delivering a concrete having a 0.50 water/cement ratio is today.

8 Research needs

As previously mentioned, the optimization of a high performance concrete is an open field in which not all the answers are known. There are still some questions that need further development particularly:

- how to evaluate simply the rheological reactivity of Portland cement and its compatibility with a given superplasticizer?
- how to evaluate simply in the laboratory and the field, the workability of a concrete having a very low water/cement ratio by means other than the slump test?
- how to diminish the rheological reactivity of a given Portland cement in the domain of low water/cement ratios?
- how to increase the efficiency of a superplasticizer in the domain of low water/cement ratios?
- how to optimize the use of supplementary cementitious materials when making low water/cement ratio concrete?
- Is it possible not to use silica fume in very low water/cement ratio concrete ?
- what is the most appropriate mixing techniques to use when making very low water/cement ratio concrete?

These are questions that demonstrate that the formulation, fabrication, and delivery of high performance concrete represent a research domain particularly rich and challenging for researchers and concrete technologists.

References

[1] AITCIN, P.-C. (novembre 1988) *Du gigapascal au nanomètre*, Séminaire ATILH-ENPC sur les bétons à hautes performances, Paris, 32 p.

[2] AITCIN, P.-C. (novembre 1988) *Les fluidifiants : des réducteurs d'eau pas comme les autres*, Séminaire ATILH-ENPC sur les bétons à hautes performances, Paris, 15 p.

[3] AITCIN, P.-C. (décembre 1988) *Comment mesurer la résistance en compression des bétons à haute performance*, Cours intensifs sur les bétons à hautes performances, Université de Sherbrooke, Sherbrooke, 15 p.

[4] KREIJGER, P. C. (1980) *Plasticizers and Dispersing Admixtures*, Admixtures, Concrete International 1980, The Construction Press, Londres, p. 1-16.

[5] JOISEL, A. (1973) *Les adjuvants du ciment*, édité par l'auteur, 3 Avenue André 95230 Soicy, France, 253 p.

[6] RIXOM, M.R. (1978) *Chemical Admixtures for Concrete*, Halstead, p. 5-89.

[7] RAMACHANDRAN, V.S. (1984) *Concrete Admixtures Handbook*, Noyes Publications, Park Ridge, N.J., p. 211-264.

[8] MALHOTRA, V.M. (1978) *Superplasticizers in Concrete*, Concrete Construction, Publications 329 Interstate Road Addison, Illinois 60101, 23 p.

[9] MEYER, L.M., PERENCHIO, W.F. (janvier 1979) *Theory of Slump Loss as Related to the Use of Chemical Admixtures*, Concrete International, p. 36-43.

[10] UCHIKAWA, H. (septembre 1986) *Effect of Blending Component on Hydration and Structure Formation*, 8th International Congress on Chemistry of Cement, Rio de Janeiro, Brésil, 77 p.

[11] RAMACHANDRAN, V.S., BEAUDOIN, J.J., SHIHVA, Z. (1989) *Control of Slump Loss in Superplasticized Concrete*, Matériaux et constructions, n° 22, p. 107-111.

[12] AITCIN, P.-C., SARKAR, S.L., REGOURD, M., VOLANT, M. (1987) *Retardation Effect of Superplasticizer on Different Cement Fraction*, Cement and Concrete Research, vol. 17, n° 6, p. 995-997.

[13] HANNA, É., LUKE, K., PERRATON, D., AITCIN, P.-C. (1989) *Rheological Behavior of Portland Cement in the Presence of a Superplasticizer*, Troisième conférence internationale sur les superplastifiants et autres adjuvants chimiques dans le béton, (sous presse)

[14] PERRATON, D., LAPLANTE, P., AITCIN, P.-C. (février 1989) *Selection of the Superplasticizer/Cement Combination for Minimizing Slump Losses When Making Very High Strength Concrete*, Réunion annuelle de l'ACI, Atlanta, 25 p.

[15] KHORAMI, J., AITCIN, P.-C. (1989) *Physicochemical Characterization of Superplasticizers*, Troisième conférence internationale sur les superplastifiants et autres adjuvants chimiques dans le béton, (sous presse)

[16] BURK, A.A., GAIDIS, J.M., ROSENBERG, A.M. (s.d.) *Adsorption of Naphthalene-based Superplasticizer on Different Cements*, communication personnelle, 21 p.

[17] HOLLAND, T.C. (juin 1989) *Working with Silica Fume in Ready-mixed Concrete - U.S.A. Experience*, Troisième conférence internationale CANMET/ACI, Trondheim, Norvège, p. 763-781.

[18] MORENO, J. (1982) *Sixteen Years of High-strength Concrete in the Chicago Area : A Historical Background*, Réunion annuelle de l'ACI, Atlanta, 14 p.

[19] AITCIN, P.-C. (1983) *Condensed Silica Fume*, Les Presses de l'Université de Sherbrooke, Sherbrooke, 52 p.

[20] RYELL, J., BICKLEY, J.A. (juin 1987) *Scotia Plaza : High Strength Concrete for Tall Buildings*, Symposium sur l'utilisation des bétons à hautes performances, Stavanger, Norvège, p. 641-653.

33

3 ULTRAFINE PARTICLES FOR MAKING VERY HIGH PERFORMANCE CONCRETES

F. de LARRARD
LCPC, Paris, France

The manufacture of very high strength concrete (28-day compressive strength higher than 80 MPa) often involves the addition of ultrafine particles together with large proportions of organic admixtures. This article compares the effectiveness of different fillers and their mixture. Silica fumes are found to be the most effective addition, and they are looked into more particularly in terms of their effect on the properties of mortars according to their proportion (optimum proportion) and quality (chemical composition).

1 Introduction

The production of concretes which are workable when fresh and have a 28-day compressive strength higher than 80 MPa is currently possible owing to superplasticizer admixtures which lead to very low water/cement ratios (less than 0.30). The use of ultrafine particles, i.e. grain size smaller than that of cement, facilitates this production, as a result of their action :

(1) on the physical level (**filler effect**, when the grains fill the voids between those of cement, reducing the water requirement)
(2) on the chemical level (for siliceous particles, **a pozzolanic effect**, reaction of silica with lime released by the cement). Depending on available products and desired concrete properties, it is important to establish quantitative and qualitative selection criteria in order to obtain the required material at the best cost.

It is consequently important to compare the performance of different compositions of binding pastes, which are designed to give the concrete a certain workability and a given strength, two

High Performance Concrete: From material to structure. Edited by Yves Malier. © 1992 Taylor & Francis.
Published by Taylor & Francis, 2 Park Square, Milton Park, Abingdon, Oxon, OX14 4RN. ISBN 0 419 17600 4.

properties which are generally contradictory. One objective way for establishing a classification is to compare at a given time compressive strengths of mortars in which **the volume and fluidity of different tested pastes have been kept constant** (fluidity being determined by a test similar to Marsh Cone Test - see Figure 1).

Figure 1 : Marsh cone flow test

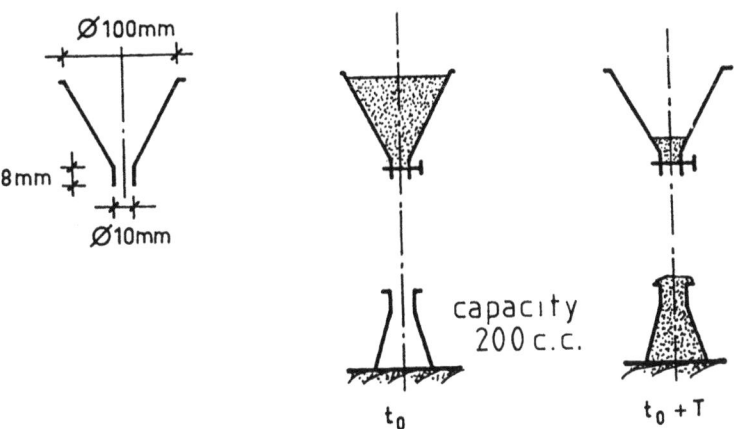

Rheological properties also depend on the amount of fluidizing admixture used. To the extent that one wishes to characterize the "mineral part" of the paste and not the organic admixture, grains are saturated with a superplasticizer so as to get the lesser water demand (see Figure 2). The choice of relative proportions of different binders then leads to a superplasticiser proportion (increasing with the specific area). The water proportion corresponds to a given flow time (5 seconds in our tests). Mechanical strengths thus obtained with mortar reflect the capacity of the cement/ultrafine mix to give the grain mixture its high strength.

This paper examines, first of all, the different types of ultrafine particle and their mixture. Then, the case of silica fumes is studied in greater detail, considering the amounts to be used and their performance in relation to their chemical composition.

2 Linear model of grain mixture packing density

A mathematical model has already been presented (1,2) designed to predict the packing density of a grain mixture based upon its particle size distribution and **specific packing density values** (packing density of each monodimensional section piled separately) using the same methods applied to the overall mixture. We applied the model to the mix design of plasticized cementitious pastes, and we showed that the following expressions gave a theoretical packing density increasing with the real packing density of the different mixtures :

$$c = \inf_{t>0} \frac{\alpha\,(t)}{1- \int_{0}^{t} f\,(t,x)\ y\,(x)\ dx\ -\ [\,1-\alpha\,(t)\,]\ \int_{t}^{+\infty} g\,(t,x)\ y\,(x)\ dx}$$

where c is the theoretical packing density of the mixture described by its particle size distribution y (with unit integral)

$\alpha\,(t) = 0.39 + 0.022\ \ln t$

where α is the specific packing density of the particles of diameter t, expressed in μ m

$f\,(t,\,x) = f\,(\,z=t/x) = (1-z)^{3.1} + 3.1\,z\,(1-z)^{2.9}$

$g\,(t,\,x) = g\,(\,z=x/t) = (1-z)^{1.6}$

Figure 2 : Cone flow time of different pastes as a function of admixture proportion (CSF: Condensed Silica Fume)

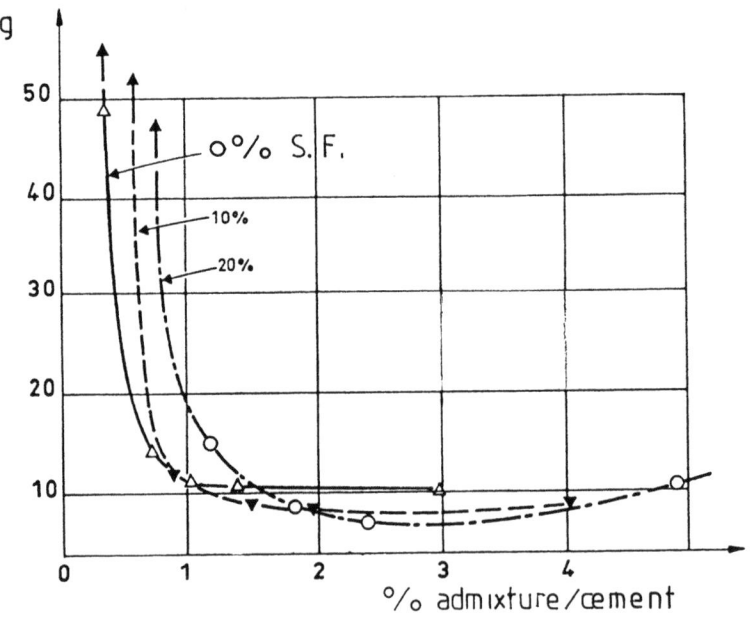

Figure 3 : Particle size distribution of different binders [1,2]

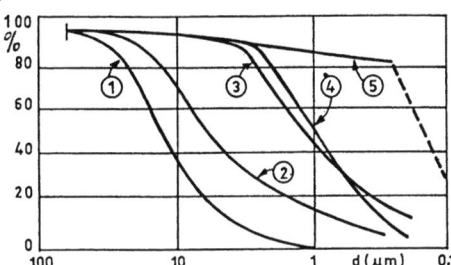

1 Portland cement
2 Limestone filler
3 Limestone ultrafine filler
4 Siliceous ultrafine filler
5 Silica fume

The particle size distributions of different cementitious materials were measured with the sedigraph (see Figure 3). The application of the model then gives the ternary diagrams shown in Figure 4, representing the locations of iso-packing density points. It is thus possible to select various cement-ultrafine mixtures of interest in different respects.

Figure 4 : Ternary diagrams predicted by the linear model of grain mixture packing density [1,2]

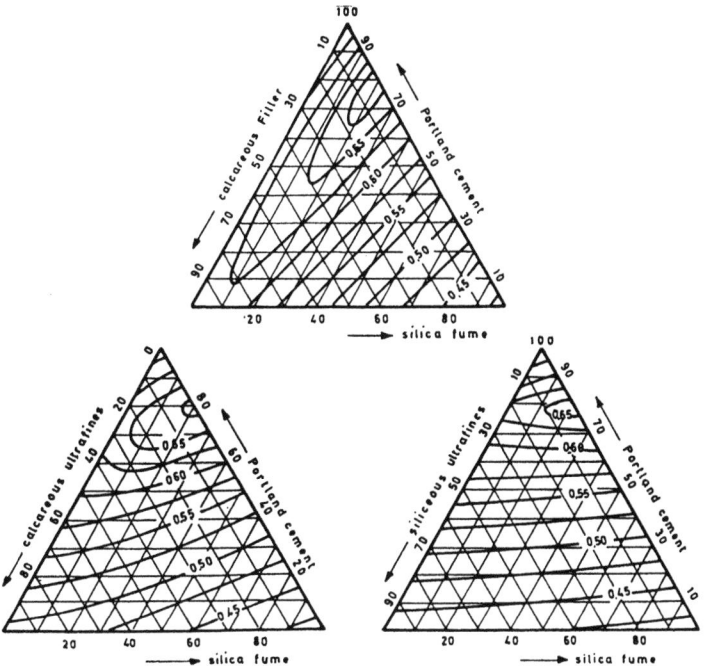

3 Hierarchy between types of ultrafines particles

The mineral binders tested were the following:

- ordinary portland cement (with a strength of 55 MPa at 28 days) with a high-silica content, chosen for its very good compatibility with superplasticizers, whose composition according to Bogue was as follows:

  - $C_3S = 63,30$ %, $C_2S = 17,72$ %, $C_3A = 1,62$ %, $C_4AF = 8,47$ %,
  - Gypsum = 4,06 %, $CaCO_3$ = 2,59 %, CaO = 0,5 %,
  - Limestone filler, with grain size between cement and ultrafine particles
  - Limestone ultrafine filler (average size 1 μm)
  - Siliceous ultrafine filler from grinding (average size 1 um).
  - Two silica fumes.

All those mixes have been prepared with naphthalene sulfonate formaldehyde.

Using these products, 12 mortars were prepared having the compositions and strengths shown in Table 1; mortar N°,1 served as a control.

It is noted first of all that compressive and bending strengths show good correlation. In the area investigated, these properties consequently appear to be controlled essentially by the **packing density of the binding phase.**

This is also the case for concrete, until one reaches a limit due to the specific strength of the aggregate, for compressive strengths of about 110/120 MPa.

Mortars N° 2, 3 and 4 have, within the limits set in the introduction, quite the optimum compositions for cement/ultrafine particles binary mixes, respectively for silica fume, calcareous fillers and siliceous fillers. The very high fineness of silica fume, combined with its pozzolanic activity, places it far ahead of grinding mill products. For the latter, siliceous fillers show a certain chemical activity in spite of their initially crystalline nature, giving them a superiority over limestone, in spite of lower efficiency regarding the filler effect. The activity of calcareous products in this type of mortar has been demonstrated by Buil et al. [3]. However, in the present case, it was not observable owing to the very low aluminate content of the cement.

In compositions N° 5 and N° 6, an attempt was made to determine the value of associating less noble products with silica fume as the price of silica fume is expected to increase in the next few years. Further, it may be reasonable to limit the amount of silica fume to about 10 to 15% achieving in the long term the maintenance of a sufficiently alkaline pH. The comparison of compositions N° 5 and 8 then shows that the composition with 10%

Table 1 : Composition and mechanical strength of mortars (1 series)

Nr	Rilem sand g	OPC g	Fillers and ultrafine particules g		Ad-mixture g	Water g	W/C	Flowing time LCL s (13)	Flexural strength MPa	Compressive strength MPa
1	1350	529	-	-	7,9	148	0,280	16	9,5	74
2	"	520	silica fume N°1 80	-	10,7	114	0,219	4	14,5	103
3	"	481	ultrafine calcareous 103	-	10,2	125	0,260	15	11,3	77
4	"	467	ultrafine siliceous filler 111	-	10,1	126	0,270	10	11,6	81
5	"	489	silica fume N°1 43	ultrafine calcareous 52	10,3	122	0,249	3	12,8	90
6	"	348	calcareous filler 149	silica fume N°1 40	9,8	132	0,379	2	9,9	72
7	"	553	industrial silica-fume 27	-	9,6	128	0,231	11	10,8	(94)
8	"	535	53	-	10,6	122	0,228	5	14,4	96
9	"	514	77	-	11,4	117	0,228	4	14,2	101
10	"	495	99	-	12,1	113	0,228	6	15,4	105
11	"	468	117	-	12,5	113	0,241	12	15,9	101
12	"	443	133	-	12,9	114	0,257	17	13,0	96

silica fume (N° 8) offers higher performances than the mix containing 10% silica fume, 10% filler and 80% cement (N° 5), although slightly lower in water. Formula N° 6 shows - by comparison with the control - that it is possible to lower the cement content of a high-strength concrete by about 34% without modifying its strength, by substituting a quasi-inert fine material for the cement, and adding a small amount of silica fume. This type of material certainly offers a lower hydration heat, as well as more stable workability, important properties for the production of large structures and for placement on the site respectively.

4 Further investigations on silica fumes

4.1 Optimum proportion

We first looked into the problem of the optimum amount of silica fume addition used in concretes to obtain maximum mechanical strength. Using additions in a mortar, with the adjustment of the admixture content, Buil et al. [4] found values between 40 and 50 % for the ratio between silica fume weight. Seki et al. [5] using the same approach, but on concrete, obtained a value of 26 %. However, **applying a constant paste content and constant fluidity, the increase in strength obtained indeed translates the original contribution of silica fume**. The improvement would not have been achievable by the addition of cement, which would have necessarily reduced workability with a constant paste content, or increased the paste volume for the same workability.

Mortar Nᵒˢ 1 and 7-12 show a variation in the water/cement ratio and compressive and bending strengths with the proportion of silica fume. Surprisingly, the cement content increases for a small proportion of ultrafine particles: this is the **"roller bearing"** effect which we described earlier [6]. Things occur as if a highly dilute suspension of silica fume in water was more fluid than pure water ! A similar result was also found by Yogendran et al. [7], who observed that a substitution of 5% silica fume for a concrete whose admixture proportion is kept constant does not cause an increase in water requirements. In fact, a small addition of silica fume prevents the sedimentation of the cement and thus facilitates its flow. This is probably to be linked with the improvement in concrete pumpability, frequently observed with low silica fume proportions.

The water/cement ratio is surprisingly stable for silica fume contents ranging from 5 to 20%. Within this range, the increase in strength is consequently essentially due to the pozzolanic activity of silica fume. It is known that this activity culminates at about 24% [8], a level beyond which all the lime released by the cement is consumed. It is also above this key level that the effectiveness of silica fume as a filler decreases (see Figure 4). **The optimum silica fume content for obtaining high strength is consequently around 20 to 25%**. It is in this proportion that ultrafine particles best fill the voids of the cement grains, with which they can then combine to form hydrates participating in the strength of the material. It is however noted that the mechanical strength gain is faster with smaller proportions. Considering the cost of silica fumes and of admixtures, the practical (economic) **optimum is located rather towards 10% silica fume in relation to cement weight** [1].

4.2 Silica fume selection criteria

Silica fume is a by-product of the manufacture of silicon and of its alloys. Depending on the composition of the alloys, on the secondary products added to the main constituents, on the manufacturing method, and so on, silica fume properties can vary considerably.

Table 2 : Mortar compositions (II series)

Mortars	RILEM Sand	Cement OPC	Silica Fume	Super-plas-ticizer	W/C	Flowing time "LCL" (13)
II Series	1 350 kg	544 g	100 g	12,2 g	0,25	depending on silica fume

An examination of Table 3 allows the following remarks :

(i) from the viewpoint of the BET specific surface, **fineness has no direct effect on rheological or pozzolanic properties**. It is hence the coarsest silica (N°1) that gives the best performance. In fact, the grain size range of silica corresponding to the specific surface (about 0.1µm for 20 m²/g of BET surface) is so fine that its granular interactions with cement are very weak (cement grains all being greater than a micro-meter). In our opinion, this effect is due to a more or less extensive aggregation of silica grains, an aggregation which can be seen on the grading curves of Figure 5, measured with the sedigraph [9].

For clarification, we carried out a series of tests on six silicas from different sources. While this does not constitute a sufficient statistical sampling, it does however make it possible to observe differences in behaviour in connection with physical and chemical properties. Silica N° 1 is a special silica from the manufacture of zirconium. Its cost makes it ill-suited to civil engineering applications, but its effectiveness and its high purity make it a particularly interesting laboratory product.

To evaluate the utilization properties of these fumes, we attempted to:

- quantity the "rheological" performance levels; for this, we measured the workability of mortars with a constant proportion;
- estimate the pozzolanic performance levels by measuring the strengths reached by the same mortars.

We then attempted to relate these utilization properties by comparing them with chemical analyses and specific area measurements. The results of these investigations are given in Table 3. The compositions of the mortars appear in Table 2.

Table 3 : Characteristics and performances of six condensed silica fumes

Silica Fumes		1	2	3	4	5	6
Specific Areas (BET) m²/g		14,2	22,3	21,6	22,2	22,2	19,3
	SiO_2	91.50%	97.35%	88.75%	92.83%	96.50%	93.30%
	Al_2O_3	5.78%	0.03%	0.08%	0.02%	0.03%	0.02%
	Fe_2O_3	0.16%	0.12%	1.60%	0.19%	0.06%	0.30%
	MgO	0.03%	0.19%	1.48%	0.54%	0.24%	0.38%
Chemical	CaO	0.08%	0.10%	0.66%	0.16%	0.01%	0.12%
composi tions	Na_2O	0.12%	0.12%	0.71%	0.24%	0.26%	0.76%
	K_2O	0.08%	0.23%	2.41%	1.76%	0.47%	1.40%
	ZrO_2	1.10%					
	alkalies	0.20%	0.35%	3.12%	2.00%	0.73%	2.16%
	carbon	--	1.06%	1.59%	2.59%	0.89%	2.75%
Mortars: (II series) flowing time L.C.L. workabilimeter (seconds)		2	5	7,5	10,5	5	9,5
Compressive strength at 28d (MPa)		123	101	87	95	94	92

Figure 5 : Grading curves of different silica fumes (Sedigraph measurement)

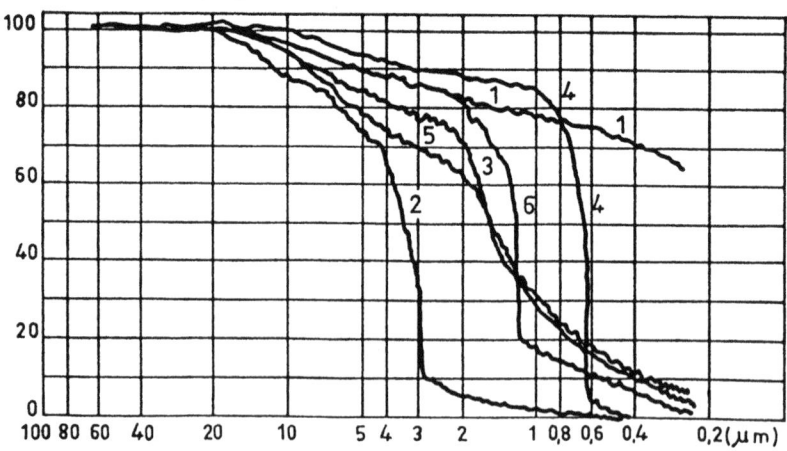

These measurements show the particularly **high degree of agglomeration** of a silica fume in suspension in a physical and chemical environment representative of that of very high strength concrete (pH 12.5, in the presence of calcium ions). The average size of the secondary grains was about one micrometer, or 100 times the average size of the basic grain! Let us however point out that these curves should just be regarded as an illustration of the "particle-size instability" of silica fume in a cement paste. It is in fact probable that sand and cement grains act as grinding agents during the mixing of the VHS concrete, so that the actual particle-size distribution of silica fume is shifted toward the small sizes in relation to the preceding figure. However, it is not excluded that silica fume is partially aggregated during its formation. This would indicate a "sintering" phenomenon whose extent depends on the chemical and thermal conditions prevailing during cooling.

(ii) **The purity of silica fume**, i.e. the percentage of $Si O_2$, **is not at first sight a decisive criterion** because, here too, it is the sample of the lowest purity (N° 1) which gives the best strengths and the best workability. However, setting aside this sample N° 1, the silicas investigated come from the manufacture of high silicon alloys. Certain authors (Regourd [10]) have examined silica fumes containing only 50 to 60% Sio2. Their strength is less sensitive to marginal fluctuations in silica content when the ultrafine content approaches (as is the case) 20% of the weight of the cement, practically exhausting all the available lime.

Figure 6 : Relationship between carbon content and workability

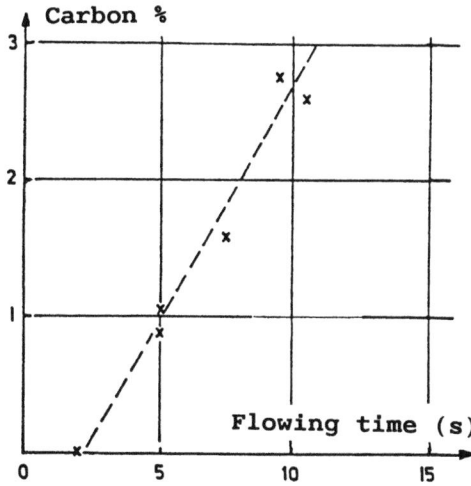

43

(iii) Carbon hold, corresponding to the more or less dark colour of silica fumes, **is related to a great extent** (in our samples) **to rheological performance levels** (see Figure 6). It is known that this carbon comes from the combustion of organic matter (coal or wood chips) added to the constituents of silicon alloys. With the optical microscope, one observes wastes which are larger than the silica grains (of the order of 10 microns in size). These particles do not appear to exercise directly any harmful rheological effect, as was corroborated by the addition of carbon black to a mortar. Its presence indicates, rather, a history of temperatures, moreover responsible for the rheological quality of the by-product, perhaps as a result of grain aggregation.

Let us point out that a similar effect was reported by Osbaeck [11] for fly ash. For this product (which, apart from its size, has many points in common with silica fume), a correlation is observed between water requirements and carbon content.

(iv) The significant chemical parameter with respect to pozzolanic performance could be the alkali proportion (see Figure 7). Based upon present knowledge regarding pozzolanic activity [12], one should expect an acceleration of the kinetics of silica attack with an increase in the proportion of alkalis. Alkalis however appear to reduce the strength of the material.

Figure 7 : Relationship between strength and alkali proportion

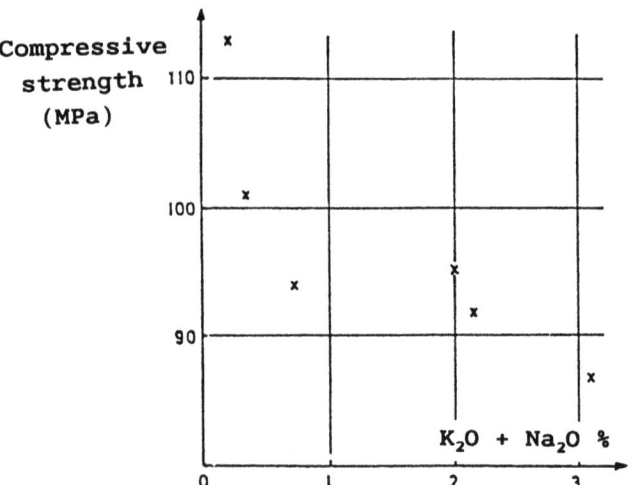

Table 4 : Compositions of mortars (III series)

MORTAR "Co"

Sand RILEM	Cement OPC 55	Silica N° 1	Superplas- ticizer	W/C
1 350 kg	544 g	100 g	12.2 g	0.25

MORTAR	C_1:	$C\sigma$ + KOH	< $K+/SiO_2$ = 1%
	C_2:	$C\sigma$ + KOH	< $K+/SiO_2$ = 2%
	C_3:	$C\sigma$ + KOH	< $K+/SiO_2$ = 3%
	C_4:	$C\sigma$ + KCl	< $K+/SiO_2$ = 1%
	C_5:	$C\sigma$ + KCl	< $K+/SiO_2$ = 2%
	C_6:	$C\sigma$ + KCl	< $K+/SiO_2$ = 3%

Figure 8 shows the compressive strength of mortars at 28 days, giving the mean values plus or minus the standard deviations of the different tests. A general decrease is in fact observed, the potash-chloride difference not being significant for the same proportion of alkalis, showing that what is involved is a direct effect of the potassium ion and not a consequence of the probable rise of the pH.

Figure 8 : Evolution of strength with increasing addition of alkalis

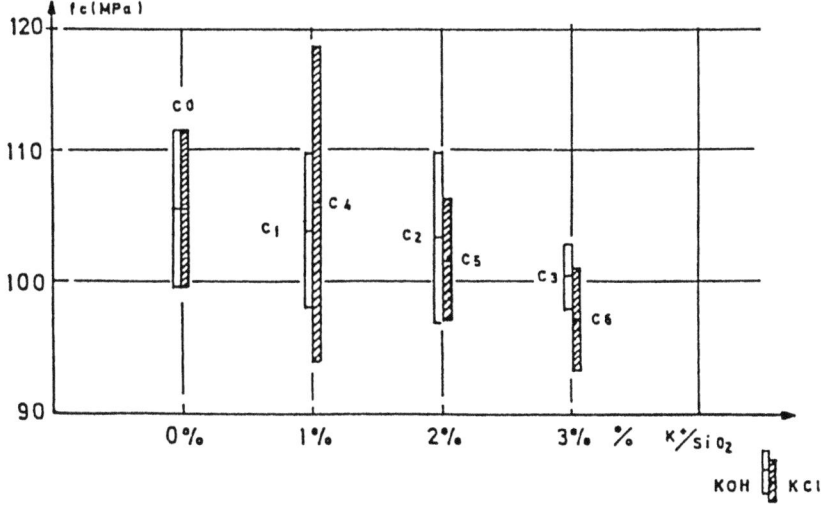

5 Conclusions

After having reviewed the conditions - which we consider to be necessary for carrying out meaningful comparisons between mineral additives for the production of very high strength concretes, we can state the following conclusions:

(a) compared with the use of a single OPC as a binder for a high strength concrete, the addition of grinding mill ultrafine particles (average size of the order of 1 micrometer) improves the strength of the material, probably with a slight superiority of siliceous products - a result to be confirmed however for cements containing aluminates.

(b) silica fume constitutes a more effective product to be used pure if what is primarily expected is strength, and mixed with less active fillers when one wishes to reduce the proportion of cement.

(c)the optimum silica fume proportion is between 20 and 25% by weight of cement. However, a proportion of half that amount of the product will lead to an economical material, easy to place and of dependable durability.

(d) the different industrial silicas available do not all have the same quality from the standpoint of their incorporation in concrete: they exhibit differences in workability and strength, these aspects not being interrelated.

(e) the water requirements of material containing silica fume increases indirectly with its carbon content, which can be evaluated visually by the colour of the by-product.

(f) binding properties of fumes sufficiently rich in silica (%SiO_2 > 85%) appear to depend primarily on the alkali content (Na_2O, K_2O), which must be as low as possible.

Additional tests are however necessary for the confirmation of these last results, and for deducing criteria enabling the classification of silica fumes into different grades.

6 References

1. De Larrard F. (1988) : Formulation et Propriétés des Bétons à Très Hautes Performances. PhD Thesis of the Ecole Nationale des Ponts et Chaussées. Rapport de Recherche L.P.C. N° 149, Paris.
2. De Larrard F. (1987) : Modèle Linéaire de Compacité des Mélanges Granulaires. "De la science des matériaux au génie des matériaux de construction", Proceedings of the First

International RILEM Congress, vol 1, Chapman & Hall, pp 325-332.

3. Buil M. , Paillère A.M. (1984) : Utilisation de fillers ultrafins dans les bétons. Bulletin de l'Association Internationale de Géologie de l'Ingénieur, 30, 13-16.

4. Buil M., Paillère A.M., Roussel B. (1984) : High Strength mortars containing condensed silica-fume. Cement & Concrete Research, 14, 693-704.

5. Seki S., Morimoto M., Yamane N. (1985) : Recherche sur l'amélioration du béton par incorporation de sous-produits industriels. Annales de l'ITBTP, 436, 15-26.

6. De Larrard F., Moreau A., Buil M., Paillère A.M. (1986) : Improvement of mortars and concretes really attributable to condensed silica-fume. Supplementary paper of 2nd International conference on fly ash, silica fume, slag and natural pozzolans in concrete. Madrid.

7. Yogendran V., Langan B.W., Haque M.N., Ward M.A. (1987) Silica-fume in high-strength concrete. ACI Materials Journal, pp124 - 129, March-April.

8. Traetteberg A. (1978) : Silica-fume as a pozzolanic material. Il cimento, 75, 3, 369-375.

9. Buil M., Witier P., De Larrard F., Detrez M., Paillère A.M. (1986) : Physico-chemical mechanisms of the action of the naphtalene-sulfonate based superplasticizer on silica-fume concretes. Proceedings of 2nd International Conference on fly ash, silica fume, slag and natural pozzolans in concrete. ACI SP 91 - Madrid.

10. Regourd M. (1983) : Pozzolanic Reactivity of Condensed Silica-Fume. Condensed Silica Fume, Edited by P.C. Aitcin, Les Editions de l'Université de Sherbrooke.

11. Osbaeck B. (1986) : Influence of residual carbon in fly ash when assessing the water requirement and pozzolanic activity of fly ash in mortar tests. Supplementary paper of 2nd International conference on fly ash, silica fume, slag and natural pozzolans in concrete. Madrid.

12. Dron R. (1986) : Private communication. Laboratoire Central des Ponts et Chaussées, Paris.

13. Baron J., Lesage R. (1973) : Proposal for a definition of workability. RILEM Seminar "Bétons frais : les propriétés importantes et leur mesure", T 1.

4 A MIX PERFORMANCE METHOD FOR HIGH PERFORMANCE CONCRETE

F. de LARRARD
LCPC, Paris, France

This paper deals with the problem of high strength concrete composition. An empirical formula is presented that allows the prediction of strength, together with a theoretical workability model. A practical mix-design method is derived, to determine a high strength concrete composition having certain specifications. In this method, it is assumed that the optimal concrete will have a low binding paste and a high admixture content. It is also supposed that the coarse aggregate is stronger than the paste. As most tests are performed on binding pastes, the number of trial batches, and the quantity of materials used during the study are minimal. Finally, the concretes obtained by this method exhibit good secondary qualities.

1 Introduction

Concretes having compressive strength at 28 days from 60 to 100 MPa will be more and more used on sites, as far as they will prove to be quite economical for a greater range of constructions (tall buildings, bridges, off-shore jackets, precast prestressed beams etc. (1). High strength concretes (HSC) can be obtained with high performance cements (i.e giving high strength on ISO mortars), admixtures and supplementary cementitious materials such as silica fume. However, due to the cost of these additives, it will be important to reach the best possible results. Furthermore, some disadvantages of HSC, such as **rapid slump loss** or excessive heat of hydration, are to be avoided.

This is why a rational experimental approach to HSC composition is needed. The aim is to control a large number of parameters, while conducting the least number of tests, and expecting this

High Performance Concrete: From material to structure. Edited by Yves Malier. © 1992 Taylor & Francis.
Published by Taylor & Francis, 2 Park Square, Milton Park, Abingdon, Oxon, OX14 4RN. ISBN 0 419 17600 4.

optimal composition to remain adequate during all the construction duration.

In this paper, two theoretical or semi-empirical mix design tools are first presented. The strength of the concrete is predicted by Feret's formula, from a limited number of mix design parameters. The workability is assumed to be closely related to the viscosity of the mix, computed thanks to the Farris model, a rheological model dealing with the viscosity of a polydispersed suspension. From these tools, the following asumptions can be made:

- the strength of a concrete made up with a given set of components is mainly controlled by the nature of the binding paste;
- the workability of a concrete, whose grading is fixed, is assumed to be the combination of two terms: the first one depending on the concentration of the binding paste, and the second depending on the fluidity of the latter.

The LCPC Experimental Method is a practical normal strength mix design method, developed in the past from previous ideas. Here, it is shown how it is possible to propose a new trend for the mix proportioning of HSC, with a similar approach. Then, this new method is applied on an example. Lastly, some particular aspects of the HSC designed with this method are pointed out.

2 Feret's formula

In the early times of concrete technology, René Féret (2) proposed an empirical formula, which is still employed, to predict the strength of mortars and concretes. Recently (3,4,5), the author proposed an extension of this formula, to be applied to HSC made up with Portland cement and silica fume.

$$f_c = \frac{Kg \cdot Rc}{\left(1 + \frac{3.1 \, w/c}{1.4 - 0.4 \exp(-11 \, s/c)}\right)^2}$$

Where :

- f_c is the cylinder compressive strength of concrete at 28 days;
- w, c, s are the masses of water, cement and condensed silica fume for a unit volume of fresh concrete;
- Kg is a parameter depending on the type of the aggregates (a value of 4.91 applies fairly commonly for river aggregates);

- Rc is the strength of the cement at 28 days, i.e. the strength of a mortar containing 3 parts of sand for one part of cement, and a half part of water (ISO Mortar).

In the range of water/cement ratios less than 0.40, the accuracy of this formula is about 5 MPa, computed from the absolute values (see Fig. 1). The formula allows us to understand the real action of silica fume in HSC: the **filler effect** shown by the lowering of the water/cement ratio, and the **pozzolanic effect**, expressed by the contribution of the s/c term.

For concretes containing various proportions of the same components, according to Féret's formula, the strength depends mainly on the nature of the binding paste. Because the most expensive components of concrete are in the binding paste, the first mix-design rule is *to make a concrete with the minimum volume* of binding paste. However, this volume of binding paste must be adjusted to give a suitable workability for casting.

Figure 1: Comparison between actual and predicted strength values according to Féret's formula, for 15 concretes mixed with the same cement, various quantities of silica fume and various gradings of the same aggregates (5).

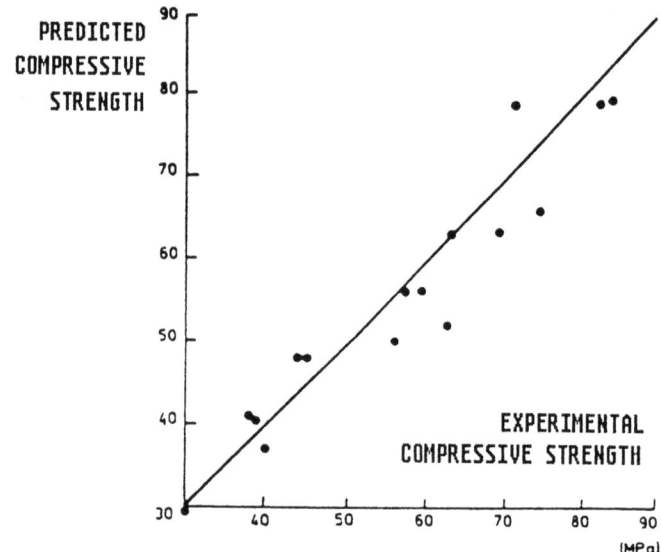

3 The Farris model (7)

The Farris model is a classical rheological model for the prediction of the viscosity of polydispersed suspensions. In a mix containing n classes of monodispersed grains of size d_i, with $d_i \gg d_i + 1$, the viscosity of the suspension is assumed to be :

$$\eta = \eta_0\, H\ \frac{\Phi 1}{\Phi 1 + ... \Phi n + \Phi_0}\ + H\ \frac{\Phi 2}{\Phi 2 + ... \Phi n + \Phi_0}\ ... + H\ \frac{\Phi n}{\Phi n + \Phi_0}$$

Where : Φ_i is the volume occupied by the ith-class in a unit volume of mix

Φ_0 is the volume and η_0 the viscosity of the liquid
the function H expresses the variation of the relative viscosity of a monodispersed suspension, as a function of its solid concentration (see Fig. 2).

The idea of the Farris model is to consider that each granular class has the same interaction with the mix of liquid plus finer classes as with an homogenous fluid. Then the different viscosities are computed by a recurrence method, starting with the finer class and adding at each step the subsequent class.

Following Petrie [8], it can be assumed that superplasticized cement binding pastes have a rheological behavior similar to purely **Newtonian** behavior. This hypothesis will be extended to high strength concretes. It means that the relation between the shear stress and the strain rate is linear, so that there is no shear

Figure 2: Relationship between viscosity and solid content of a monodispersed suspension (6).

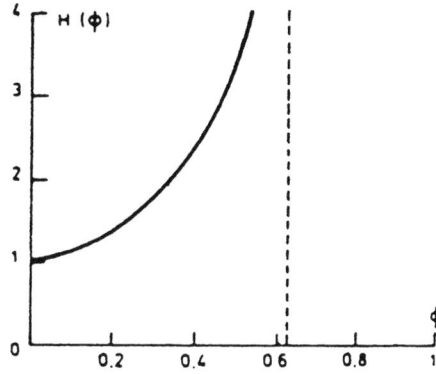

threshold. The viscosity describes the whole behavior of the fresh HSC, and the Abrams cone indicates a fluid consistency (slump greater than 200 mm).

Using the Farris model for the composition of HSC, it will be considered that the three solid components (coarse aggregate, sand and cement) represent an interlocking of three classes of grains, in accordance with the previous hypothesis.

With regard to the HSC composition, the aim is to proportion a workable concrete, i.e. with a certain viscosity, whose strength will depend on the water/cement ratio. It is now easy to demonstrate [4] the three following results :

(i)   the maximum strength will be theoretically achieved with a pure cement binding paste (without sand or aggregates);

(ii) for a given strength, and a given workability, there is an **optimal sand/aggregate ratio** which minimizes the quantity of binding paste;

(iii) this optimal ratio will remain unchanged if the quantity of binding paste is replaced by an equal volume of a different binding paste having the same viscosity.

Although the Farris model does not apply quantitatively to concrete (because it does not take into account the real grading of granular components), it is a good comprehensive tool in understanding the general problem of concrete mix design.

4 LCPC mix-design experimental method
(Normal Strength Concrete)

Fifteen years ago, Baron and Lesage [9] developed an experimental mix design method, based on the above-mentioned ideas. The LCL apparatus [10] is convenient to measure the flowing time of concrete submitted to vibration. This measurement of the workability is more reliable and more representative than the slump test. For a given application, a **critical workability** and a 28 day compressive strength of concrete are fixed beforehand. The different steps of the method are the following:

- an arbitrary composition (w/c ratio) and a volume of binding paste per cubic meter of concrete are fixed;
- these parameters are kept constant, and the optimum sand/aggregates ratio is searched to obtain the best workability (here, the proportions between coarse and medium size aggregates, if any, are fixed taking regularity and economy into account);

- with the same grading, concretes are manufactured with various amounts of cement, and the water content is adjusted to obtain the critical workability;
- compressive tests are performed at 28 days, and the optimum concrete is determined by a linear interpolation between the mix design parameters.

In fact, the optimum value of the workability does not lead to the exact strength optimum, because the compressive strength is slightly improved when the concrete contains more coarse aggregates. However, the concretes obtained with this approach are:

- **denser** (the minimum water amount is used for the desired properties) ;
- more **regular** (variations of the workability are minimal, because the proportions of the various kinds of aggregates are near the optimal values) ;
- more **stable** (the risk of segregation is avoided because the exact amount of fine elements is available to conveniently fill the voids in the coarse element clusters).

This method needs much experimental work, compared with classical methods based on the concept of a reference curve (e.g. **Fuller**'s curve). However, it appeared to be useful for important job sites (when casting large volumes of concrete), and/or when high quality concrete is required. This is the case in most high strength concrete applications.

5 Application to high strength concrete mix design

5.1. Stages of mix design
Compared with normal strength concretes (NSC), the HSC mix-design is a more complex job, because of the greater number of parameters involved. Up to four additional components may be used with these materials, including superplasticizers, retarding agents, silica-fumes, and fly ashes (or slag). A direct optimization of all of them would lead to several hundreds of trial batches! The main idea of the method presented herein is to perform most tests on **model materials**: grout for rheological tests and mortar for mechanical tests.
The process which allows us to obtain an optimal HSC, based on a NSC composition, is presented in the following. The fluidity of the different binding pastes (or grouts) are measured with the **Marsh Cone** (which allows a flowing time measurement, see Fig. 4 and Appendix).
1st phase: proportioning a **control concrete** containing a large amount of superplasticizer together with the amount of cement

Figure 3: L.C.L. apparatus for the measurement of workability of concrete.

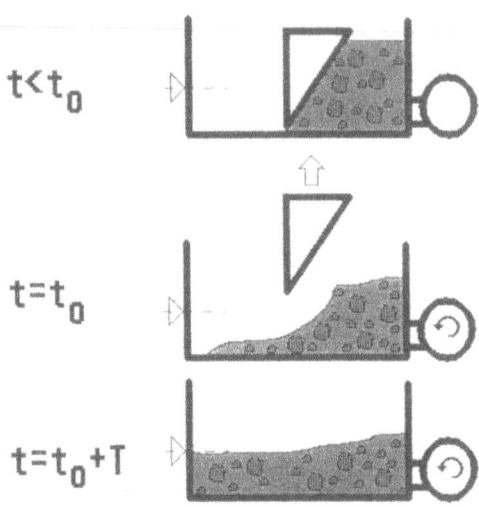

$t < t_0$

$t = t_0$

$t = t_0 + T$

Figure 4: The Marsh cone (all dimensions in mm). For this particular use, the classical test must be modified. Gravity is often stet enough to empty the cone as most concrete binding pastes are more sticky than usual grouts. The flowing time is considered as the time necessary to fill a 200 cm^3-bottle.

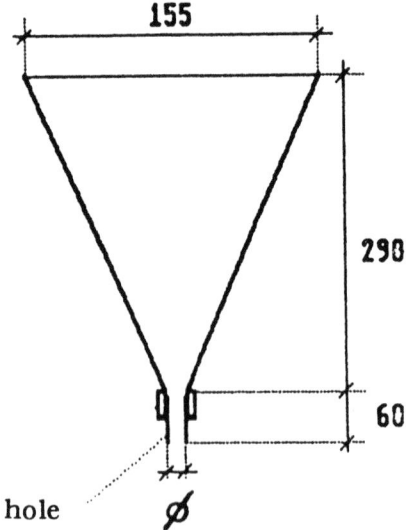

155

298

60

hole

∅

corresponding to the least water demand. For the usual aggregate gradings, this quantity of cement is about 425 kg/m^3. The water content of this control concrete must be adjusted to obtain the right workability, as measured with a dynamic apparatus, such as the LCL apparatus or the VB test apparatus. Previously, the granular skeleton, taken from a regional normal strength concrete composition, is supposed to have been well-proportioned (following the LCPC experimental mix design method, or at least in accordance with any graphic method involving a reference grading curve).

2nd phase: measuring the **flowing time** of the binding paste of this control concrete. The water/cement ratio of this binding paste must be computed taking the **moistening of aggregates** into account: when fresh concrete is flowing, a part of its water content is absorbed by the granular skeleton, and does not lubricate the mix (see Fig. 5). For non-porous good quality aggregates, this amount varies from 10 to 20 liters per cubic meter.

3rd phase: arbitrarily choosing the **percentages of binders** for several grouts (for example: 90% OPC/10% silica fume, or 75% OPC/20% fly ash/5% silica fume etc). Penttala [11] gave some interesting data about the choice of binding paste components. The author has also given advise on choosing the ultrafine particles to be used in combination with cement [4, 12].

4th phase: for each grout, adding a small amount of super plasticizer, adjusting the water content to obtain a sticky binding paste (although passing through the cone, e.g. with a flowing time aproximately equal to 20 s). With the water/cement ratio temporarily fixed, adding superplasticizer until the flowing time does not decrease anymore. This amount of super plasticizer represents the **saturation value** (see Fig.6), and remains fixed once and for all.

Figure 5: the various states of water in fresh concrete.

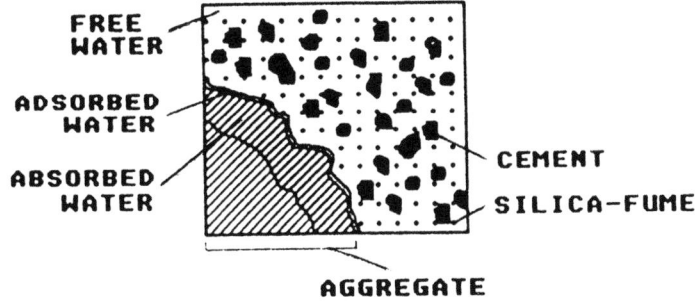

Figure 6: Determination of the saturation value of superplasticizer, for a certain grout. The figure shows the Marsh cone flowing time versus the amount of superplasticizer, for the binding paste formula #2. Each point represents an experimental value obtained for a grout whose w/c ratio is fixed.

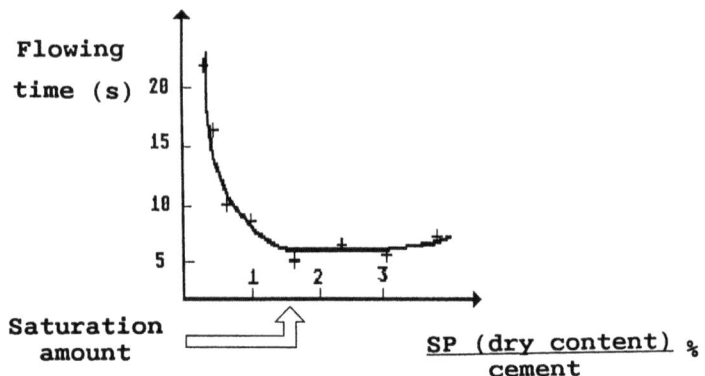

5th phase: adjusting the water content to obtain the same flowing time as the control grout. In consequence, the water/cement ratio is now fixed for each binding paste.

6th phase: measuring the evolution of the flowing time during the supposed practical use duration of HSC (this is to prevent **slump loss** hazards). If the flowing time increases too much, a retarding agent must be added to maintain it under the reference value. Hot weather concreting can be also simulated by heating the components before mixing and keeping the binding paste in a temperature-controlled container.

7th phase: predicting the strength of HSC made with the differents grouts. This prediction can be made with the generalized Feret's formula. A better precision will be achieved with compression tests performed on the different mortars. Each mortar must be proportioned with the same volume of binding paste, to exhibit the same workability.

8th phase: manufacturing the HSC, using the same granular skeleton and the same volume of binding paste as for the control concrete, but with the modified binding paste (plus the moistening water). Following the Farris model, the HSC and the control concrete will have the same workability. Subsequently, one **checks** that the properties (consistency and strength) are convenient. If the strength is too high or too low, one returns to the 7th phase and chooses another binding paste formula.

5.2 Some remarks

In this HSC mix design method, it is assumed, as deduced from the Farris model, that the optimal proportions between the skeleton components are the same with any binding paste. Fig. 7 shows the variation of the workability as a function of the sand/aggregate ratios, for NSC, HSC, and VHSC (Very High-Strength Concrete) made with the same aggregates. The corresponding optimal ratios lie very close, which proves the validity of keeping the same granular skeleton for the control concrete and for the HSC.

Figure 7: Flowing time obtained with the LCL workabilimeter versus the aggregate/sand ratio, for three concretes. A is the control concrete; B has the same composition but a lower water content due to the addition of a superplasticizer; C contains silica fume, more superplasticizer and less water [4] .

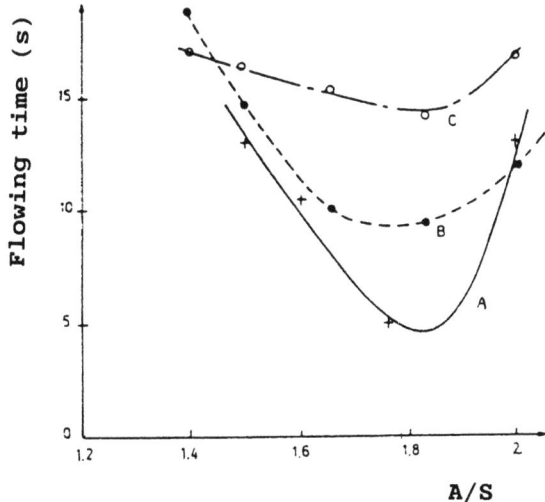

The way of proportioning the control concrete from which HSC is determined leads to a moderate volume proportion of binding paste, but to an unusually large amount of superplasticizer. The cheapest concrete achieving the desired performances (workability and strength) will not always be a concrete of this type. Depending on the comparative price of cementitious materials and organic admixtures, it will in some cases be more advantageous to use more binding paste - which means more binders - but less admixtures.

However, this approach ensures the following qualities (compared with other mix design methods):

- a **moderate** proportion of cement, which reduces the thermal cracking hazards (particularly for thick elements)
- a **low segregation**
- a **regular workability**: casual errors in weighing the aggregates or the admixture will have little effect on concrete consistency [3,4]
- a **high Young's modulus**
- a **low creeep** [13] and a **low total shrinkage**.

Considering the **slump loss** hazards, one may obtain a high strength concrete with a rapidly decreasing workability over time, although its binding paste has a quite constant flowing time during the concrete casting period. In such a case, it can be concluded that the slump loss is due to the **absorption of water by the aggregates**. An addition of a retarding agent will not cancel the phenomenon. Here, a preliminary wetting of the aggregates is probably the best solution.

6 Example

The aim is to make a very high strength concrete, with a fluid consistency, and a 90 MPa design strength at 28 days, from the following normal strength concrete composition (Table 1).

The control concrete has been obtained from the NSC, using 425 kg/m³ of cement (strength class: 55 MPa at 28 days) superplasticizer and an adjusted amount of water in order to get a 200 mm slump (Table 2).

The Marsh cone test, performed with the control binding paste, exhibits a flowing time of 5 seconds (Table 3).

Three trial binding pastes are prepared (Table 4), with an increasing percentage of silica fume (5, 10 and 15% respectively). The saturation proportions of superplasticizer are then determined, (see Fig. 6) and the water amounts are fixed to give the same flowing time as for the control binding paste. It is shown than the saturation amount of superplasticizer increases when more silica fume is added. This is due to the high specific area of the silica fume [4].

Table 1 : Composition of the Normal Strength Concrete (kg/m³).

Crushed limestone aggregates (sieve grading in mm)			River sand	OPC	Silica fume	Super plast.	Water
12.5/20	5/12.5	0/5	0/5				
826	398	315	315	410	-	-	181

Table 2 : Composition of the Control High Strength Concrete (kg/m^3).

Crushed limestone aggregates (sieve grading in mm)			River sand	OPC	Silica fume	Super plast.	Water
12.5/20	5/12.5	0/5	0/5			(powder)	
855	412	326	326	425	-	6.4	129

Table 3 : Composition of the Control Binding Paste (kg/m^3)

(total volume: 257.6 l) Moistening water: 10 l/m^3

OPC	Silica fume	Super plasticizer	water	Marsh cone flowing time
425	-	6.4	119	5 s

The compressive strengths are predicted using Féret's formula. A mean value of 100 MPa is assumed to correspond to the 90 MPa design strength; the formula # 2 is chosen at this step. This leads to the composition given in Table 5.

Once the first batch is made, the very high strength concrete appears to be a little sticky. Thus, a small amount of water is added to the original composition (Table 6).

The characteristics of the concrete obtained are the following:

Slump: 200 mm.
Mean compressive strength at 28 days
(160x320 mm cylinders):   101 MPa.

In this example, the desired characteristics have been reached after the second trial. In a more current field case, a certain number of batches (between one and ten) will be necessary to obtain the required concrete.

One may then ask about the strategy to follow if all the predicted strengths had been significantly lower than 100 MPa. In this case, it is no longer possible to keep the same volume of binding paste. The binding paste composition giving the highest strength must be chosen, and the water/cement ratio should be lowered sufficiently to ensure a suitable strength, as predicted by Féret's formula. When making the VHSC, the binding paste volume should be increased until a good workability is obtained for the resulting concrete. Sometimes, this will lead to excessive amounts of binders; the choice of components must then be re-examined.

Table 4 : Compositions and predicted corresponding strengths of the High Strength Binding Pastes

Total volume (l)	OPC (kg)	Silica fume (kg)	Super plast. (kg)	Water (l)	Marsh fl. time (s)	W/C	S/C	Compr.* strength (MPa)
# 1 257.6	444	5% 21.6	7.6	103	5	0.25	0.05	95
# 2 257.6	428	10% 42.3	8.5	97	5	0.25	0.10	102
# 3 257.6	411	15% 61.5	9.1	93	5	0.25	0.15	106

* *Theoretical compressive mean strength of corresponding concrete at 28 days according to Feret's formula with kg = 4.91 and Rc = 55 MPa.*

Table 5 : Theoretical composition of the Very High Strength Concrete (kg/m^3) from previous trials #2

Crushed limestone aggregates (sieve grading in mm)			River sand	OPC	Silica fume	Super plast.	Water
12.5/20	5/12.5	0/5	0/5				
855	412	326	326	428	42.3	8.5	108

Table 6 : Actual composition of the Very High Strength Concrete (kg/m^3).

Crushed limestone aggregates (sieve grading in mm)			River sand	OPC	Silica fume	Super plast.	Water
12.5/20	5/12.5	0/5	0/5				
854	411	326	326	421	42.1	7.59	112

7 Conclusion

Optimising the composition of HSC needs a rigorous approach, because of the increased number of parameters. A theoretical model and an empirical formula have been presented, from which a practical mix design method is proposed. It allows the composition of a concrete having a given strength and a given workability to be determined, with a minimal number of trial concrete batches. The tests used in this method are conventional, simple and cheap, and may be performed in any laboratory. The main assumption of the method concerning the use of the saturation amount of super pasticizer, leads to concretes having good secondary properties, even if they may be more expensive than others containing more binder and less admixture.

8 References

1. Utilization of high-strength concrete". Conference held in Stavanger (Norway). Tapir Ed. , Trondheim (1987).
2. Feret R. : "Compacité des mortiers hydrauliques" (Packing densities of mortars). Annales des Ponts et Chaussées, Paris (1892) (in French).
3. De Larrard F. , Acker P., Malier Y.: "Very High-Strength Concrete: from the laboratory to the construction site". in [1] (1987).
4. De Larrard F. : "Formulation et propriétés des bétons à très hautes performances" (Mix-design and properties of very high-strength concretes). PhD Thesis presented to the Ecole Nationale des Ponts et Chaussées. Also, Rapport de Recherche des LPC N° 149 (1987) (in French).
5. De Larrard F. : "Formulation et propriétés des bétons à très hautes performances; la méthode des coulis" (Mix-Design of High-Strength Concrete: the Grout Method). Bulletin de Liaison des laboratoires des Ponts et Chaussées, Nr. 161, pp. 75-83, May-June (1989) (in French).
6. Allou F., Charif H., Jaccoud J.P.: "Bétons à hautes performances". Chantiers Suisse, Vol. 19, N° 9, pp. 725-730 (1988) (in French).
7. FARRIS J.; "Prediction of the Viscosity of Multimodal Suspension from Unimodal Viscosity Data". Trans. Soc. Rheol., 12 (2), 281-301 (1968).
8. Petrie E.M. : "Effect of Surfactant on the Viscosity of Portland Cement-Water Dispersions". Ind. & Eng. Chem. Products Research and Development, 15, p. 242 (1976).
9. Baron J., Lesage R. ; "La composition du béton hydraulique, du laboratoire au chantier" (The composition of hydraulic concrete, from the laboratory to the construction site). Rapport de Recherche LPC N° 64 (1976) (in French).

10. Baron J., LesageE R. : "Proposition pour une définition de la maniabilité" (Proposal for a definition of the workability). RILEM Conference "Fresch concrete : important properties and measurements". Vol 1, Leeds Univ. (1973) (in French).
11. Pentalla V. : "Compatibility of Binder and Superplasticizer in High-Strength Concrete". Nordic Concrete Research, N° 5, 117-128 (1986).
12. De Larrard F. : "Ultrafine Particles for the Making of Very High-Strength Concrete". Cement and Concrete Research, Vol. 19, N° 2, pp. 161-172, 1989.
13. De Larrard F. : "Creep and Shrinkage of High-Strength Field Concretes". 2nd International Conference on the Utilization of High-Strength Concrete, Berkeley, ACI SP 121, May 1990.

Appendix. The modified Marsh cone test experimental procedure

The fluidity of a cement paste is assumed to be characterized by the time necessary to fill a flask placed below the cone.

The aim is to determine the proportion of each grout so that the flowing time be equal to that of the reference binding paste. As the latter may vary in a wide range from one concrete to another, the operator must choose the hole diameter, and the flask volume - for instance 8 mm and 200 cc. - to get a reference flowing time between 5 and 15 seconds. Then, the note and the flask content will remain the same during all the mix design process.

1) Mixing the binding paste (with a conventional two speed mortar mixer)

If the silica-fume is provided in bulk form, or if it is densified, add 50% of water by weight to the silica fume, and mix at high speed for 3 minutes to get a dispersing effect. One must then obtain a smooth slurry, without lumps.

Add the rest of the water, one third of the total amount of super-plasticizer, and the total amount of retarder (if any), and mix at low speed for 3 minutes.

Add the cement and the other binders (if any) and mix for 1 minute at low speed, then 1 minute at high speed.

Add the rest of the super plasticizer amount and mix for 2 minutes at high speed.

2) Testing the binding paste
As soon as the mixing is finished, measure the flowing time of the binding paste. When it is desired to investigate the evolution of the flowing time vs. time (6th phase), remix the paste at high speed for 1 minute (because that kind of suspension generally exhibits thixotropic behavior).

5 HIGH PERFORMANCE CONCRETE SUPPLIED BY A NETWORK OF READY-MIX CONCRETE PLANTS

C. LEVY, Lafarge Coppée Recherche
J-P. LE BOULICAUT, Bétons Granulats Lafarge, France

1 Introduction

Following the example of countries such as the Unites States, Canada and Norway, high performance concrete (HPC), is now developing in France. The first pilot works have shown and demonstrated the numerous technical and economic qualities of this new material. In proving the feasibility of HPC projects, the first stage of which consisted of moving from the laboratory to the construction site, has been passed. The second stage is to leave the experimental and individual stage of HPC construction to reach a national and industrial dimension. In order to do this, clients and contractors must, among other things, be able to count on a reliable and available partner for the manufacture and delivery of high performance concrete.

The purpose of this paper is to show that a ready-mix concrete company Lafarge Béton Granulats, is today in a position to meet these needs.

First of all, we shall present the experience of the supply of B 60 concrete for the construction of the Joigny bridge by one of its subsidiaries, Béton Chantiers Bourgogne. We shall then deal with the development of HPC at national level within the Lafarge Coppée group.

Lastly, we shall give a few examples of the development of regional HPC formulations, illustrating the possibility of supplying high performance concrete ready for use, virtually anywhere on French territory.

2 Manufacture of HPC for the Joigny Bridge

2.1 Conditions of the project

2.1.1 Objectives of the concrete producer

High Performance Concrete: From material to structure. Edited by Yves Malier. © 1992 Taylor & Francis.
Published by Taylor & Francis, 2 Park Square, Milton Park, Abingdon, Oxon, OX14 4RN. ISBN 0 419 17600 4.

At the end of 1988, more than 1000 m^3 of HPC B 60 had been cast to make up the deck of the bridge crossing the River Yonne at Joigny (89- Yonne)).

The main objectives to be followed by the concrete producer were the following:

1) Use of the network's ready-mix concrete plants, near the site, in order to establish the reproducibility of this experiment:

- equipment of plants: standard,
- manufacturing procedure: current,
- human and physical resources: traditional in ready-mix concrete.
2) Use of local materials. No silica fume.
3) Casting of the deck in a single phase. Regularity of the production of B 60 over more than 1 000 m^3.
4) Use of means of transport between the manufacturing sites and the utilization location: with traditional ready-mix mixer trucks delivering at a regular rate, to avoid any interruption of the casting, concrete with good working properties, with proper maintenance of handling over time.

2.1.2 Béton Chantiers Bourgogne

The ready mix concrete company chosen for the supply of the Joigny bridge concrete was Béton Chantiers Bourgogne, a subsidiary of the Compagnie des Sablières de la Seine, which is part of Lafarge Bétons Granulats.

The B.C.B. company comprises:

5 plants in the department of Yonne delivering,
70,000 m^3 of concrete per year operating,
17 mixer trucks 6 m^3 capacity,
30 personnel, and
1 monitoring laboratory.

The construction site was supplied from two ready-mix plants, located respectively 15 km (Passy) and 26 km (Gurgy) from the structure. This is quite a normal configuration in France, since, taking into account the whole of the national network of ready-mix companies, no site is more than 30 km distance from a concrete plant. The two concrete plants are of different design:

* Passy plant: weighing of aggregates under a hopper with a weighing belt; mixer of 1.5 m^3 electro-mechanical type BETP automation; volumetric measurement of additives.
* Gurgy plant: tower plant; mixer of 2 m^3; Betomatic III automation; weight measurement of additives.

2.2 Methods

2.2.1 Mix design of the B 60 concrete
The main requirement for the development of this HPC was the use of local materials. It was worthwhile proving that it was possible to manufacture a B 60 without silica fume at a given site.

As soon as the contract was awarded, a true partnership was setup between the contractor, the client and concrete supplier. A working group (D.D.E. of Yonne - Bouygues - Lafarge) met regularly in order to develop the final mix proposed for suitability tests.

Selection thus related to:

- cement CPA HP from Cormeilles (Ciments Lafarge) at
  450 kg/m³,
- correction natural sand from Etampes (C.S.S.),
- natural sand 0/4 from Passy (C.S.S.),
- crushed gravel 6/20 from Balloy (C.S.S.),
- a g/s ratio of 1.3 - 1.4,
- a w/c ratio of approximately 0.37,
- Melment 40 superplasticizer,
- Melretard retarder.

The introduction of a quarter of the superplasticizer and all of the retarder took place with the mixing water, the remaining three quarters of the superplasticizer being introduced after 30 seconds of prior mixing. The total mixing time was 90 seconds.

This mix resulted in an uniform concrete, of good workability (slump of approximately 20 cm for one hour) and pumpable (over 120 m horizontally).

It should be noted that these tests for the development of the composition of B 60 were carried out in the plant, and thus under construction site conditions.

2.2.2 Manufacture of B 60 concrete
The very early use (when the formulation was being settled), of the construction site conditions with simulation of transport in a transit mixer, enabled the whole of the manufacturing process to be well tried and to be optimized with respect to:

- the volumes and rates,
- the quality and regularity.

The conditions for introducing additives, the conditions of their handling on the plant and the mixing time led us to propose and choose the stand-by plant planned as a second operational plant for implementation of the pouring of B 60. The objective of this was:

- firstly, having concreting rates compatible with the requirements of the construction site,
- secondly, ensuring concreting continuity in the event of a major mechanical incident.

This decision also had consequences as part of one of the objectives of the experimental site: to make the use of ready-mix plants more credible for the manufacturer of HPC since, as we have previously seen, the two plants are of different design and technical characteristics.

The two plants were naturally supplied with the same components necessary for the manufacture of B 60; that is the same formula but different processes.

The setting up of the Quality Assurance Plan made allowance for the characteristics of each of the two plants.

The choice of the date of pouring had been set at 20th of december 1988, with all the procedures relative to the Quality Assurance Plan having been set up before 4 p.m. on 19th of december 1988.

Background to concreting:

- 1 m³ of 'slop' (that is concrete not subsequently used) was manufactured at Passy from 1 a.m.
- the Gurgy plant started up at 1 a.m. on 20th of december 1988. The first mixer truck going to the site was released at 1:10 a.m.
- the Passy concrete was to start up at 2 a.m. On the third loading the cable of the level probe in the hopper holding the stock of buffer aggregates broke ans was repaired by intervention of the maintenance structure planned in the Quality Assurance Plan; start up again at 3 a.m.
- Gurgy produced 554 m³ of B 60 from 1 a.m. to 11:13 p.m. on 20 december, i.e. 25 m³/hour on average.
- Passy produced 517 m³ of B 60 from 3 a.m. on the morning of 20 december to 0:20 a.m. on 21 december, i.e. 24 m³/hour on average.
- production peaks were 30 m³/h at each of the plants.

2.2.3 Manufacturing controls
These were carried out at several stages, in accordance with the Quality Assurance Plan:

- manufacture, supply and reception of materials,
- conformity of equipment,
- conformity of the concrete manufacturing process,
- conformity of the quality of the fresh concrete.

The test conditions on this type of concrete are essential for acquiring the knowledge that we wish to have of the quality of HPC structures: this requires the necessity for adapting some operating modes. The checks carried out by BCB during studies and supply to the site showed a certain incidence of the sample taking conditions and control tests. Attention was particularly paid to the production of test pieces, placing these by vibration (although the slump was greater than 10 cm) and on the surface quality of the test pieces or their grinding, etc.

2.3 Results obtained

- A very few cubic metres of concrete were refused for reasons of incorrect workability (less than the required value).
- Production variations of each of the weighings (water, cement, aggregates, additives) remained appreciably lower than the standardized limits.
- At 28 days, the compression strength on 16 x 32 cm test cylinders were the following :

Mean value	= 78 MPa
Variation coefficient	= 8.3% (on 2 plants)
Max value	= 94 MPa
Min value	= 65 MPa
Density	= 2.42 kg/m^3

In conclusion, this site, on which high performance concrete was supplied by a ready-mix company for the first time in France, was crowned with success. The reasons for this were the utilization of:

- a study program for the concrete to adapt the available resources to the objectives sought,
- mastery of the equipment,
- stringent manufacturing and checking procedures,
- awareness and responsible behavior by the concrete producer.

3 The development of HPC in France with Lafarge Béton granulats

3.1 Meeting a need
This new material, high performance concrete, is undergoing full development. In France, laboratories are increasingly mastering its characteristics and its behavior: contractors and design offices have codes available (BAEL 90, BPEL 90) that permit them to design structures using B 60 concrete; clients and architects are aware of the economic and an esthetic advantages of the use of the material; companies are increasingly proposing HPC variants.

Particularly in the context of small or medium sized construction sites, a deep need has occurred in the field of building and public works for the production of high performance concrete. This is why the ready-mix concrete profession had to effectively meet this demand.

Based on the experience of the Joigny bridge, the Lafarge Béton Granulats Company obtained the resources in men, equipment and technical means, for proposing, at any time and over a large part of the national territory, high performance concrete.

3.2 The Lafarge Performance Concrete Network

A working group was set up in order to start, expand and manage the HPC development process within Lafarge Béton Granulats. This was undertaken involving all the resources of the Lafarge Coppée Group:

- Lafarge Béton Granulat,
- Ciments Lafarge,
- Lafarge Nouveaux Matériaux (Chryso admixtures),
- Lafarge Coppée Recherhce (Research Center).

Mastery of all the technical knowledge of each of the components necessary for making HPC (aggregates, cement, admixtures), a national manufacturing and distribution network of ready-mix high quality concrete, as well as experience with HPC by the Laboratoire Central de Recherche (France, North America) are the main assets of this assignment to develop HPC.

Today, the Lafarge Bétons Granulats has set up a "Lafarge Performance Concrete Network", capable of supplying B 60 concretes or even B 80 concretes in the major part of France. At the end on 1989, some 60 concrete plants belonged to this network, which continues to expand. Most of the formulations proposed are tested and checked in the Regional Ponts et Chaussées laboratories and other recognized laboratories. Certain subsidiaries, such as the Béton de Paris company, are even in a position to supply B 100 concrete.

HPC supplied by a ready-mix network has thus become a reality in France.

4 Examples of regional HPC formulae

In order to illustrate the regional feasibility of HPC, we shall present three studies carried out by Lafarge Coppée Recherche, in the Lyon, Nice and Toulouse regions. In each of the cases, the objective was to make a high performance concrete with local materials. The laboratory study was followed by tests at the ready-mix plant. The specifications are as a general rule, the following:

- characteristic strength at 28 days of 60 MPa, which leads to obtaining at least 70 - 75 MPa at 28 days on cylinders in the laboratory,
- high fluidity of the concrete,
- good maintenance of workability over time,
- permitted admixtures with the NF mark,
- no use of silica fume.

4.1 HPC Lyon

4.1.1 Materials

Cement
Preliminary tests have enabled us to determine, at regional level, the best mix of cement and superplasticizer. Certain cements exhibited more or less speedy setting and thus did not offer the necessary maintenance of workability.

We therefore chose CPA HPR cement from Val d'Azergues (Ciments Lafarge) which among other things, has the following advantages:

- HPR cements are the top of the range, and so are well suited for HPC,
- the cement is on the COPLA list of materials suitable for pre-stressing (worthwhile for the development of HPC in structures),
- its content of C_3A is extremely low (too high a content would have a tendency to increase the water demand and disturb the maintenance over time of the concrete's handling qualities).

Aggregates
The aggregates were chosen in between the Lyon area quarries. Frequently used and having a good quality, we took the aggregates from Millery: sand 0/5 mm, per gravel 5/12 and coarse aggregate 12.5/20.

These silico calcareous aggregates are hardly porous, rather strong and having a good cubic form, so they are well-adapted for a HPC formulation.

Admixtures
With an equivalent dry extract content, the superplasticizer giving us the best performance (largest reduction in water, very high concrete fluidity without segregation) is Durciplast (Chryso), which is based on modified melamine formaldehyde sufonate.

In order to retain very good handling qualities of the concrete for 1 hour, a retarder is used, Chrytard (Chryso). For more effectiveness, one-third of the Superplasticizer and the retarder are introduced into the mixing water; the remaining two-thirds of the superplasticizer are only poured into the mixer after preliminary mixing of the concrete.

4.1.2 Formulation of the concrete

The cement content was fixed at 425 kg/m³ (normal content value of CPA HPR for an HPC). The quantity of superplasticizer was determined so as to obtain a sufficient water reduction, i.e. a W/C ratio = 0.35 (W total/C = 0.365).

Initially, the composition of the concrete was determined by the Faury method. However, optimisation of the fine gravel/coarse gravel ratios and fine + coarse gravels/sand was carried out with the CEBTP.

Thus, we have adopted the following formulation:

coarse gravel 12.5/20	852	kg
fine gravel 5/12.5	267	kg
sand 0/5	765	kg
CPA HPR Val d'Azergues	425	kg
water	150	liters
Durciplast (1.5% of cement weight)	6.37	kg(1/3+2/3)
Chrytard (0.4% of cement weight)	1.7	kg

This composition applies to dry aggregates and leads to 1 m³ of concrete in position. Slump in an Abrams come was 25 cm.

4.1.3 Control of the concrete characteristics

Mixing

The concrete were made up in an Eirich brand vertical axis mixer of , in batches of 40 liters (at 20°C). The mixing cycle was the following:

- dry mixing of constituents for 1 minute,
- introduction of mixing water with the retarder and 1/3 superplasticizer,
- mixing for 2 minutes,
- introduction of the remaining 2/3 of superplasticizer,
- mixing for 30 seconds.

Settling of the concrete and maintenance of workability

We monitored for 1 hour the trend in the slump of the concrete in the Abrams cone (concrete at rest and re-homogenization every 15 minutes before testing):

Time	0	15 min	30 min	45 min	1 h
Slump	25 cm	25 cm	25 cm	20 cm	18 cm

In addition, we followed the development of the mixed concrete for 30 minutes, in an inclined axis concrete mixer revolving constantly at three revolutions/min (which is a very severe simulation of the conservation of concrete in a mixer truck). The slump was still 25 cm after a 15 minutes ; it changed to 20 cm at 30 minutes.

All these results are quite satisfactory and indicate very good rheological behavior over time, necessary for pouring the concrete on the site.

Measurement of entrained air
Measured on fresh concrete with an aerometer, the quantity of entrained air amounts to 1 %.

Mechanical strength
The mechanical strength was checked on cylinders (diam = 16 cm, h = 32 cm) in compression at periods of 1,7, 28 and 90 days and in tension by splitting at 28 days and 100 % relative humidity. The strengths obtained were the following:

Time	1 day	7 days	28 days	90 days
Compressive strength MPa)	17.8	60.6	74.0	82.5
Splitting tensile strength (MPa)			5.3	

These laboratory results completely satisfy the required demands both from the viewpoint of handling and that of strength and, with this formulation, permit the production of a B 60 concrete at 28 days in a concrete plant.

4.2 HPC Nice

4.2.1 Materials

Cement
For producing a high performance concrete, in the Nice region, we used a "top of the range" cement manufactured locally, CPA HPR cement from Contes (Ciments Lafarge).

The very good development of these compressive strengths enables us to envisage concretes whose mechanical characteristics when young would be high, and which maintain a satisfactory trend of strength over time.

Aggregates
The aggregates have been selected from among the best products of quarries and ballast centers in the region.

We selected the crushed limestone materials from Spada, to which we have added, for a supplementary concrete study, a rolled crushed sand from Var.

Crushed limestone sand 0.1/2.5 mm Spada: its granular distribution extends from 0 to 4 mm, with 11 % passing through an 80 μm screen. The fineness modulus is equal to 2.33. Its sand equivalent is correct (ESV = 88) and there is no organic matter.

Sand from Var, rolled and crushed 0/5 mm from the St-Isidore ballast pit: this sand is of silica origin and has a granular distribution between 0 and 5 mm with 2 % passing through the 80 μm screen. As for the preceding sand, satisfactory characteristics are noted:

. fineness modulus   = 2.39
. correct sand equivalent : ESV = 91
. no organic matter.

Crusched limestone fine aggregates 6/14 mm Spada: its granular metric curve extends from 4 to 16 mm with 3 % passing the 6.3 mm screen and 6 % retained at the 12.5 mm screen. These aggregates are hard and not very porous.

Crushed limestone coarse aggregates 10/20 mm Spada: its granular distribution extends from 0 to 4 mm, with 11 % passing through an 80 μm screen. The fineness modulus is equal to 2.33. Its sand equivalent is correct (ESV = 88) and there is no organic matter.

Sand from Var, rolled and crushed 0/5 mm from the St-Isidore ballast pit: this sand is of silica origin and has a granular distribution between 0 and 5 mm with 2 % passing through the 80 μm screen. As for the preceding sand, satisfactory characteristics are noted:

- fineness modulus   = 2.39
- correct sand equivalent : ESV = 91
- no organic matter.

Crushed limestone fine aggregates 6/14 mm Spada: its granular metric curve extends from 4 to 16 mm with 3 % passing the 6.3 mm screen and 6 % retained at the 12.5 mm screen. These granulates are hard and not very porous.

Crushed limeston coarse aggregates 10/20 mm Spada: its granular distribution lies between 6.3 mm and 25 mm with 2 % passing the 10 mm screen and 2 % residue at the 12.5 mm screen. Same geological origin as the fine aggregates.

Admixtures

In order to effectively reduce the water/cement ratio of the concretes to be made, we used the superplasticizer Durciplast (Chryso Company). After a few preliminary tests, we decided to add this additive at 2 % of the weight of the cement. This proportion is appreciably greater than that of the Lyon HPC. This is explained by different behavior of the product depending on the cements used and also by the presence of much more considerable sand fines content in the case of the Nice HPC.

In order to obtain good workability retention, we used retarder Chrytard (Chryso Company).

4.2.2 Concrete mix design

In all possible cases, we use the Faury method for the theoretical composition of concrete, aiming at a high percentage of quartz aggregates.

As in the case of Lyon, we laid down a cement content of 425 kg/m³. The first formulation, based solely on the use of Spada limestone aggregates, was as follows (1 m³ of concrete on dry aggregates):

- Limestone Spada 10-20 mm	763 kg
- Limestone Spada 6-14 mm	465 kg
- Limestone sand Spada 0.1-2.5 mm	709 kg
- CPA HPR Contes	425 kg
- Water	165 liters
- Durciplast (2 % of cement weight)	8.5 kg (1/2 + 1/2)
- Chrytard (0.4 % of cement weight)	1.7 kg

The superplasticizer percentage is thus greater than the Lyon HPC case, and even so, the quantity of water added is nevertheless greater (due to absorption of water by the limestone fines). In addition, the following table shows that we had to introduce half of the superplasticizer in the mixing water (although a smaller proportion would have been more effective) :

Quantity of Durciplast		Water Quantity	Slump	
in mixing water	in mixer (30 s before end of cycle)		at time 0 h	at time 1 h Extractor simulation
1 %	1 %	160 l	5 cm	5 cm
1 %	1 %	165 l	15 cm	13 cm
0.66 %	1.34 %	165 l	Concrete too dry. Does not fluidify.	

We then carried out tests that replaced the sand by crushed rolled sand from Var, which contained a greater proportion of fines. The formulation then became (per m³ on dry aggregates):

- Limestone Spada 10-20 mm   698 kg
- Limestone Spada 6-14 mm   465 kg
- Sand from Var   738 kg
- CPA HPR Contes   425 kg
- Water   160 liters
- Durciplast (2 %)   8.5 kg (1/2 + 1/2)
- Chrytard (0.4 %)   1.7 kg

The quantity of water and the method of introducing the superplasticizer were determined by tests, the results of which are shown in this table:

Quantity of Durciplast			Slump	
in mixing water	in mixer (30 s before end of cycle)	Water Quantity	at time 0 h	at time 1 h Extractor simulation
1 %	1 %	150 l	2 cm	1 cm
1 %	1 %	155 l	7 cm	7 cm
1 %	1 %	160 l	14 cm	11 cm
1 %	1 %	165 l	18 cm	18 cm
0.66 %	1.34 %	160 l	Concrete too dry. Does not fluidify.	

We observe that:

- the replacement of the Spada limestone sand which is rich in fines, by a silica sand from Var which is low in fines, leads to a reduction of 5 liters of mixing water to obtain an identical slump,
- the 1/3 + 2/3 mixture of Durciplast is still not suitable.

4.2.3 Checking the characteristics of the concrete

Mixing
  - Batch of 40 liters at 20° C
  - Cycle : . dry mixing   = 1 minute
             . water + retarder + half superplasticizer
             . mixing   = 2 minutes

. introducing of other half of superplasticizer
. mixing = 30 sec.

Settling of concrete - Maintenance of workability

We checked the development of the settling of the concrete over time in an Abrams cone, with the concrete being conserved in a "simulation mixer" i.e. in a concrete mixer with an inclined axis turning at 3 revolutions/min. The results of the slumps (in cm) are the following:

Time (min)	0	15	30	45	60
Concrete with Spada limestone sand	15	20	20	18	13
Concrete with Var silica sand	14	19	19	17	11

The two concretes have a satisfactory rheological behavior. An increase in the slump is noted in the first half-hour (these limestones absorb a lot of water, which they then release over time, especially when dry materials are used).

Air content

Measurements of entrained air are performed at 20° C, on fresh concrete using an aerometer. The following is obtained:

Concrete with Spada limestone sand : 1.4 %
Concrete with Var silica sand : 1.7 %

Mechanical strengths

Conditions were identical to the Lyon HPC (see section 2.3.4),and used 3 test cylinders 16 x 32 cm per test - standardized cuting conditions. The results appear in the following table (MPa):

		Times				
Concrete type	Test	1d	2d	7d	28d	90d
Concrete with Spada limestone sand	Compression	36.2	51.8	68.3	75.9	81.5
	Splitting				4.5	
Concrete with Var silica sand	Compression	37.1	51.3	69.5	76.1	87.3
	Splitting				4.3	

The compression results obtained at 28 days are very good and compatible with the requirements of a B 60 concrete.

All of these tests thus prove the feasibility of a B 60 concrete in the Nice region, using solely local materials.

4.3 HPC Toulouse

4.3.1 Materials

Cement
In the Toulouse region, we have the choice between two cements of grades meeting the requirements of HPC :

- CAP HP from Boussens (Ciments Lafarge)
- CAP HP from Lexos (Ciments Lafarge).

We chose the first, and then performed supplementary tests with the second (its content of tricalcium aluminate is slightly lower than that of HP Boussens).

Aggregates
The choice of aggregates is a major problem in the Toulouse region for the development of an HPC. As we will see later, we have undertaken various supplementary studies with variations in the origins and qualities of aggregates. Initially, we chose rolled silica limestone aggregates from the Sablières de Garonne.

La Garonne sand rolled 0-2 mm: its granular metric distribution lies between 0 and 4 mm with 3 % passing at 80 µm and 12 % residue with the 2 mm screen. Its fineness modulus of 2.38 is satisfactory. Its sand equivalent is good (ESV = 89.6) and we did not discover any organic matter.

La Garonne natural coarse sand 2.5 - 6.3 mm: its granulometric curve lies between 1.25 mm and 10 mm with 4 % passing the 2.5 mm screen and 12 % residue at the 6.3 mm screen.

La Garonne natural fine gravel 6 - 10 mm: its granulometric distribution lies between 2.5 and 12.5 mm with 12 % passing the 6.3 mm screen and 3 % residue at the 10 mm screen.

La Garonne natural coarse gravel 10 - 20 mm: its granulometric curve lies between 4 mm and 25 mm with 18 % passing the 10 mm screen and 6 % residue at the 20 mm screen.

All these aggregates, of the same origin, are clean. Their specific weight lies between 2.66 and 2.71.

Admixtures
As in the previous cases of the Nice and Lyon HPC, we have used the following additives: superplasticizer Durciplast and retarder Chrytard (Chryso Company).

4.3.2 Concrete formulation

The composition of the concrete has been based on the Faury method, starting from a cement content of 425 kg/m³ and 2 % Durciplast (in view of previous experiments). We thus obtain the following formulation: (by m³ of dry aggregates):

- La Garonne gravel 10/20	875 kg
- La Garonne gravel 6/20	137 kg
- La Garonne sand 2.5/6.3	299 kg
- La Garonne sand 0/2	608 kg
- CPA HP Boussens	425 kg
- Water	150 liters
- Durciplast (2 % of cement weight))	8.5 kg (1/2 + 1/2)
- Chrytard (0.4 % of cement weight))	1.7 kg

The quantity of mixing water as well as the method of introducing the superplasticizer have been designed so that the concrete should be fluid after manufacture and so that its slump should remain high for an hour.

The following table shows that 150 liters of water are necessary to obtain the required workability (W/C = 0.35). It will be observed that the best distribution of the superplasticizer (introduction in the mixing water + introduction of the remainder in the mixer) turns out to be 1/2 + 1/2.

In the case of 2/3 + 1/3, there is a loss of handling capacity over time and in the case of 1/3 + 2/3, the concrete is too dry and does not fluidify well.

Quantity of Durciplast			Slump	
in mixing water	in mixer (30 s before end of cycle)	Water Quantity	at time 0 h	at time 1 h Extractor simulation
1 %	1 %	145 l	20 cm	3 cm
1 %	1 %	150 l	25 cm	17 cm
1.34 %	0.66 %	150 l	25 cm	11 cm
0.66 %	1.34 %	150 l	Concrete too dry. Does not fluidify.	

4.3.3 Checking the characteristics of the concrete

Mixing
Identical to section 4.2.3. Nice HPC.

Concrete slump - Maintenance of workability
We checked the development over time of the concrete slump with the Abrams cone, with the concrete being conserved for an hour in an inclined axis concrete mixer rotating at 3 revolutions/mn. This permits simulation of conservation of the concrete in a mixer truck.

Time (min)	0	15	30	45	60
Slump	25 cm	20 cm	20 cm	17 cm	17 cm

The rheological behavior of this concrete is satisfactory and corresponds properly to the requirements of an HPC site.

Air content
Measurement of entrained air on fresh concrete at 20° C with aerometer = 1 %.

Mechanical strength
Conditions were identical to the two preceding HPC studies (see section 4.1.3.), with 3 test cylinders 16 x 32 cm per test - standardized curing cylinders.

Times

Time	1d	2d	7d	28d	90d
Compressive strength (MPa)	29.3	43.7	53.9	59.6	69
Splitting strength (MPa)				4.5	

Compression strengths are inadequate (it should be close to 60 MPa at 7 days at 70 MPa at 28 days). In spite of the relatively high cement and superplasticizer contents, this concrete does not exhibit the characteristics of the B 60 concrete.

4.3.4 Supplementary studies
In view of the inadequate results obtained with the basic formulation, we modified several parameters:

- Superplasticizer Content

- Sand fines
- Gravel washing
- Maximum aggregate size
- Change of cement
- Crushing of the gravel
- Utilization of a filler
- Mineralogical nature of the gravel.

Superplasticizer
We performed tests for increasing the content of Durciplast to 3 % (1.5 % in the mixing water + 1.5 % in the mix), the quantity of water was reduced by another 10 l (W/C = 0.33). The concrete exhibited a segregation phenomenon quite clearly, which explains a reduction in the strength obtained at 7 days and 28 days:

Compression strength (tp. 16 x 32, in MPa)	7 days	28 days
Control cement with 2 % Durciplast	53.9	59.6
3 % Durciplast	47.8	53.2

Test with fillerised sand
In view of the low fines content of the rolled sand, used in the main study, we added (to the extent of 50 % of the weight of the total sand) a fillerised crushed sand of the same mineralogical origin. This was 0/2.5 mm with 18 % passing the 80 µm screen and 3 % residue at the 2.5 mm screen.

Sand equivalent ESV   = 72.6
No organic matter
Fineness modulus   = 2.13

The formulation thus became (per 1 m³ of dry aggregates):

- Rolled La Garonne gravel 10/20   875 kg
- Rolled La Garonne gravel 6/20   137 kg
- Rolled La Garonne chippings 2.5/6.3   299 kg
- Rolled La Garonne sand 0/2   304 kg
- Fillerised crushed sand 0/2.5   304 kg
- CPA HP Boussens   425 kg
- Water   150 liters
- Durciplast (2.5 % of cement weight)   10.6 kg (1/2 + 1/2)
- Chrytard (0.4 % of cement weight)   1.7 kg

This is because 2.5 % of Durciplast was necessary to retain the same quantity of water (150 l) as previously :

Quantity of Durciplast			Slump	
in mixing water	in mixer (30 s before end of cycle)	Water Quantity	at time 0 h	at time 1 h Truck-mixer simulation
1 %	1 %	150 l	10 cm	0 cm
1.25 %	1.25 %	150 l	23 cm	15 cm
1.5 %	1.5 %	150 l	25 cm	20 cm

The results obtained with this concrete are as follows:

Time (min)	0	15	30	45	60
Slump (cm)	23	19	18	16	15

Measurement of entrained air = 0.9 %

Age (days)	1	2	7	28	90
Compressive strengths (MPa)	27	39.2	49.1	56.8	65

Splitting strength				4.1	

These results are thus a little lower than those of the main formulation without fillerised sand. This addition thus brings no improvement.

Maximum aggregate size
Certain studies on HPC and VHPC propose limiting the value of the maximum diameter D of the coarsest aggregate. With this in mind, we mixed concretes starting from the basic formulation but eliminating the 10/20 mm gravel. The following is the chosen composition (calculated by the Faury method):

- Rolled La Garonne gravel 6/10   687 kg
- Rolled La Garonne coarse sand 2.5/6.3   528 kg
- Rolled La Garonne sand 0/2   700 kg
- CPA HP Boussens   425 kg

- Water   150 liters
- Durciplast (2 % of cement weight)   8.5 kg (1/2 + 1/2)
- Chrytard (0.4 % of cement weight)   1.7 kg

The results then progressed a little compared with the control concrete (main formulation) :

Compression strength (MPa)	7 d	28 d
Control concrete	53.9	59.6
Concrete 0/10	55.1	63.0

But this improvement is not enough to obtain a B 60 concrete.

Gravel washing
Hoping to increase adherence at the level of the paste - aggregate interface, we washed the 6/10 and 10/20 gravel. However, no strength gain was observed:

Compression strength (MPa)	7 d	28 d
Control concrete	53.9	59.6
Concrete with washed 6/10 and 10/20 gravel	52.4	57.0

New series of tests
We performed a new series of tests starting on a new basis:

a) Replacement of the CPA HP cement from Boussens by CPA HPR from Lexos (Ciments Lafarge). This cement is also manufactured in the region of Toulouse and has a more reliable content of C_3A, which should admit more efficient action of the superplasticizers.

This sampling of HPR Lexos cement exhibited mechanical strengths close to that of the Boussens HP previously used:

Compression strength on mortar (MPa)	7 d	28 d
HP Boussens	50.9	65.8
HPR Lexos	52.9	64.1

b) The use of only one gravel, 5/16, of the same origin (Roques-Garonne), but semi-crushed so as to attempt to grind the aggregates consisting of arenaceous granite that enters into the composition of the Garonne gravel and which are certainly a source of the weakening of the concrete.

c) After several tests on pouring, we observed that the Durciplast superplasticizer is highly suitable for the new cement and that higher contents could be envisaged.

d) We thus retained the rolled sand 0/2.5 ; the gravel 2.5/6.3 and the Chrytard as retarder.

The formulation of the concrete was then the following (Dreux method): per m³ of dry granulates:

- La Garonne sand 0/2.5   630   kg
- La Garonne coarse sand 2.5/6.3   217   kg
- La Garonne semi-crushed gravel 5/16   955   kg
- Fillerised crushed sand 0/2.5   304   kg
- HPR Lexos   425   kg
- Water   160   liters
- Durciplast (3 % of cement weight)   12.75   kg (1/2 + 1/2)
- Chrytard (0.4 % of cement weight)   1.7   kg

The following are the characteristics:

Time (min)	0	15	30	45	60
Slump (cm)	23	23	22	21	20

Time (days)	7	28
Compressive strength (16x32 - MPa)	56.0	63.9

In spite of a larger quantity of water than in the initial concrete, the strengths are a little higher, but still somewhat low.

e) We then made concretes that had higher contents of fines (to compensate for the low sand bottom of the curve) :

- Cement overdose at 450 kg/m³

and/or

- Use of limestone filler from Baixas at 40 kg/m³.

Demand, without bringing any gain in strength :

(3 % Durciplast, 0.4 % Chrytard)

HPR Lexos (kg)	Baixas Filler (kg)	Water (l)	Slump (cm) at time (min)					Strengths (MPa)	
			0	15	30	45	60	7 d	28 d
425	40	160	23	23	23	22	21	55.6	63.4
450	0	170	25	23	23	22	22	52.2	59.9
450	40	170	23	23	23	23	23	55.7	63.0

f) As a last resort, we were led to change the gravel. This was because all the tests in this supplementary study led us to explain the rather tricky performance of our concretes by the poor quality of the aggregates and in particular the gravel (crushing did not reduce the quantity of arenaceous granite enough, and in addition, increased the gravel's shape coefficient).

Consequently, we made the same concrete as in d), but replacing the 5/16 Garonne semi-crushed by a 6/14 crushed limestone from Bruniquel (Montauban). This quarry is not far from Toulouse and offers harder aggregates that are better adapted to our needs. By retaining the same other parameters (3 % Durciplast, 0.4 % Chrytard, 160 liters of water, etc), we obtained the following results:

Time (min)	0	15	30	45	60
Slump (cm)	21	21	21	21	20
Time (days)		7		28	
Compressive strength (MPa)		57.7		67.2	

There is thus an appreciable gain in strength thanks to the use of this 6/14 limestone.

In conclusion, it was much more difficult for us to develop a high performance concrete in the Toulouse region than for Lyon or Nice. The main reason was certainly the unsuitable quality of La Garonne materials. We were able to obtain concrete exhibiting worthwhile strengths using a crushed limestone from the Montauban region.

5 Conclusion

We have attempted to demonstrate, in this article, the feasibility of high quality and high performance concrete on the construction site, manufactured and delivered by ready-mix plants.

In addition to the example of 1 000 m³ of B 60 concretes supplied by Béton Chantiers Bourgogne (Lafarge Group), we have presented a few examples of regional HPC formulations with the purpose of demonstrating the possibility of making, in virtually any region of France, an HPC composition with local materials.

Lafarge Bétons Granulats, with the assistance of the Lafarge Coppée Recherche and the Ciments Lafarge Division, has set up a "Concrete Performance" network, which groups together very many ready-mix plants, distributed roughly all over French territory and which are in a position to manufacture high performance concrete industrially and reliably.

6 ENGINEERING PROPERTIES OF VERY HIGH PERFORMANCE CONCRETES

F. de LARRARD, L.C.P.C. Paris, France
Y. MALIER, E.N.S. Cachan, France

This article sums up current knowledge of very high peformance concretes (VHPCs) with a view to their use in civil engineering structures. From fabrication to behaviour in service and fracture, the various stages of the life of the material are described, with its measured properties related as much as possible to the composition and microstructure of the concrete. A few important questions that are sometimes obstacles to its use (feasibility, fragile breaking behiaiviour, cost) are then discussed. Finally, the durability of VHPCs is emphasized, as one of the major features making them attractive for many applications.

1 Introduction

At a time when 'high performance concretes' (HPCs) are being used more and more routinely, this article deals with the next generation of structural materials, still hardly out of the laboratory stage except for a few skyscrapers in North America [39]. By convention, very high performance concretes (VHPCs) are hydraulic concretes having a characteristic compressive strength greater than 80 MPa at 28 days. They are in fact the top of the line in field-placeable concretes, with common aggregates and a binding paste improved by the use of a few special products such as superplasticizers and silica fume.

Our aim is to give an overview of these materials, starting from the logic of their composition and relating their main properties to their microstructure. We shall then attempt to discuss a few important questions, correcting some received opinions in the light of recent experiments.

High Performance Concrete: From material to structure. Edited by Yves Malier. © 1992 Taylor & Francis.
Published by Taylor & Francis, 2 Park Square, Milton Park, Abingdon, Oxon, OX14 4RN. ISBN 0 419 17600 4.

2 Principle of composition

We should in fact talk about 'very-high-packing-density concretes', since that is the main distinguishing feature of their constitution.

The idea behind concrete as a material is to attempt to reconstitute a solid rock from elements having complementary gradings. The packing density of the mixture so constituted will be determined by its grading range, e.g. by the ratio of the largest and smallest sizes. At the coarse end, placement requirements routinely limit the maximum aggregate size to some 20 or 25 mm. At the small end, the limit is determined by the physics of surfaces that causes natural flocculation of the grains of cement.

This led at first to the idea of using the deflocculating properties of certain organic substances [2] to restore the original grading of the cement in suspension in water (between 1 and 80 μm). But the spectrum of the mixture can be extended even further by adding an ultrafine, chemically reactive product to fill part of the pores of the packing that are beyond the reach of grains of cement. Silica fume, a byproduct of the electrometallurgy of silicon and its alloys, is the best product yet found to perform this function [14].

The application of these simple principles leads to formulations like the one presented in Table 1. The non-specific components (aggregates, Portland cement) must be of good quality, with rigorous selection being especially necessary if the aim is to exceed a mean strength of 100 MPa at 28 days. The cement content may range between 400 and 550 kg/m^3, with a silica fume content that is commonly between 5 and 15% of the weight of cement and a water/cement ratio of less than 0.35. It is moreover by lowering this last parameter that we advance from HPCs (f_c 2 8 between 50 and 80 Mpa) to VHPCs.

Table 1 : Composition (kg/m^3) of a VHP concrete (11)

Crushed limestone aggregates			River sand	Ordinary Portland cement	Silica Fume	Superplast. (dry powder)	Water	W/C
12.5/20	5/12.5	0/5						
854	411	326	326	421	42.1	7.59	112.3	0.27

3 Behaviour in the fresh state

3.1 Workability

In spite of water contents that are much lower than usual, these concretes most often exhibit a slump of approximately 200 mm in the Abrams cone as they leave the mixer. The separation of the particles made possible by superplasticizers in effect strongly reduces the shear threshold of the fresh cement paste, giving a concrete that flows under its own weight alone, with, however, a velocity that depends on the viscosity of the mixture, e.g. the degree of loosening of the granular packing by the mixing water. With water/cement ratios of less than 0.30, the consistency, while fluid, is often viscous and 'sticky'. The concrete can be placed properly, but requires vibration comparable to that of common concretes of 'plastic' consistency. At form removal, there is occasionally some formation of bubbles, a consequence of the viscosity of the fresh material.

The workability may be of short duration, if this problem has not been anticipated at the time of mix design [15]. It is however quite possible to obtain a practical duration of workability greater than one hour (see Figure 1), using a set retarder if necessary, when, for example, the cement contains a large proportion of aluminates. We should point out that conventional means are unsuitable for evaluating the rheological behaviour of VHPCs in the fresh condition, as their workability is outside the limits of normal concretes.

3.2 Concrete in the 'dormant' period

What we call the 'dormant' period is the time that elapses between the placement of the concrete and the start of setting. This period varies considerably from one VHPC to another. Since the hydration reaction is self-activated, a very small temperature rise is enough to initiate it. Conversely, cold weather and/or a large proportion of

Figure 1 : Evolutions of flow time (measured with the LCL workabilimeter) and slump of VHSC Vs time after during one hour (11).

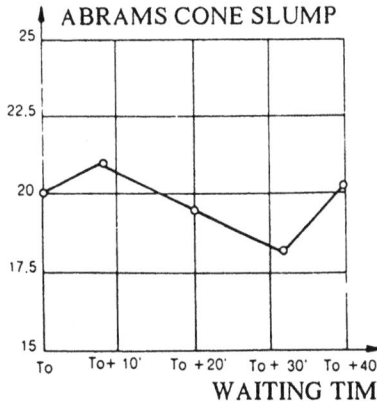 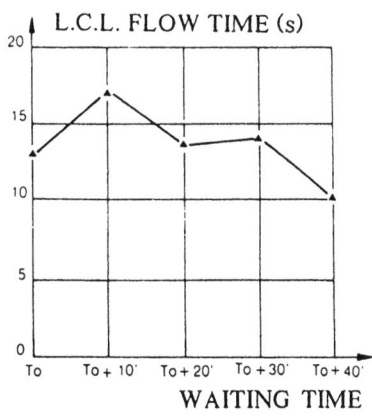

retarder can delay setting by 48 or even 72 hours.

The presence of ultrafine particles, if there are enough of them (in practice, about 7% of the weight of cement), allows stabilization of concrete in the fresh condition. This phenomenon should be considered in relation to the thixotropic behaviour of the grout, which 'congeals' and takes on the appearance of a gel some time after mixing, then recovers its fluidity after agitation (if the superplasticizer has not lost its effectiveness). Conventional cement pastes compact their granular structure and expel excess water (bleeding) under the effect of gravity. By contrast, it follows that there is no bleeding of water to the free surfaces of these concretes.

For the same reasons, segregation is practically unknown in VHPCs : the separation of a grading class from the mixture would involve its sedimentation through the skeleton, and failure of the bonds between grains of binder: bonds that are probably strong, as indicated by the stiffness of the material after compaction and vibration.

However, this stability has two less favourable aspects. First of all, since there is no bleeding water to protect the concrete from early desiccation, it will require careful curing, without which there is a risk of cracking by plastic shrinkage (shrinkage occurring before the concrete sets). It will also be difficult to subject the concrete in place to shear stresses aimed at changing its shape - as for example when a large slab is refloated to level it to its final position.

On the other hand, the impossibility of migrations of water and fines in the fresh concrete guarantees a remarkably homogeneous material, with interfaces that are free of defects (cf. Section 5).

4 Hardening

4.1 Heat given off during setting
The first parameter one thinks of is naturally the total proportion of binders, which varies significantly from one formulation to another. But it is in particular the exothermic reaction of the cement that will be decisive [9] (as it is, moreover, in conventional structural concretes). Silica fume seems to act as a catalyst at the start of the reaction [5], so that the initial temperature rise is rather 'steep'. The result is a substantial temperature rise even in thin sections (we have measured temperature rises of as much as 8°C in cylinders 160 mm in diameter [3]). Laplante et al. [26] report temperature rises of 40°C in columns 1 m on a side.

In such massive structures, the risks of cracking of thermal origin do not seem to be especially high, because of the rapid development of tensile strength [9]. Delicate cases can be investigated by numerical simulation of the temperature and blocked strain fields in the parts during hardening [11].

4.2 Hydration shrinkage

This is the shrinkage related to the degree of hydration of the binder occurring in the absence of exchanges of moisture and under isothermal conditions. In practice, while very massive elements fully satisfy the first condition, it is practically impossible, even in the laboratory, to maintain a constant temperature in the specimen, so that this endogenous shrinkage, or even self-desiccation shrinkage, is nearly always found superimposed on some degree of thermal shrinkage.

We have recently presented [3] shrinkage measurements on various ordinary, HP and VHP concretes, measurements made between the time of setting and the time of mould release of the specimens. We evaluated this early strain at approximately 50×10^{-6} for the VHPCs, against 30×10^{-6} for the ordinary concrete tested. Figure 2 presents some curves showing the endogenous shrinkage (from the time of mould release) of the VHP concrete of which the formula is given in table 2, together with that of a control concrete made with the same cement and the same aggregates, but without admixtures or ultrafines. Two coats of resin, separated by aluminium foil, had been applied to the specimens, a technique that guarantees practically no drying (as shown by the cessation of shrinkage after 90 days, (while the same specimens, allowed to dry (cf. Section 6.4), continued to become shorter).

The VHPC therefore exhibits substantially more endogenous shrinkage than the control. One way to explain this phenomenon is to reflect on the quantity of free water present in the hardened concrete:

- at the outset, the VHPC is formulated with a quantity of water of the same order as the mean stoichiometric proportion of the hydration reactions.
- the pore space of the hardened VHP concrete, with a very extensive grading range, contains water free only in its smallest pores. These account for a proportionally larger volume than in the ordinary concrete (cf. Section5), because of the original fineness of the silica fume. These small pores become the locus of especially large capillary (or other) forces that result in strains of the paste, limited by the stiffness of the aggregates.

Table 2 : Composition (kg/m^3) of a control concrete

Crushed limestone aggregates			Ordinary Portland cement	Superplast (liquid)	Water	W/C
12.5/25	4/12.5	0/5				
890	208	725	375	3.75	190	0.51

Figure 2 : Autogenous shrinkage of VHPC, compared with the one of control concrete (measured on 160 x 1,000 mm cylinder, reference length 500 mm, at 20°C and 50% R.H.).

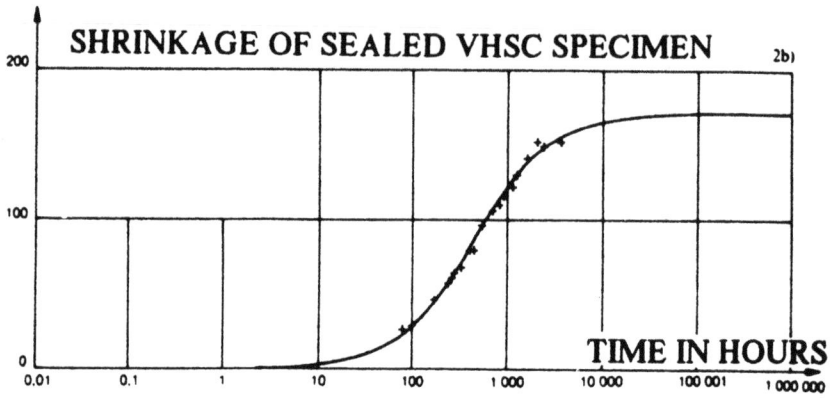

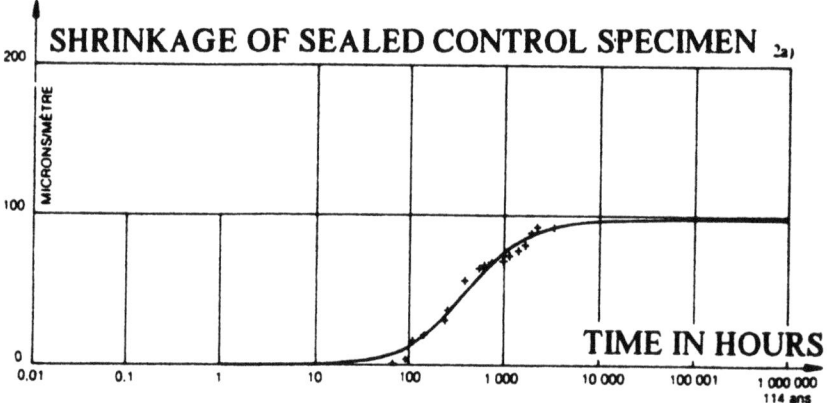

5 Microstructure [33,34]

This can be imagined thanks to the principles that governed the formulation of the material, with the help of what we know of the mechanisms of hydration and the contribution of some experimental techniques.

When VHP concretes are formulated, we saw (Section 2) that the water content is reduced by spreading out the fine part of their grading. This initial state explains to a large extent the microstructure of the hardened concrete.

If we consider the non-solid (liquid or gaseous) volumes of the hardened concrete as we gradually refine the scale of observation, we find the following 'voids':

- Occluded air bubbles, entrained during the mixing of the concrete, in a small quantity because of the fluidity of the VHP concrete (less than 2% of the total volume [45]).
- The paste-aggregate interfaces, favoured positions for the accumulation of water by internal bleeding. But we have seen that there is practically no such bleeding because of the presence of silica fume (section 32).
- The voids constituted by the vestiges of the intergranular spaces of the fresh paste. These pores are incompletely filled by the hydrates because of 'Le Chatelier' contraction (volume loss approximately 10% of the hydration reaction of the cement) and failure of a part of the mixing water to combine; this water may be lost by evaporation. But there is a very limited quantity of free water in VHPC, because of the low initial water content. Moreover, in this range of w/c ratios, the quantity of hydrates is probably smaller than in an ordinary concrete having the same proportion of cement.
- Finally, on the scale of the nanometer, there are the voids found in the very structure of the hydrates.

Now, if we observe a few fragments of hardened VHPC with the scanning electron microscope (SEM), we find that the microstructure is very dense, on the whole amorphous, and that it contains an unusual volume of anhydrous grains, cement remains that are uncombined because of a lack of available water. Moreover, the paste/aggregate interfaces (the "auréole de transition" [21]) are of remarkably low porosity and do not exhibit the usual accumulation of lime crystals. This is understandable in view of the fact that the activity of the silica fume results from the pozzolanic reaction between the silica and the lime released by the cement during its hydration. Mercury porosimetry, a technique rich in information although difficult to interpret because of its partially destructive character, shows a refined structure with practically no capillary porosity. Finally, the ambient humidity in the pores of the concrete [6] can be measured versus the age of the material. Whereas it is always 100% for the usual concretes (in the absence of exchanges with the surrounding atmosphere), it falls to 75% at 28 days of age for VHPCs

Finally, these various approaches make it possible to describe the microstructure of VHPCs as follows:

- a reduced proportion of paste, since the anhydrous grains must be counted as part of the granular skeleton of the hardened concrete,
- a paste of low total porosity,
- smaller pore sizes (pore grading),
- very little free water, since only the smallest pores are saturated with water,

- paste-aggregate interfaces that are no different from the core of the paste, eliminating a traditional zone of weakness of hydraulic concretes,
- a low free lime content,
- a state of self-stresses illustrated macroscopically by the endogenous shrinkage, which undoubtedly leads to a powerful confinement of the aggregates.

6 Main mechanical properties and modeling

6.1 Compressive strength

This is the most important of the material's functional properties. It is also the one that is most spectacularly improved: it has been possible to produce in the laboratory [4], by making optimum use of the components mentioned in Section 1, concretes having a crushing strength exceeding 200 MPa! However, if the aim is to produce workable concretes from ordinary aggregates, one must settle, in the current state of the technology, for mean strength values between 100 and 120 MPa at 28 days (measured on a cylinder), which is even so a giant step when compared to the usual concretes.

The kinetics of strength increase are notably faster than that of conventional concretes [44] (see Table 3). This results from the initial proximity of the grains of cement in the fresh concrete and from the accelerating action of silica fume mentioned above. How early the strength values appear will depend in practice on the nature (aluminates content, fineness of grinding) and proportion of cement, on the proportion of set retarder, if any, and of course on the temperature of the concrete.

To explain these high strength values, we have given [12] a physical description of the fracture of concrete in compression, which would seem to occur by successive buckling of the microstructure of the hardened cement paste, then on a larger scale, of the elementary granular edifice called a small column. Using a structure made up of bars to model the matrix of the material, it is easy to show that the compressive strength should

Table 3 : Evolution of the mechanical properties of the VHP concrete of Table 1.

Age (days)	1	3	7	14	28	90	360
Mean compressive strength MPa	27.2	72.2	85.6	92.6	101.0	109.6	114.1
Splitting strength MPa	2.2	5.4	6.4	6.1	6.5	-	-
E⁻modulus PMa	34.9	48.7	51.2	52.4	53.4	53.6	56.8

increase with the square of the compactness of the hardened cement paste - defined as the proportion of the solid phase by volume - while the tensile strength increases only in proportion to the first power of this value (Féret's law). This is roughly what is found in practice [11].

6.2 Tensile strength

This increases significantly, even if to a smaller extent than the compressive strength (for the reasons stated above). The ratio f_t/f_c therefore decreases to as little as 1/20 in the strongest concretes. However, splitting strength values greater than 6 MPa can be attained, an advantage structural engineers can easily turn to account (especially in prestressed concrete). A law of correlation in the square root of the compressive strength (like the one used in the American regulations) would be consistent with the considerations of the previous paragraph. However, the experimental results are in rather good agreement with the equation of the French regulations:

$$f_{t\,j} = 0.6 + 0.06\, f_{c\,j}$$

where $f_{t\,j}$ and $f_{c\,j}$ are the mean splitting strength and the characteristic compressive strength at j days, stated in MPa, respectively.

The tensile strength values are attained even faster than the compressive strength values (see Table 3). The densification of the matrix and of the paste-aggregate interface explain the improvement of the tensile strength. During splitting tests, the fracture faces are systematically transgranular (even with siliceous aggregates), proving the mechanical homogeneity of the material. It may however be wondered why the tensile strength ceases to increase at approximately 14 days of age, unlike the compressive strength, which may increase by a further 10 to 20%. The state of tension of the matrix, resulting from blocking of its shrinkage by the aggregates, may explain this phenomenon. It may be also a structural effect on specimens preserved in water: the skin of which would tend to swell, causing internal tensile self-stresses that increase as time passes.

6.3 Young's modulus

Hashin [22] has proposed mathematical expressions derived from homogenization theories that illustrate the influence of three key parameters - the volume of aggregate (g), the modulus of the paste (E_p), and the modulus of the aggregate (E_a) - on the modulus of the concrete (E).

$$E = \frac{E_p\,[(1-g)E_p + (1+g)E_a]}{[(1+g)E_p + (1-g)E_a]}$$

The modulus of the paste will be governed by its packing density, with an exponent that is logically close to 1 (as it is for tensile strength)*. The terms g and E_a do not differ much between a conventional concrete and a VHPC. This leads to a rather moderate increase in the modulus, which ranges from 40 GPa (high-paste VHP concrete with a rather flexible aggregate) to approximately 55 GPa (VHP concrete with a skeleton that is densely packed - such as that the one shown in Table 1 - and rigid).

Note: for the VHPC shown, the theoretical modulus of the paste can be calculated using Hashin's formula and a modulus assumed to be 65 MPa for the aggregates (hard limestone). The correlation between the modulus of the paste and the tensile strength (cf. Figure 3) shows the soundness of the assumption in that these values are proportional to the packing density of the paste, which here corresponds to its degree of hydration.

The growth of the Young's modulus with time roughly follows that of the tensile strength, without however showing the same tendency to level off from 14 days. Here again, the empirical formula of the regulations, relating modulus and characteristic compressive strength, can be extended without difficulty :

$E_{ij} = 11,000 \, (f_{c\,j})^{1/3}$ (all quantities in Mpa)

Figure 3 : Correlation between paste modulu (calculated with Hashin formula) and splitting strength, for several ages of VHPC.

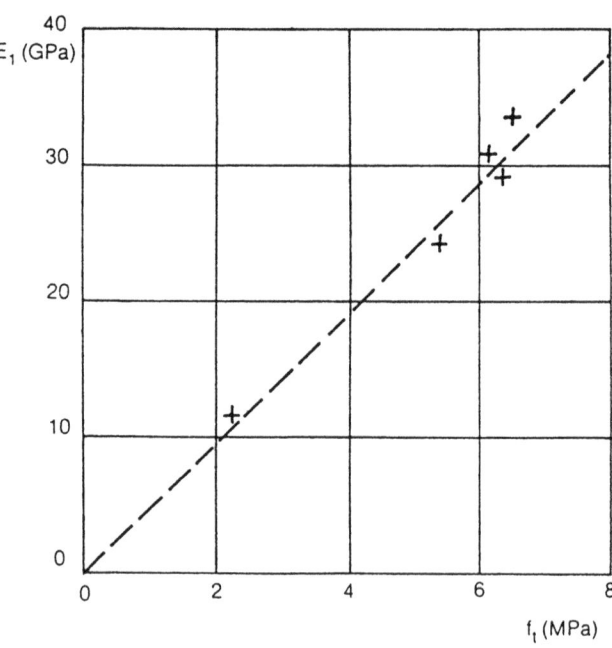

6.4 Shrinkage

It is drying shrinkage that is of interest to us here. Figure 4 shows the shrinkage of a specimen removed from the mould at 24 hours, having dried in a room in which the relative humidity (50±10%) and temperature (20±1°C) were controlled. The desiccation

Figure 4 : Drying shrinkage of VHPC, compared with the one of control concrete (measured on 160 x 1,000 mm cylinder, refrence length 500 mm, at 20°C and 50% R.H.).

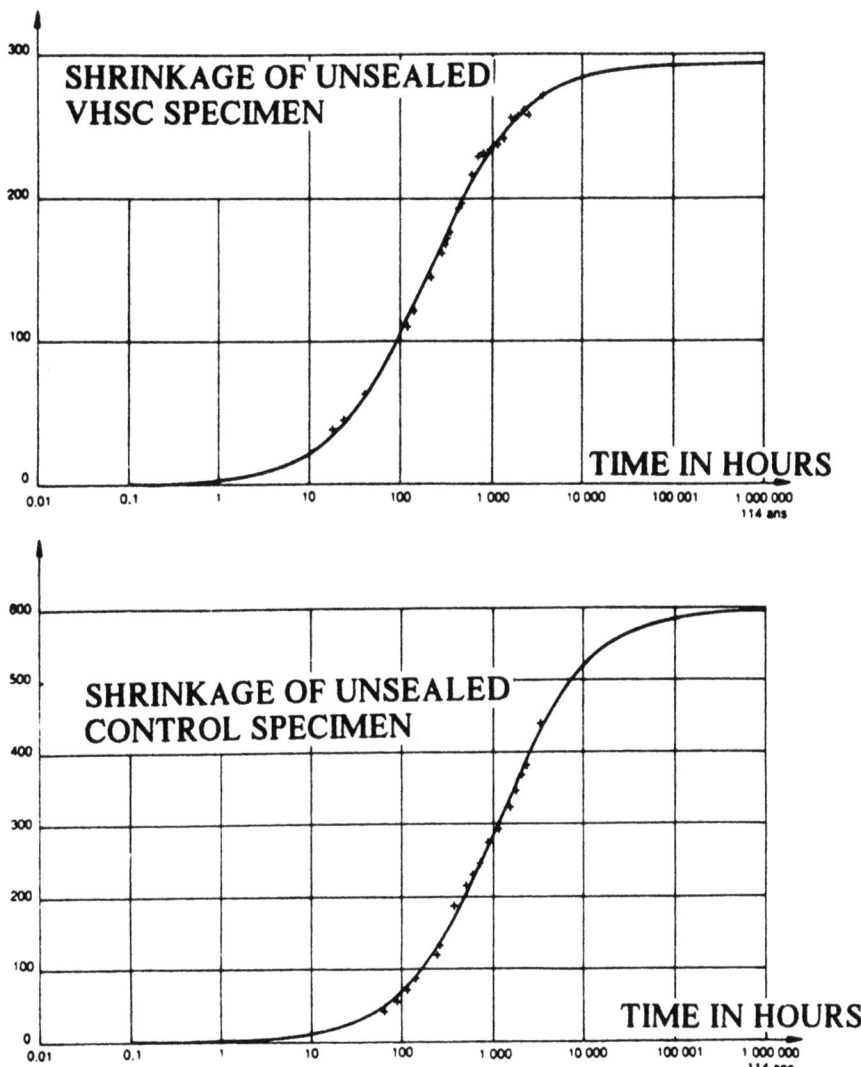

shrinkage is by convention taken to be equal to the difference between the total shrinkage and the shrinkage of the same specimen with no loss of water (cf. Section 4.2). Table 4 gives the shrinkages at the end of the experiment and in the long term (extrapolations).

While the final endogenous shrinkage is roughly doubled, the drying shrinkage is greatly reduced, since the material contains very little free water after hydration. The total shrinkage of the VHPC, measured on specimens 160 mm in diameter, is approximately half that of the same specimens made of the control concrete. The especially rapid kinetics of the shrinkages of the VHPC, which could lead to errors if comparisons were made on the basis of short-term tests, should be noted.

Is there any reason to fear the effects of the endogenous shrinkage of VHPC on the scale of structures ? For bridges, most of this strain will occur after form removal, and its effects will then be similar to those of a homogeneous strain of thermal origin. The points of contact of the structures with their substructure are designed so that this type of strain will not cause any particular damage.

For pavements, and other slabs cast on the ground, the only case in which there would really be a penalty with respect to normal concretes would be that of a very thick unreinforced slab strongly bonded to its substrate. Such a case arose in one of the first recorded applications of VHPC [24]. Otherwise, there is either a thin slab (the construction of which would be quite justified for a wearing course, given the greater abrasion resistance - cf. Section 6.7), in which drying shrinkage is predominant, or a reinforced slab in which the passive reinforcements distribute the cracking. But the drying shrinkage is much smaller, and the adherence greater (cf. Section 6.6) with VHPC. Even so, it should be noted that the higher tensile strength of VHPC then entails a larger percentage of reinforcements (determined by the condition of non-fragility).

Table 4 : Comparative numerical values of shrinkages (in μm/m) measured on the control (Table 2) and VHP (table 3) concretes. Allowance is made for the initial hydration shrinkages before mould release.

	Control concrete	VHP concrete
Total shrinkage	470	320
- extrapolation	650	340
Endogenous shrinkage	120	200
- extrapolation	120	220
Desiccation shrinkage	350	120
- extrapolation	530	120

6.5 Creep

This has already been discussed in previous articles [3,11,13]. The strains versus time (Figure 5) of specimens identical to those used to measure the shrinkage can be seen (cf. 4.2 and 6.4); the VHPC has the same composition as that of Table 1, while the control is different (Table 5). The basic creep is in principle determined as the difference between the strain of the loaded specimen coated with resin and that of the equivalent specimen with no load. With the uncoated specimens (preserved under a polyane film for 28 days, then exposed to a controlled atmosphere (cf. 6.4)), the same calculation gives the 'total' creep: the sum of the basic creep and of an additional strain conventionally called desiccation creep.

In these experiments, the moisture barrier was not as effective as in the shrinkage measurements, because there was no aluminium foil between the two coats of resin. The basic creep of the control is therefore probably slightly overestimated. However, the VHPC exhibits the same creep on both specimens. It is therefore unaffected by the foregoing artefact.

The creep of the VHPC is characterized, finally, by:

- rapid kinetics (at 7 days' loading, 67% of the strain at one year is already attained, against only 41% for the control),
- a very low amplitude ($K_{c\ r} \leq 0.6$, against nearly 2 for the control; this point is perhaps less favourable for loadings at a very early age: we found in a previous study [3] $K_{c\ r}$ near 4 for 1 day old VHPC loading),
- independence of the effects of moisture and of the geometry of the structures, a veritable boon for the structural engineer, who, with these materials, will at last be able to believe in the validity of his calculations!

6.6 Bond with reinforcement

The literature on this subject is still sparse. Rosenberg et al. [37] give some results of pull-out tests (smooth-walled tube embedded in a cylinder of HPC) on two concretes, with and without silica fume. The mean adherence increases by 40%, with an increase in compressive strength of approximately 50%. Bürge [7] has compared the adherence of materials at a constant water/cement ratio, again with and without silica fume. The adherence varies in

Table 5 : Composition (kg/m^3) of another control concrete (11)

Crushed limestone aggregates			River sand	Ordinary Portland cement	Water	W/C
12.5/25	4/12.5	0/5				
826	398	315	315	410	181	.044

Figure 5 : Shrinkage and creep of VHPC (Table 1) and control concrete (Table 5) from the age of 28 days. Between the casting adn the beginning of measurements, the unsealed specimens have been cured under plastic sheets. After 28 days, the specimens have been air-cured (20°C and 50% R.H.).

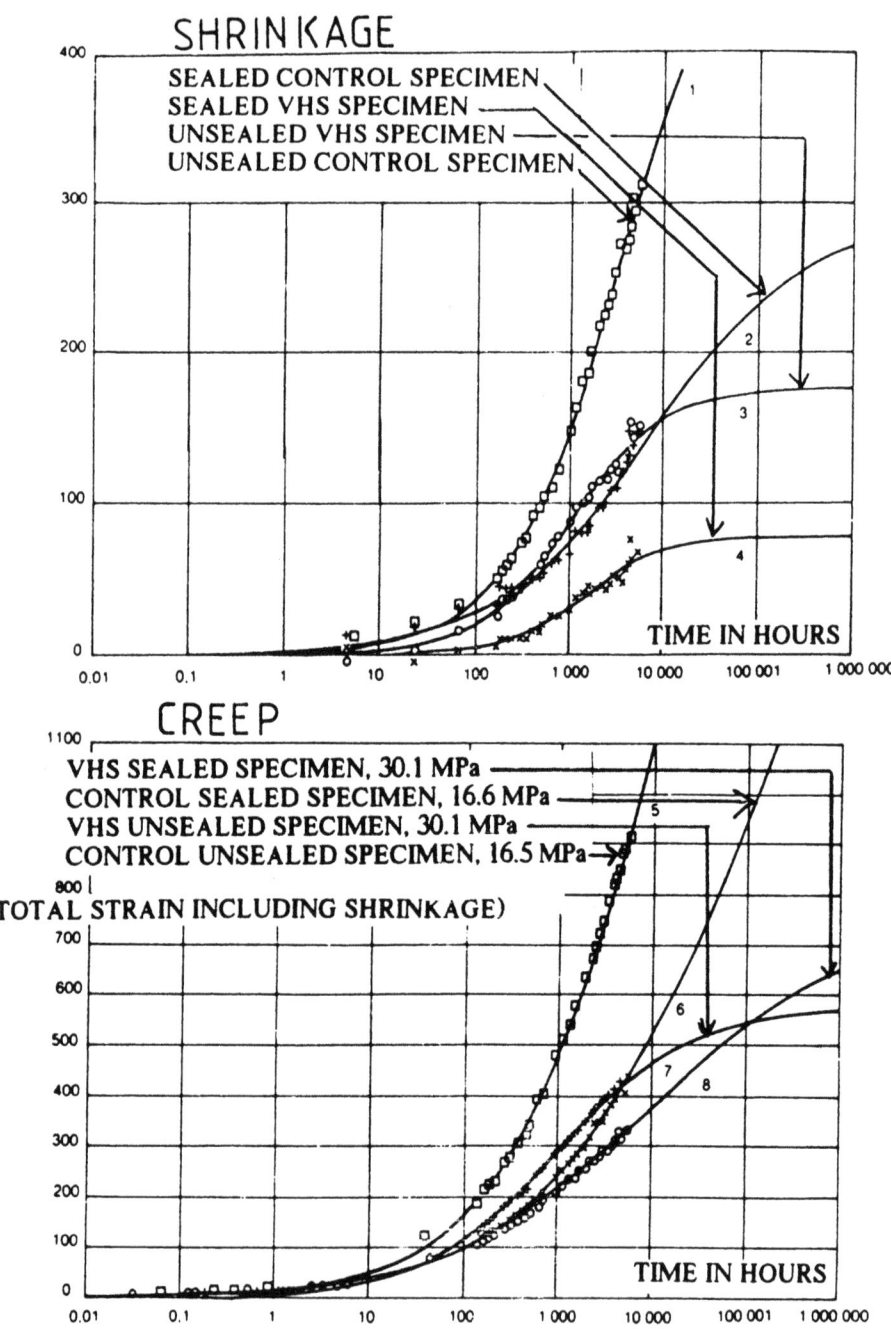

98

the ratios 3, 2 and 1.5, respectively, in tests on pure paste (w/c+s = 0.20), on mortar (w/c+s = 0.30), and on concrete (w/c+s = 0.35). Wecharatana et al. [42] have also carried out pull-out tests on an HPC having a mean strength lying between 75 and 80 MPa, but with no control concrete. By comparison with the tests reported in the literature, they note a more brittle behaviour of the bond - e.g. smaller slippages before the force of adherence drops. Finally, Lorrain et al. [28] have written what seems to be the most complete paper on the subject. Two types of test were used: the pull-out test on a coated bar from a specimen cast in a metallic cylinder used as a lost form, and the pulling test, in which one pulls shows a good correlation with the tensile strength of the concrete. The maximum force, stiffness, and reversibility parameters evolve in a positive direction as the age of the concrete, the length of the concrete-reinforcement contact, and the binder/water ratio increase.

Beam tests, the principle of which is shown by Figure 6, have recently been conducted at the Nancy Regional Public Works Laboratory [30]. These are conventional tests, from which it is impossible to derive intrinsic mechanical parameters directly. But they have the advantage of involving a mode of loading of the interface that is fairly representative of what takes place in a beam in bending. The presence of transverse reinforcements serves notably to confine the concrete 'duct' surrounding the reinforcement.

Figures 7 and 8 show the mean shear/slippage curves of the reinforcement (on the non-loaded side). The shear stresses reached in VHPC beams are far greater than those measured on the control. They exhibit a rather large scale effect, probably explained by the endogenous shrinkage, the confinement induced which is greater the smaller the percentage of steel in the beam. This thesis is moreover consistent with the results of Bürge, who has shown that the adherence increases more on a pure paste than on mortar, which itself adheres better than concrete. Now, the shrinkage decreases rapidly as the cement paste is filled with sand (mortar), then with gravel (concrete).

Figure 6 : Principle of the beam-test.

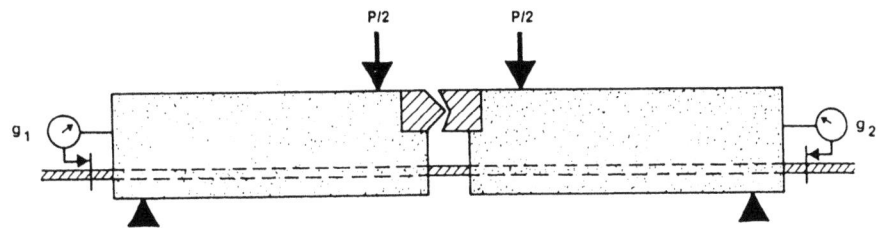

Figure 7 : Load vs bond slippage for both concretes (10 mm and 16 mm diameters).

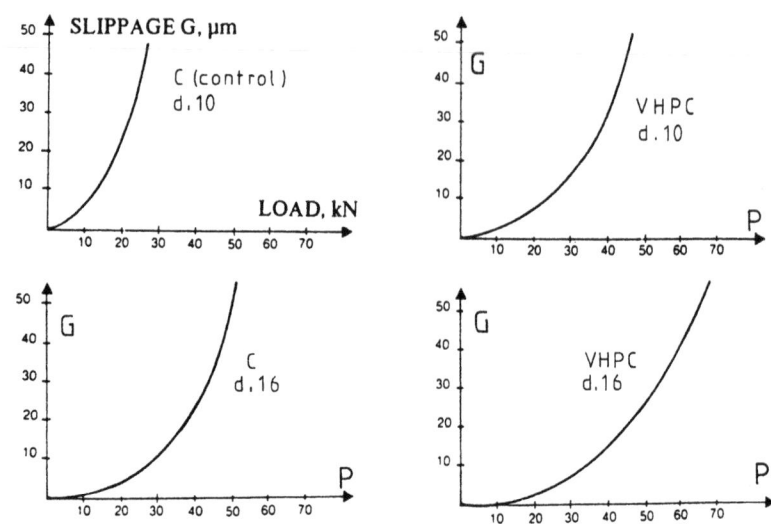

Figure 8 : Load vs bond slippage for both concretes (25 mm diameters).

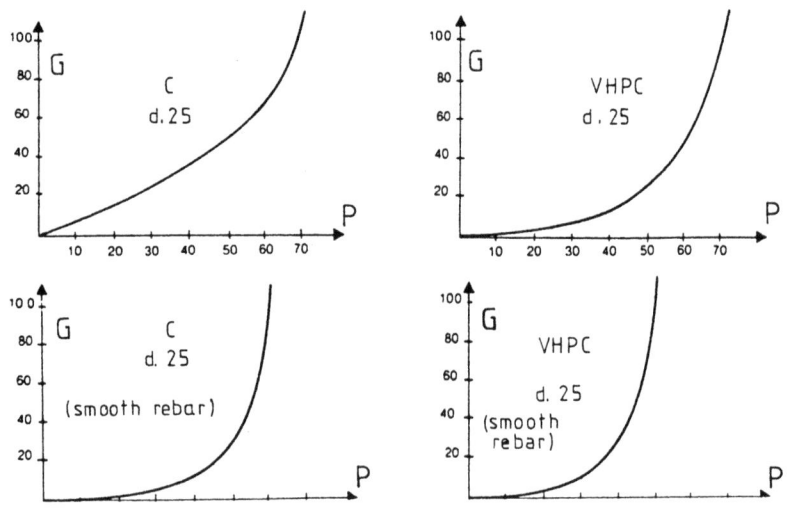

Large reinforcements therefore have less adherence than small ones; it may however be noted as a first approximation that the ratio of the mean shear stresses (mean of the stresses for slippages of 10 and 100 µm, for all diameters tested) is the same as for the tensile strength values. This is again a normal result for common concretes that can be extrapolated to VHPCs.

It should be noted that these beam tests clear the VHPC/passive reinforcement bond of the suspicion of brittleness. Wecharatana's result can doubtless be explained by the nature of his test ('pull-out test') in which an increase in the pull-out force leads to splitting of the specimen (in the absence of transverse reinforcements).

One immediate consequence of the improved adherence is the corresponding reduction of anchorage lengths. Moreover, there is a positive effect for the design of bent reinforced concrete beams when cracking is judged harmful or very harmful (according to the terminology of the current regulations): in such cases, the structural engineer must diminish the working stress of the reinforcement so as to limit the crack opening width in ordinary concrete. A comparative calculation [11] shows in this case that, in a slab bent in one direction, designed to withstand a given load, it is possible, with the VHPC slab, to have the reinforcements work at their maximum mechanical capacity and still obtain a theoretical crack opening width smaller than in the same slab made of ordinary concrete. In an unexpected manner, recourse to VHPC results both in a thinner slab and in a reduction of the section of the reinforcements, and so in a lower cost of materials.

6.7 Other properties

Abrasion resistance has been the subject of a few articles describing research aimed at improving the resistance of concrete surfaces to localized mechanical actions. For example, Holland [24] did research on a material for dam stilling basins (Kinzua Dam, USA), which, when there is flooding, are subjected to very violent flows of water laden with alluvial matter and various debris. Since fibre-reinforced concretes failed to give better performance than ordinary concrete, it was finally a VHP concrete that was chosen; according to the operator, it has been completely satisfactory. Gjorv [19] was interested in this material for pavement surfacings, and has tested a series of compositions of various strength values on a fatigue test track. With the thickness of material removed from the structure versus number of cycles as quality criterion, the (inverse) correlation of this factor with compressive strength is rather good: with a 150 MPa VHPC, the behaviour is comparable to that of solid granite, thanks to the better quality of the paste-aggregate bond (cf. Section 5).

As we have seen in [46], structures made of VHPC should logically be comparable, in shape, light weight, and flexibility, to structures with metal frameworks. It may well be that fatigue problems would then appear. A few articles are available on this subject [25,27], which has been investigated primarily for offshore construction applications. Both in tension and in compression, the resistance of the VHPC does not seem to be different from that of conventional concretes at comparable working stress/fracture stress ratios.

7 Burning questions

Here we shall attempt to take stock of the grounds for some ideas concerning high-strength materials in general and VHPCs in particular, on the basis of which the civil engineering profession has continued to use the same materials for structural works for 30 years.

7.1 Is the fabrication of VHPCs 'critical' ?

With the usual components of concretes, everyone knows that it is easy to obtain characteristic strength values ($f_{c\ 2\ 8}$) of 15 to 25 MPa, and possible with some difficulty to obtain values of 30 to 40 MPa. "Top-of-the-line" strength values can be maintained for the whole duration of a project only by rigorous quality control. Even then, in such cases, there is a risk of an accidental deterioration of the characteristics of a component.

How, under these conditions, can we hope to pour thousands of cubic metres of VHP concrete worthy of the name, with materials some of which do not offer any prima facie guarantees of regularity (silica fume is a byproduct, and sales of it are pratically negligible when compared to those of silicon and its alloys)? But then, how are the Americans managing to place a 130 MPa concrete (mean strength) in the field (a skycraper recently built in Seattle)?

In fact, we must realize that, when you decide to build with VHPC, you acquire new weapons, moving forward to a new generation of materials in which, in the limit, B80 concrete corresponds to the current B25 concrete of the building industry and B120 concrete to the B35 concrete of our bridges.

To quantify these considerations, we have carried out in the laboratory a simulation of the conditions of industrial fabrication [10]. We imposed on two basic compositions (one a control concrete, the other a VHPC), distortions corresponding to the errors of proportions that might occur in a ready-mix plant.

It was found that while the VHPC is, in terms of strength, relatively more sensitive to excess water, it is on the other hand relatively unaffected by a deficit of cement, of plasticizer (if the basic formula includes enough of it [15]), or of silica fume. Combining the harmful errors results in strength losses of the same order in both materials (10 to 15%).

Fluctuations in the sand/aggregate ratio have a limited effect on this concrete's workability, which may change if the components change (carbon content and degree of agglomeration of the silica fume, fineness of the cement, temperature), but will remain compatible with high-quality placement if the precaution has been taken at the outset of aiming for 200 mm slump concretes. It should be recalled that this fluidity, formerly forbidden in bridge concretes, no longer poses the conventional problems of bleeding and segregation (Section 3.2).

VHP concretes should therefore provide satisfactory reliability under industrial conditions. However, their fabrication does impose a few additional requirements, such as more prolonged mixing.

7.2 Are VHPCs brittle ?

With metals, in particular steel, increasing strength often goes hand in hand with greater brittleness. This is reflected by characteristic fracture faces, by reduced fracture toughness, and by lower elongations at failure. Let us briefly examine these three aspects of very high performance concrete.

Fracture faces

The failure surfaces of VHPCs are typical of the material: as we have said, cracks pass through paste and aggregate alike (see figure 9). There is therefore here some similarity with the 'cleavage' faces of brittle metals.

Fracture toughness

The same does not hold for the fracture toughness, or critical stress intensity factor. We have measured [11] this parameter on three concretes, namely an 'ordinary' concrete, an HPC without silica fume, and a VHPC. The values found are 2.16, 2.55 and 2.85 $MPa.m^{1/2}$, respectively, while the fracture energies are 131, 135 and 152 J/m^2. This means that a crack of a given length and environment will require a loading a third greater to propagate in the VHPC than the same common crack in an ordinary concrete. Here again, the main cause lies in the densification of the paste and in the improvement of the matrix/inclusion bond.

Figure 9 : Fracture faces of VHPC after splitting.

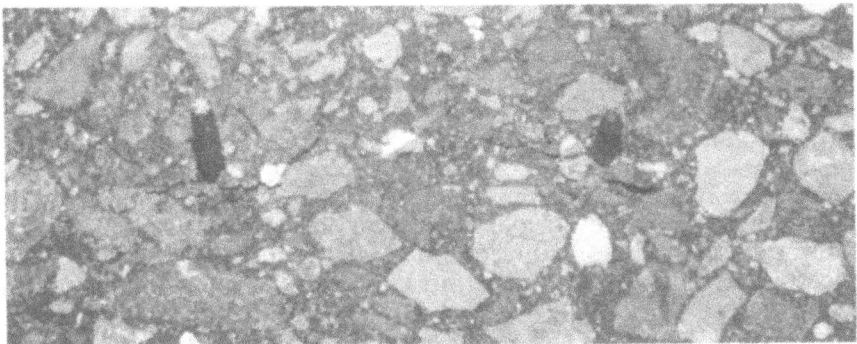

It will, however, be noted that the fracture toughness increases even more slowly than the other 'secondary' parameters (tensile strength and modulus). To understand this phenomenon, we may call on the help of the following considerations. The classic expression, taken from Fracture Mechanics and valid in stress planes, is:

$$K_{Ic} = (EG_{1c})^{1/2}$$

(K_{1c} fracture toughness, G_{1c} specific energy of fracture, and E Young's modulus).

Let us assume that the fracture energy G_{1c} is the sum of the energy necessary to break the paste (proportional to its packing density c_p) and the fracture energy of the aggregate G_{1ca}, weighted by the volumes. We then have:

$$G_{1c} = (1-g) G_{1cp} + g G_{1ca}$$

(where g is the partial volume of aggregate).

We have measured the fracture toughness of a VHP paste [11]; the order of magnitude was 0.2 MPa $m^{1/2}$. Working from the estimate of the modulus of the paste made in section 6.3, we find

$$G_{1cp} = 1.2 \text{ J/m}^2$$

against

$$G_{1ca} = 152 \text{ J/m}^2$$

for the concrete the matrix of which consists of this paste.

The 'paste' term is therefore practically negligible with respect to the 'aggregate' term in the equation for the fracture energy of concrete (this is why the control concrete and the HPC have roughly the same fracture energy). Whence the law:

$$K_{1c} = (g G_{1c} g E)^{1/2}$$

At a constant aggregate content, the fracture toughness therefore increases only with the sixth root of the compressive strength, a somewhat pessimistic prediction because of the improvement of the paste-aggregate bond, for which no allowance is made in the foregoing calculations.

Behaviour in uniaxial compression
This is obviously the normal working mode of the material. Before fracture, the stress-strain relation is highly linear (see Figure 10). At the stress peak, the plastic strain is only 15% of the total strain, against 29% in the control. According to the definition proposed by

Figure 10 : Stress-strain curve of VHPC under compressive stress.

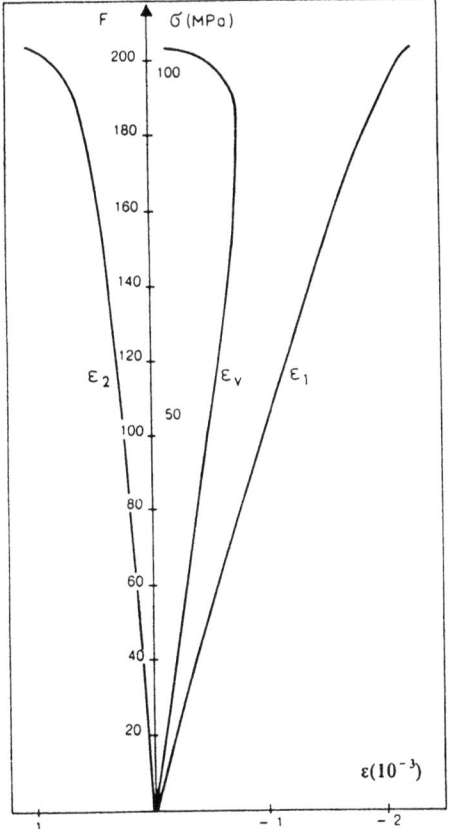

Rossi [38], VHPC would therefore be more fragile as a material. The elongation at the stress peak is slightly greater than for an ordinary concrete (here 2.1 against $1.8.10^3$). The transverse strains are qualitatively of the same type as those of the control, with, however, a smaller increase.

We have recently [43] been able to control the compression testing of the VHPC in its 'post-peak' part, using a device that prevents rotation of the press plates and controlling the test by a linear combination of the force and the displacement between the plates, chosen so as to be always increasing during the test [36] (cf. Figure 11). It is found that, in these experiments, the load-displacement curve is nearly perfectly elastic-fragile. In the absence of the device preventing rotation of the ends of the specimen, the curve even shows snap-back.

This behaviour can be explained in at least two ways:

Figure 11 : Total load-displacement curve under compressive stress (110x220 mm specimen), with a special apparatus avoiding any rotation of the testing machine plattens.

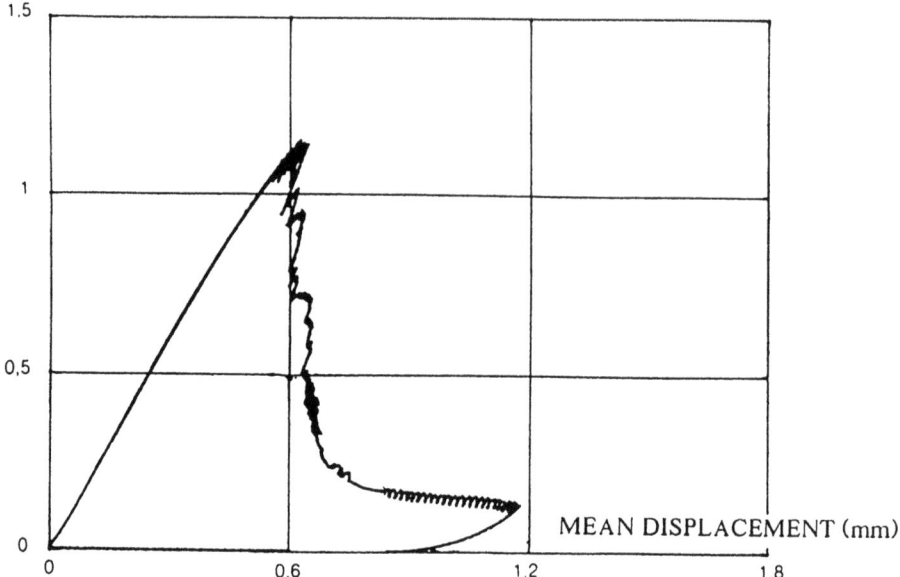

FORCE IN MN

MEAN DISPLACEMENT (mm)

- according to a global energetic approach, we have seen that the fracture energy of the material increases very little with its compressive strength. It follows that the area under the curve cannot be proportional to the maximum stress. The 'post-peak' part is therefore necessarily increasingly reduced to the bare minimum.
- according to a local approach, we refer once again to the model of buckling on two scales (cf. [12] and Section 6.1). The stress peak would here correspond to the buckling threshold of the small columns formed following the longitudinal cracking of the material. After this bifurcation, we distinguish the flow phase, during which slippages occur along the oblique cracks of the small columns, comparable to shearing cracks. At the moment of fracture, the tension is maximum in direction q, the angle of these cracks varying with the direction of compression. Classical considerations in the Mohr plane give:

$$\theta = (1/2) \text{ Arctg } 2(f_t/f_c)^{1/2}$$

(f_t and f_c are the strength values of the material in tension and compression, respectively).

Since the tensile strength increases more slowly than the compressive strength, q decreases for large values of f_c. On the other hand, the fracture faces are very smooth, and these flow cracks can no longer take out a force other than by friction, via a transverse stress condition. Whence, in the VHPC in uniaxial compression, a sudden drop in the force as soon as the bifurcation of the small columns is reached.

But is this post-peak behaviour actually representative of the fracture of concrete in a real structure? Surely not in the case of bending: experimenters who have carried out fracture tests on over-reinforced VHPC beams [23,29] - e.g. beams that fail by crushing of the concrete - report that the same models of behaviour, similar except for an affinity on the stress axis (and in the ratio of the strength values), correctly predict the time of fracture of both the VHPC and the control concrete beams.

The strains measured on the compressed fibre exceed 0.4% at the end of the test, and yet the concrete does not disintegrate - except for the coating, which flakes, by a failure of confinement due to the transverse reinforcements. Moreover, the beams exhibit a ductile behaviour; with under-reinforced beams, the VHPC even provides additional ductility. The improved adherence, by which a greater length of reinforcement would yield in the VHPC, may explain this phenomenon.

The disagreements between the two types of test - cylinder in compression and beam in bending - probably have two causes. The first lies in the confinement exerted by the rest of the structure on the zone that has exceeded the strain limit in compression. Fracture is extremely localized in the cylindrical specimen [43], and the redescending part of the curve of Figure 11 reflects a progressive slippage of two blocks of concrete one with respect to the other. Now, this sloping fracture grows with difficulty in the beam; for it to lead to unloading, slippages must occur in the compressed zone. The only possible diagram is then the one given in Figure 12.

It is here that the second cause of disagreement, namely (of course) the transverse reinforcements, intervenes. Figures 13 and

Figure 12 : Failure of an over-reinforced concrete beam under flexural moment.

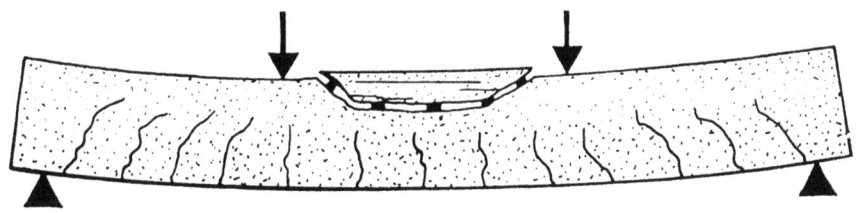

Figure 13 : Total load-displacement curve of control concrete, with several percentages of transverse reinforcement.

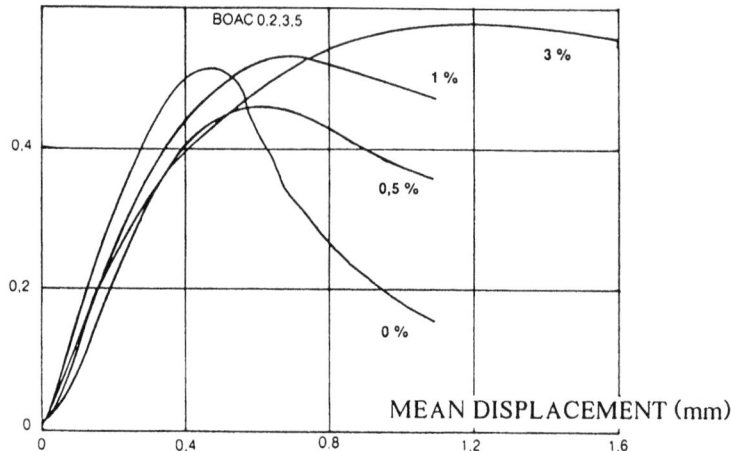

Figure 14 : Total load-displacement curve of VHPC, with several percentages of transverse reinforcement.

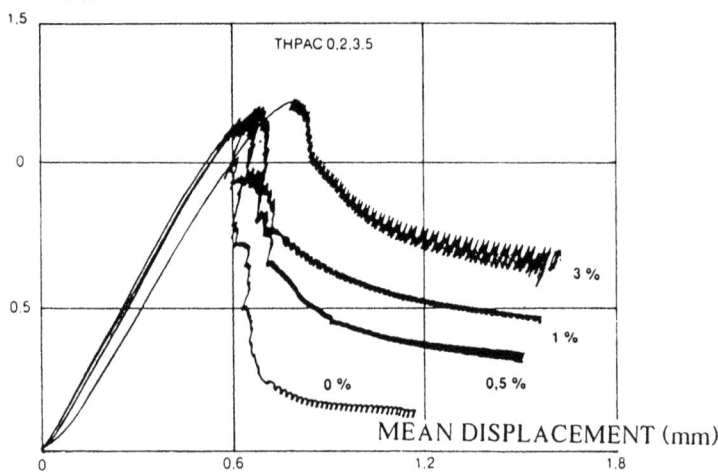

14 show the evolution of the load-displacement curve in the presence of helical reinforcements, in increasing percentages by volume. The drop in force then declines as the transverse reinforcements increase.

By way of a conclusion, we may put forward the following ideas :

- for bending calculations, the classical approach consisting of taking an experimental (stress-strain) curve with a softening branch, obtained on a specimen in simple compression, as the law of behaviour of the concrete is physically meaningless. It is necessary to reverse this approach, and choose conventional laws that agree with the macroscopic results of bending tests. In this respect, it would seem, according to the literature, that a trapezoidal curve, like the one proposed in Section 6.8, may be suitable, provided, however, that the element contains at least a minimum of transverse reinforcements (a minimum that remains to be determined). It should also be noted that because of the bulk of the passive reinforcements, over-reinforced sections will be very rare with these materials,
- for fracture calculations in slightly eccentric compression, new calculation methods must be developed, probably based on solid mechanics rather than on continuum mechanics. For the moment, it is possible to stay on the safe side by using diagrams with no redescending portion, in other words the pre-peak part of the trapezoidal diagram.

7.3. Are VHPCs expensive ?

This question is a difficult one to answer from a research laboratory and for a material that still has very few applications. We can, however, attempt to estimate some of the extra costs in connection with the use of VHPC.

Superplasticizers are most often sold in solution, and the concentrations of dry extract vary by a factor or more than two to one (from 20 to 45 %). However, although their effectiveness at the same concentration of dry extract varies from one product to another, the needs of competition keep the normal prices of these admixtures in a rather narrow range, around 20 French Francs/kg dry extract (price not including tax, for large quantities). Unlike admixtures, silica fume has only very recently appeared on the civil engineering materials market. The equilibrium prices have therefore not yet been reached, and the costs may vary rapidly, given the possible arrival in France of foreign silica fume. A price of 1,000 F/t not including taxes may be given as an order of magnitude.

Per cubic metre of concrete, the extra cost of 'materials', not including taxes, for a VHPC having a formula close to the one given in Table 1, is about 200 F/m^3, against 120 F/m^3 for an HPC.

To this extra cost for the components of the concrete, other secondary costs must obviously be added. At the fabrication level, special silos are needed to store the silica fume - unless the ultrafine material is put into the mixer manually, which may be the simplest and most reliable approach for occasional use of this byproduct. The mixing must in most cases be prolonged (total duration at least 1'30") for fractionated introduction of the plasticizer [31] and good dispersion of the phases present. Finally, the necessity of careful curing entails further extra costs.

On the other hand, there are savings under other headings - placement made easier by the fluidity of the concrete, and the possibility of rapid form release, thanks to the early appearance of strength - to such an extent that it was recently claimed that "HSCs cost less than ordinary concretes" [35].

Table 6 : Summary table of the main mechanical properties of the VHPC, determined in conformity with the spirit of current French regualtions.

Compressive strength and
E modulus (MPa) at :

1 d. : $f_{c\,1}$	**20**
$f_{t\,1}$	**2.0**
$E_{1\,1}$	**35,000**
7 d. : $f_{c\,7}$	**75**
$f_{t\,7}$	**5.5**
$E_{1\,7}$	**51,000**
28 d. : $f_{c\,28}$	**90**
$f_{t\,28}$	**6.0**
$E_{1\,28}$	**53,000**

Behaviour under uni-axial
compressive stress :

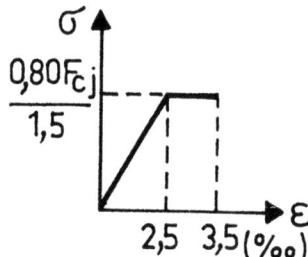

Shrinkage :
ε_s **2 or 3.10^{-4}** (depending on the local climatic conditions)

Creep :
$K_{e\,r}$ **0.8** (ratio between creep strain and elastic strain, when the load is applied at 28 days)

8 Conclusion

We have attempted to give a consistent overview of very high performance concretes, starting from the logic of their composition and from it deducing their microstructure and the mechanical properties to which it leads. Then we took stock of a few burning questions raised by this material.

On the theoretical level, neglecting possible difficulties of construction, VHPC manifestly makes it possible to attain higher performance (spans, slenderness ratios [11], etc,) and to achieve an overall reduction of the materials used, in both quantity and total cost.

There are already structures (columns of skyscrapers, beams prestressed by pre-tensioning wires, lattice structures, etc.) in which the use of the VHPC will certainly yield significant short-term savings with no change of design. But *it is by redesigning the structures on the basis of the mechanical functions to be performed, the properties of the material, and the techniques of construction that the largest savings will be achieved,* justifying the current investment in research and innovation.

We have not mentioned *durability* in connection with VHPCs. And yet, most of the physico-chemical mechanisms of deterioration of concrete change in a direction favourable to its preservation [6]: reduced permeability to water and air, low mobility of chloride ions, capacity of the hydrates to trap foreign ions, high resistivity, greatly reduced depth of carbonation, excellent frost resistance, and all this with no air entraining agent - as soon as the material has matured enough to have practically consumed its free water - etc. There also seems to be no skin cracking by drying of the concrete - neglecting plastic shrinkage -, and 'mechanical' cracks close better, because of the more regular shape of the fracture faces. Only resistance to fire might be a problem [17], and further research is currently being done on this point.

Even though it has not yet been confirmed at full scale, there is every reason to think that the durability of VHPCs should be such as to make the lives of civil engineering structures longer: in many cases, this argument by itself will lead to greater use of this new generation of materials.

Bibliography

1. Acker P. Comportement mécanique des bétons : apports de l'approche physico-chimique. Doctoral thesis, ENPC, L.P.C. research report, N°. 152, 1988.
2. Aitcin P.C., Albinger J. Les bétons à hautes performances. Expériences Nord-Américaine et Française. Annales de l'ITBTP, N°. 473, March-April 1989.

3. Auperin M., De Larrard F., Richard P., Acker P. Retrait et fluage de bétons à hautes performances - Influence de l'âge du chargement. Annales de l'ITBTP, N°. 474, pp. 50.75, May-June 1989.
4. Bache H.H. Densified cement-ultrafine particle-based materials. 2nd International Conference on Superplasticizers in Concrete, Ottawa, pp. 185-213, June 1981.
5. Buil M., Paillère A.M., Roussel B. High-strength mortars containing silica fume. Cement & Concrete Research Vol.14, N°. 5, pp.693-704, 1984.
6. Buil M. Des matériaux à très hautes performances pour le Génie Civil grâce aux fumées de silice. Travaux, N°. 624, pp 8-12, September 87.
7. Burge T.A. Densified matrix improves bond with reinforced steel. MRS Symposium, Boston, 1987.
8. Chardin G. Private communication. PPB-SARET company, 1988.
9. De Larrard F., Acker P., Lejeune D., Le Maou F. Contraintes d'origine thermique dans le bétons H.P. Rapport M.R.T. F.O. 379, 1985.
10. De Larrard F., Acker P., Malier Y. Very-high-Strength Concrete: Froom the Laboratory to the Construction Site. Conference on "Utilization of high-Strength concrete" , Stavanger, June 1987.
11. De Larrard F. Formulation et propriétés des bétons à très hautes performances. Doctoral thesis, ENPC - L.P.C. research report N°. 149. March 88.
12. De Larrard F., Torrenti J.M., Rossi P. Le flambement à deux échelles dans la rupture du bétons en compression. Bulletin de liaison des Laboratoires des Ponts et Chaussées, N°. 154, pp. 51-55, March-April 1988.
13. De Larrard F., Acker P., Attolou A., Malier Y. Fluage des bétons à très hautes performances. IABSE congress, Helsinki, 1988.
14. De Larrard F. Ultrafine Particles for the Making of Very-High-Strength Concrete. Cement and Concrete Research,v. 19 Jan.,1989
15. De Larrard F. A Method for Proportioning High-Strength Concrete Mixes. This book. Also presented at the 3rd International Cconference on fly ash, Silica Fume, Slag and Natural Pozzolans in Concrete, Trondheim, June 1989, and published in Cement, Concrete and Aggregates, Summer Issue, 1990.
16. Several author Un pont à hautes performances: l'ouvrage expérimental de Joigny. Travaux, N°. 642, pp. 57-65, April 89.
17. Diederichs U., Jumppanen U.M., Penttala V. Material Properties of High-Strength Concrete at Elevated Temperatures. IABSE congress, Helsinki, June 1988.
18. Fouré B. Etude expérimentale de la résistance du béton sous contrainte soutenue. Annales de l'Institut Technique du Bâtiment et des Travaux Publics, N°. 435, June 1985.

19. Gjorv O.E., Baerland T., Ronning H.R. High-strength concrete for highway pavements and bridge decks. Conference on "Utilization of high-strength concrete", Stavanger, June 1987.
20. Grattesat G. et al. Ponts de France. Presses de l'Ecole Nationale des Ponts et Chaussées, Paris, 1983.
21. Hanna B. Contribution à l'étude de la structuration des mortiers de ciment Portland contenant des particules ultra-fines. Doctoral thesis, INSAT, June 1987.
22. Hashin The moduli of heterogeneous materials. Journal of Applied Mechanics, March 1962.
23. HellandD S., Einstabland T., Hoff A. High-strength concrete The Norwegian Concrete Society, Norsk betongdag, 1983.
24. Holland T.C., Krysa A., Luther M.D., Liu T.C. Use of silica-fume concrete to repair the abrasion-erosion damage in the Kinzua dam stilling basin. 2nd International conference on fly ash, silica fume, slags and pozzolans in concretes, Madrid, ACI SP 91, April 1986.
25. Lambotte H., Taerwe L. Fatigue of plain high-strength concrete subjected to flexural tensile stresses. Conference on "Utilization of high-strength concrete", Stavanger, June 1987.
26. Laplante P., Aitcin P.C. Field study of creep and shrinkage of a very high-strength concrete. 4th International Congress of the RILEM, Z.P. Bazant Ed., Evanston, August 1986.
27. Lenschow R. Fatigue of high-strength concrete. Conference on "Utilization of high-strength concrete", Stavanger, June 1987.
28. Lorrain M., Khelafi H. Sur la résistance de la liaison armature-béton de haute résistance. Annales de l'ITBTP", N° 470, Décembre 1988.
29. Maro P. Bending an shear tests up to failure of beams made with high-strength concrete. Congress on "Utilization of high-strength concrete", Stavanger, June 1987.
30. Maton R. Adhérence acier-béton: propriétés des bétons à hautes et très hautes performances. Degree thesis, Université de Nancy I, December 1988.
31. Paillère A.M., Serrano J.J., Buil M. Optimisation de la mise en œuvre du couple ciment-adjuvant. Recherches sur les bétons de plus hautes performances, Rapport Ministère de l'Urbanisme et du Logement. Convention N° 8371093, July 1984.
32. Pliskin L. Ouvrage prototype en béton à hautes performances. Engineering memo, Entreprise SGE, May 1985.
33. Regourd M., Mortureux B., Aitcin P.C., Pinsonneault P. Micro-structure of field concretes containing silica-fume. 4th International conference on Cement Microscopy, Las Vegas, pp 249-260, April 1982.
34. Regourd-Moranville M. Microstructures des bétons à hautes performances. Formation continue ENPC, March 1989.
35. Richard P. Journée ATHIL-ITBTP.

36. Rokugo K., Ohno S., Koyanagi W. Automatical measuring system of load-displacement curves including post-failure regions of concrete specimens. Lausanne Congress, 1987.
37. Rosenberg A.M., Gaidis J.M. A new mineral admixture for high-strength concrete - Proposed mechanism for strength enhancement. Second International Conference on the Use of Fly Ash, Silica Fume, Slag and Natural Pozzolans in Concrete, Supplementary paper, Madrid, 1986.
38. Rossi P. Fragilité et ductilité des matériaux et des structures de Génie Civil. Bulletin de Liaison des Laboratoires des Ponts et Chaussées, N° 1988.
39. Russell H.G. High-strength concrete in North-America. Conference on "Utilization of high-strength concrete", Stavanger, June 1987.
40. Setra Ponts mixtes acier-béton bipoutres - Guide de conception. October 1985.
41. Trouillet P. Private communication, 1989.
42. Wecharatana M., Chimamphant S. Bond strength of deformed bars and steel fibers in high-strength concrete. MRS Symposium, Boston, 1987.
43. Weber L. Etude de la ductilité du béton à très hautes performances armé en compression. Degree project, LCPC/ENPC, June 1988.
44. Wolsiefer Ultra high-strength field placeable concrete in the range of 10.000 to 18.000 psi (69 to 124 MPa). Concrete International, vol. 6, N° 4, pp.25-31, 1984.
45. Yogendran V., Langan B.W., Haque M.N., Ward M.A. Silica-fume in high-strength concrete. ACI Materials Journal, pp 124-129, March-April 1987.
46. De Larrard F., Malier Y. Propriétés constructives des bétons à très hautes performances - De la micro- à la macro-structure. Annales de l'ITBTP, December 1990.

7 CREEP IN HIGH AND VERY HIGH PERFORMANCE CONCRETE

F. de LARRARD and P. ACKER
L.C.P.C., Paris, France

We introduce here the results of creep tests carried out H P concretes of the bridges of Joigny (60 MPa without silica fume) and Pertuiset (65 MPa with silica fume), an H P concrete specially designed to limit cracking in a nuclear reactor containment, as well as on a V H P concrete (90 MPa) designed for a bridge application. The results are analyzed in terms of present knowledge regarding creep mechanisms. The most important results are kinetics and ageing which are clearly different from current concretes and, for silica-fume H P concretes, the absence of drying creep and hence of a size effect.

1 Review of creep mechanisms

If the loading of a specimen is maintained for a certain time (creep test), its deformation doubles, if it is an ordinary concrete, in a few weeks, triples after several months and, under extreme conditions, can even quintuple in a few years. The same phenomenon has been observed with the same intensity, under tension, torsion, etc. It depends on many parameters : nature of concrete, age at loading and, especially, ambient conditions.

When loading is removed, a clear instantaneous reduction in deformation is observed (very close in absolute value to that of a control specimen loaded for that age) followed by a delayed reduction, called recovery. This is however much smaller in absolute value than the corresponding creep and stabilizes after a few weeks.

Creep depends essentially on the following:

Applied load: under moderate loads, it may be considered to be proportional to the applied load and hence to the instantaneous deformation ; beyond 50 % of the ultimate load, it however increases faster than the stress,

High Performance Concrete: From material to structure. Edited by Yves Malier. © 1992 Taylor & Francis.
Published by Taylor & Francis, 2 Park Square, Milton Park, Abingdon, Oxon, OX14 4RN. ISBN 0 419 17600 4.

The nature of the concrete: in the same way as instantaneous deformation, except for special concretes whose characteristics specific to drying kinetics are different. This is the case for lightweight aggregates concretes in which the porous aggregates constitute a water reserve ; they generally exhibit less creep than an ordinary concrete of the same strength.

Ambient conditions: when there is no exchange of water with the exterior, creep (then called basic creep) is roughly proportional to the evaporate water content, and a concrete previously dried at 105°C practically does not creep. However, in practice, a concrete undergoes a more or less rapid loss of water depending on the ambient climate, and this variation leads to significant creep, two to three times greater than the basic creep (Figure 1). This drying creep may be explained by a structural effect related to drying shrinkage : in an unloaded specimen, drying leads to free deformations which are faster and greater at the surface than in depth, and this results in tensions on the surface and cracking. In a specimen loaded under compression, this cracking decreases and drying results in greater macroscopic deformation. It has not been possible yet to completely quantify this effect, but it surely accounts for a large part of the drying creep ; further, it also provides a very good explanation of the size effect because, in thick pieces, drying is limited to the surface and conditions of basic creep are approached, with tensions and cracking of the surface.

Figure 1 : Creep deformations (delayed deformation less shrinkage) of ordinary concrete under different humidity conditions : creep in concrete which dries is not within the envelope, whereas it goes continually from the uniformly humid state to the uniformly dry state.

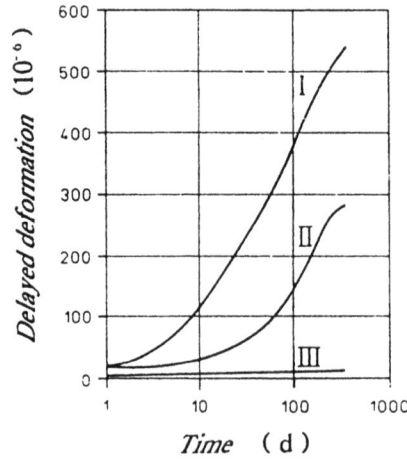

I   specimen drying normally,

II   specimen whose surface is coated with an impervious resin upon mould removal,

III   specimen previously dried at 40° C for 35 days then coated with an impervious resin.

2 Experimental results

2.1. Apparatus

The tests are carried out on cylinders 160 mm in diameter and 1 m long. The sections, located 250 mm from the ends, are provided with three inserts which are previously fixed inside the mould. They are arranged at 120° on the periphery of the section. The measurements after mould stripping are thus carried out on a length of 500 mm in which the Navier-Bernoulli principle applies. The symmetrical arrangement of the inserts also makes it possible to avoid spurious flexural effects.

Figure 2 : LCPC creep apparatus

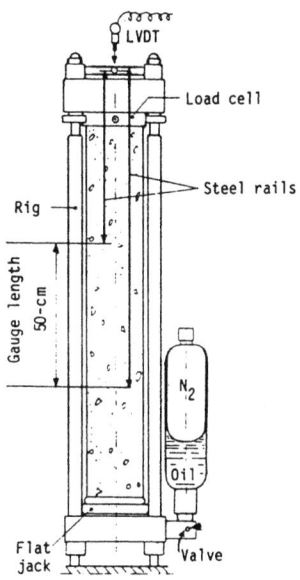

For the measurement of deformations, metal rods are fixed on the inserts, and two plates are placed on their upper ends (one corresponding to the lower section, the other to the upper section). With a mechanical comparator, the distance between these two plates is measured, giving exactly the average deformation of the concrete on the 500 mm of basic length.

The loading of the concrete is carried out by means of a flat Freyssinet type ram on which the specimen is placed. Pressurizing first takes place in a bottle containing oil and nitrogen, which are placed instantaneously in contact with the ram by means of a 3⁻way valve. Loading thus takes place in a few seconds, making it possible to obtain early creep in the material. This system moreover guarantees excellent maintenance of the applied load, owing to the high compressibility of nitrogen.

2.2. Concrete of Joigny bridge
The mix for the Joigny bridge concrete is given in Table 1.

Table 1 : Composition of Joigny bridge concrete

Aggregates, 5/20 mm	1027 kg/m^3
Sand of the Yonne, 0/4 mm	648
Fine sand, 0/1 mm	105
Cormeilles "HP" OPC cement	450
Water	158
Melment superplasticizer	11.25
(40 % dry extract)	
Melretard retarder	4.5
Slump	230 mm

Creep results (less shringkage) were smoothed with a law of the form :

$$\text{where :} \qquad \mathcal{E}_v = K_{f1}(t_o) \cdot \frac{\sigma_i}{E} \cdot f(t-t_o) \qquad\qquad f(t-t_o) = \frac{\sqrt{t-t_o}}{B + \sqrt{t-t_o}}$$

The values of K_{f1}, the creep coefficient and of B (parameter indicative of the kinetics of the phenomenon) are given in Table 2.

Table 2 : Results of mechanical tests on Joigny bridge concrete.
  1) S   = concrete poured on the site
     L_1 and L_2 = concrete tested in the lab.
  2) P   = specimen protected from drying
     U   = unprotected specimen

N°	1	2	3	4	5	6	7	8	9	10	12
Batch[1]	S	L_2	L_2	L_1	L_1	L_1	S	L_2	L_2	L_1	L_1
Cure[2]	P	P	P	P	P	P	U	U	U	U	U
f_c (MPa)	78	40	40	?	58	70	78	40	40	58	70
t_o (d)	28	3	3	5	7	28	28	3	3	7	28
σ (MPa)	?	14.5	8.0	8.0	14.0	22.6	21.0	14.0	9.0	14.0	23.0
ε_e (10^{-6})	140	260	150	115	220	270	560	470	240	310	620
ε_v (10^{-6})	446	390	240	182	314	490	490	350	240	90	544
K_{f1}	?	1.58	1.67	1.26	1.37	1.05	2.34	2.96	2.38	1.96	2.33
B ($d^{1/2}$)	4	1.7	1.7	5	4.5	3.7	11	4	5	7	9

By way of comparison, the corresponding values proposed by Appendix 2 ("Deformations in concrete") of BPEL specifications (French prestressed concrete design code) are shown in Table 3.

Table 3 : Creep in ordinary concrete according to BPEL 83 regulation (French Code)

Type	P	P	P	P	U	U	U	U
t_o (d)	3	5	7	28	3	5	7	28
K_{fl}	1.04	1.03	1.02	0.92	4.01	3.03	3.86	3.30
B $(d^{1/2})$	10	10	10	10	10	10	10	10

It is noted that the basic creep in slightly greater than that of ordinary concrete, particularly at early age. As concerns drying concrete, it is clearly reduced (by at least 50 %) and does not appear to be highly dependent on the loading age (which would appear to be obvious considering the slowness of the drying phenomenon), contrary to what is specified by the BPEL. Kinetic effects are also accelerated, in any case for the basic creep.

2.3 Pertuiset Bridge concrete

With the opportunity of the construction of the Pertuiset bridge (concrete G_1), several concrete mix designs were tested in order to analyze the influence of the different compositional parameters, Table 4.

Table 4 : Concrete mix designs tested during construction of the Pertuiset Bridge

Type	G_0	G_1	G_2	H_1	H_2
Seine aggregates, 5/20 mm	1020	1015	1015	1018	1022
Seine aggregates, 0/5 mm	657	651	650	647	650
Fine Fontainebleau sand 0/1 mm	41	42	43	43	43
Cormeilles HP OPC cement	456	453	453	453	455
Silica fume	36	36	36	-	-
Melment superplasticizer	35	33	48	20	12
Melretard retarder	2.27	2.27	2.27	2.27	2.27
Water	121	147	172	156	175

Table 5 : Concrete mix designs tested during construction of the Pertuiset Bridge: mechanical properties

Type	G_0	G_1	G_2	H_1	H_2
Compressive strengths					
$f_{c\,1}$ (at 24 hours)	-	-	27.6	26.3	-
$f_{c\,14}$	88.5	77.0	67.7	63.3	59.3
$f_{c\,28}$	94.5	83.3	73.8	72.5	64.0
Tensile strength (splitting)					
$f_{t\,14}$	-	5.1	4.7	4.5	4.0
$f_{t\,28}$	5.8	5.9	4.9	5.0	4.2

The main creep results are given in Figures 3 and 4. A more thorough analysis is given by Auperin and de Larrard (ref. 5).

Figure 3 : Reduced deformations (total deformation, less shrinkage, divided by instantaneous deformation at 28 days) measured on concrete G_1, without drying, at different loading ages.

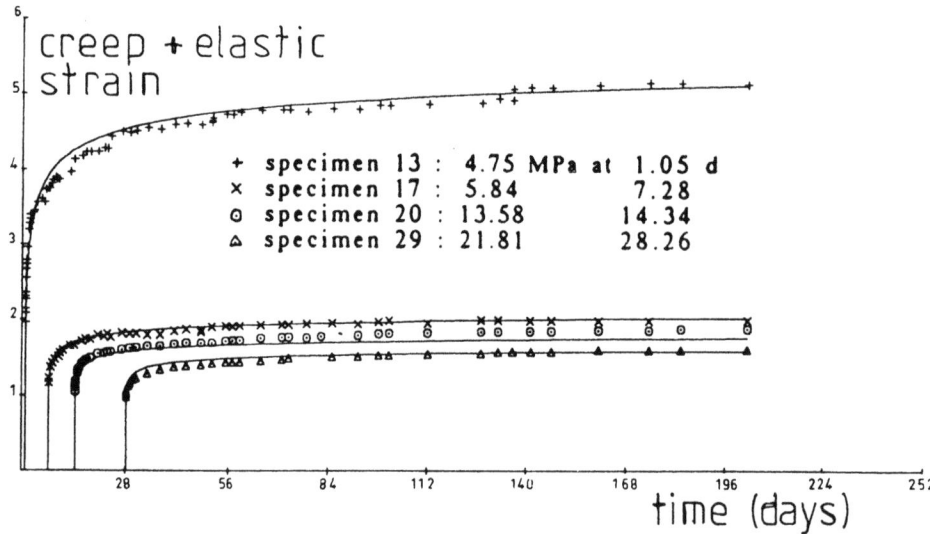

120

Figure 4 : Reduced deformations measured on concrete G_1, with drying and compared with the interpolated curves of the preceding results without drying (continous curves). It is noted that there is no, or practically no, drying creep.

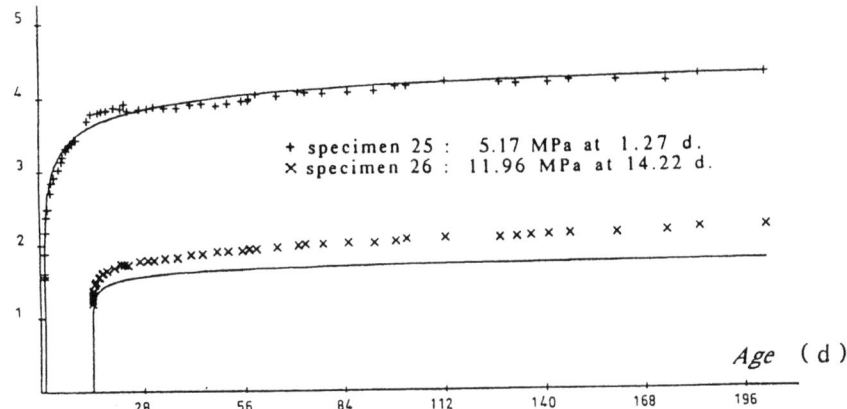

An examination of these results shows that, for this type of concrete containing silica fume:

- creep is proportional to the applied stress for loadings at the same age,
- the creep coefficient K_{fl}, varies very quickly with the age of the concrete under load, much faster than predicted by the formula of the french regulation (BPEL, Appendix 2) (See Fig. 5)..

Figure 5 : creep coefficient (creep deformation divided by instantaneous deformation at 28 days) of concrete G_1 as a function of the loading age and compared with values provided by the french regulation BPEL with and without drying.

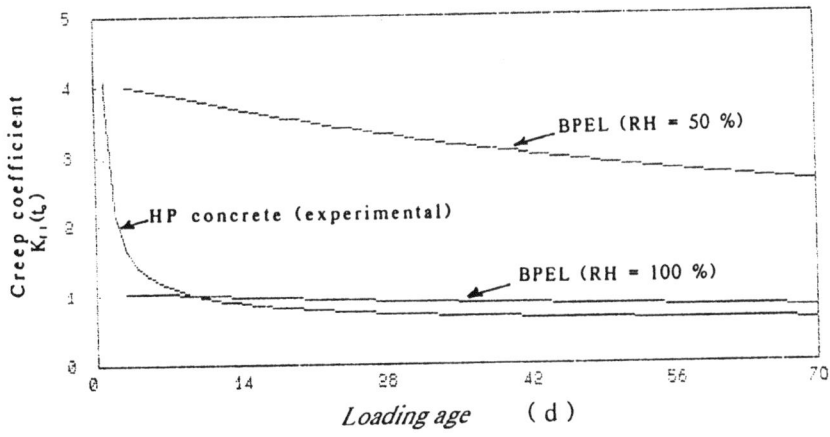

- drying creep is practically nonexistent ; there is consequently no longer any size effect on creep,
- these deformations occur with much faster kinetics than for ordinary concrete (29 % in 24 hours !).

2.4. Concrete for the Civaux power plant containment

As part of the construction of the nuclear power stations at Civaux, an HP concrete mix design study was carried out by the LCPC with EDF (Electricity of France) and the Fougerolle-France company. The objective was to improve the imperviousness of the internal containment, while complying with the construction requirements under current jobsite conditions.

Considering the importance, in this type of structure, of cracks of thermal origin during construction, conventional high performance concretes, in spite of their intrinsic sealing, are penalized by their hydration heat and their endogeneous shrinkage. On the other hand, by substituting slightly exothermic mineral additives (fillers or silica fume) for part of the cement, it was possible to obtain a HP concrete (about 70 MPa at 28 days) which is compact and not highly subject to cracking.

The mix design of the HP concrete proposed for the Civaux containment in Table 6.

Table 6 : Formula for HP concrete with low hydration heat proposed for the construction of the internal containments (further study led to a slight subsequent modification) and main mechanical properties.

Arlaut aggregates,	12.5/25 mm	791 kg/m³
	4/12.5 mm	309
	0/5 mm	786
Airvault blended Portland cement		266
Limestone filler		87
Silica fume		40.3
Water		161
Rheobuild 1000		9.08
Compressive strength f_{c28}		67 MPa
Splitting strength		4.1 MPa
Modulus		36 000 MPa
1 - day bending strength		2.3 MPa
1 - day maximum deformation		120.10^{-6}
28 - days bending strength		5.5 MPa
28 - days maximum deformation		160.10^{-6}

Creep deformations with and without drying are given in Figure 6 in which they are compared with those of the control concrete (40 MPa of conventional mix design obtained with the same cement and the same aggregates). Creep specimens were loaded at 28 days.

The load was about 30 % of their instantaneous ultimate load. Drying specimens began to dry at demoulding (24 hours), at the same time as the corresponding shrinkage specimens.

The extrapolated values after 90 days are given in Table 7.

To the extent that the average radius of the tested specimens is very small and the drying conditions (20° C, 50 % RH) are very severe, it may be said that the two values of the creep coefficient K_{fl} given in this table constitute, experimentally, bounds for this coefficient for each concrete. Based on these results, the following observations may be made :

- this HP concrete exhibits negligible drying creep,
- as a direct consequence, there is no size effect on the creep, and the basic creep measured represents quite accurately the behaviour of the material in a structure,

Figure 6 : Delayed deformations (creep + shrinkage) of HP concrete of Civaux and of its control concrete with and without drying.

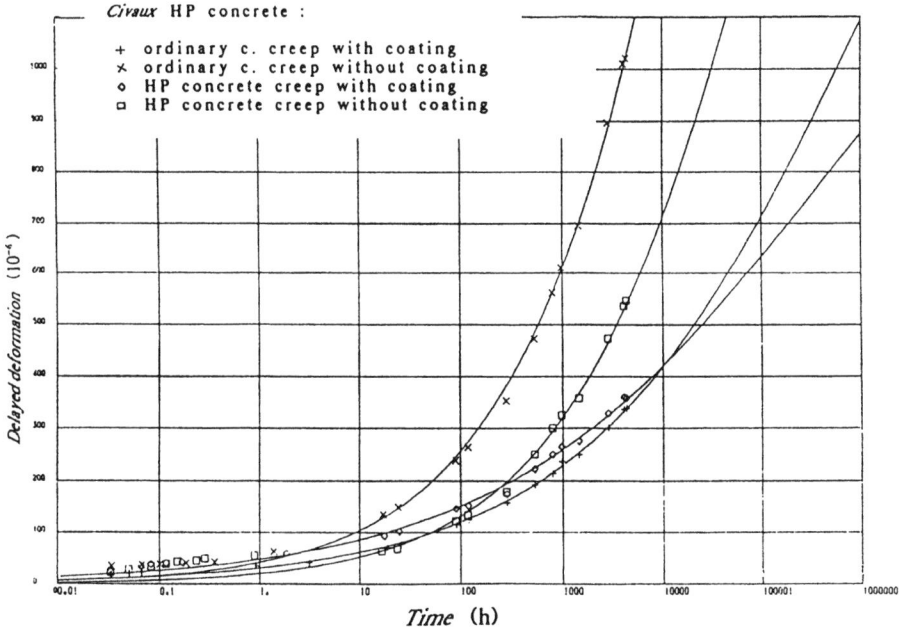

Table 7 : Values of shrinkage and creep (in μm/m) for HP concrete of Civaux and its control (specimens loaded at 28 days, values after 90 days loading).

	Control	HP concrete
Applied stress	12 MPa	20 MPa
Protected specimens		
Shrinkage between 28 and 90 days	20	10
Shrinkage + creep	250	280
Basic creep	230	270
Basic creep for 1 MPa	19.2	13.5
K_{fl} without drying	0.65	0.49
Drying specimens		
Shrinkage between 28 and 90 days	150	80
Shrinkage + creep	700	360
Drying creep	320	10
Drying creep for 1 MPa	26.5	0.5
K_{fl} with maximum drying	1.56	0.50

- the reduction of creep due to the substitution of HP concrete for ordinary concrete is thus between 30 and 70 %, depending on the extent of the drying effect ; thus, for the internal containment, in which drying affects no more than one-third of the section, a reduction of 30 to 40 % must be expected.

2.5 Very high performance concrete
In his paper, de Larrard investigated creep in a V H P concrete whose mix design (given in Table 8) complied with current industrial requirements, particularly with regard to placement under jobsite conditions.

Creep deformations (given in Figure 7, without subtracting shrinkage) were obtained with and without drying under the same conditions as the preceding series.

Table 8 : Composition of VHP concrete and main mechanical
properties

Boulonnais limestone aggregates, 5/20 mm	1 265 kg/m³
0/5 mm	326
Seine sand	326
HTS 55 OPC cement	421
Silica fume	42.1
Plasticizer	7.59
Water	112
Slump	20 cm
28-days compressive strength	101 MPa
Modulus	53 400 MPa

Figure 7 : Delayed deformations of VHP concrete and of its control
concrete with and without drying.

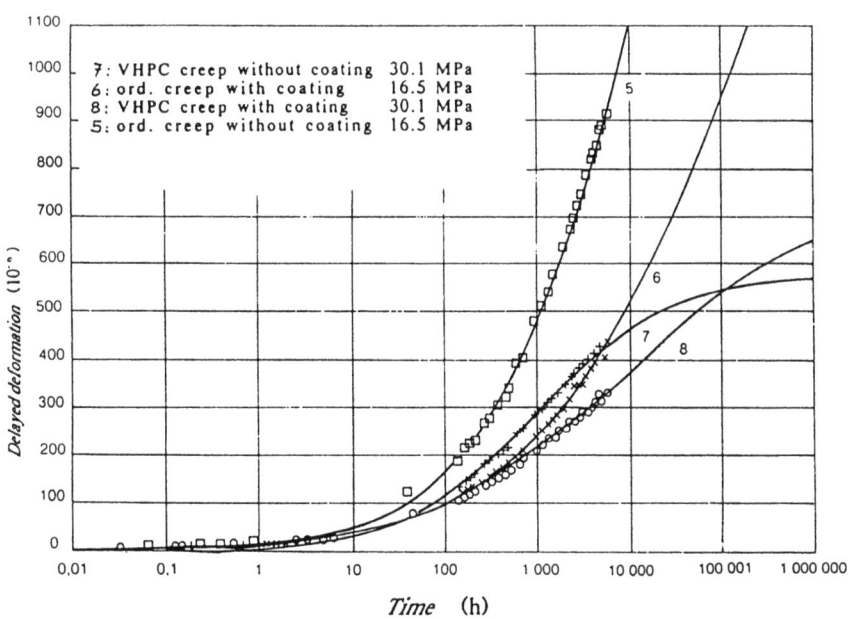

125

Compared with a control concrete having the same components but without additives, creep is divided by two (it drops from 23.3 to 10.5 10^{-6} for a unit stress of 1 MPa). De Larrard explains this decrease quite simply by the reduction in the volume of hydrates formed in the matrix of the V H P concrete, as well as by the reduction in internal moisture.

As regards drying creep, it is eliminated whereas drying shrinkage is barely divided by two. This disappearance would thus be due both to the reduction in the drying intensity (because there is much less evaporable water) and to the increase in the tensile strength of the VHP concrete.

3 Conclusions

The results obtained on HP and VHP concretes quite different in their mix designs give a quite complete view of the various aspects of the delayed behaviour of these materials, a view which appears today to be consistent with the state of knowledge regarding the physical and chemical mechanisms that control creep, and particularly :

- the very great decrease in *drying creep* - and its elimination when silica fume is used - and, what is important for the design engineer, the absence of any resulting *size effect*,
- a modification in the kinetics of *basic creep*, stabilizing much faster than that of ordinary concrete,
- note should however be made of a much more marked *"ageing"* which results, at a very early age (1 to 3 days) of loading, in total deformations of the same order as those of an ordinary concrete under extreme drying conditions.

References

1. Acker P., "Comportement mécanique des bétons : apports de l'approche physico-chimique", Rapport de recherche des LPC, **1 5 2**, 1988
2. De Larrard F., "Formulation et propriétés des bétons à très hautes performances", Rapport de recherche des LPC, **1 4 9**, 1988
3. De Larrard F., Acker P., Attolou A., Malier Y., "Fluage des bétons à très hautes performances", IABSE Congress, Helsinki, 1988
4. De Larrard F., Malier Y., "Propriétés constructives des bétons à très hautes performances : de la micro à la macrostructure", Annales de l'ITBTP, **4 7 9**, Dec. 1989
5. Auperin M., Richard P., De Larrard F., Acker P., "Retrait et fluage de bétons à hautes performances. Influence de l'âge au chargement", Annales de l'ITBTP, **4 7 4**, May 1989.

8 BOND PROPERTIES OF HIGH PERFORMANCE CONCRETE

M. LORRAIN
I.N.S.A. de Toulouse, France

1 Introduction

A considerable amount of research has been carried out on the bond between reinforcement and concrete. Since Ward in 1883, every angle of bond in reinforced concrete has been studied and many attempts have been made, however without complete success, to give it a numerical model in order to describe in detail the mechanical behaviour of reinforced concrete. The new concrete technologies, especially high strength concretes, brought forward once more the problem of bond strength.

This paper presents the first results we have obtained on this subject, in collaboration with Dr. H. Khelafi (1) (2) (3), between 1984 and 1988. In the first section, we explain the experimental procedure used to evaluate the bond strength. The second section is devoted to the presentation of the results we got in comparisons with the values reached in the case of ordinary concrete. The conclusions are given in the third section.

2 Experimental method

2.1. Testing of specimens

The tests carried out consisted of pull-out tests performed on cylindrical specimens. The testing device, schematically represented in Figure 1, has been developed in our laboratory according to R. L'Hermite's idea (4). The distinctive feature of this test is the permanent presence of a metallic mould allowing the distribution of the pull-reaction.

2.2. Measurements

During the tests, the displacement of the free edge of the embedded bar and the intensity of the pulling force have been

High Performance Concrete: From material to structure. Edited by Yves Malier. © 1992 Taylor & Francis.
Published by Taylor & Francis, 2 Park Square, Milton Park, Abingdon, Oxon, OX14 4RN. ISBN 0 419 17600 4.

measured. The displacement measurements have been performed with the aid of an inductive transducer (range + 1 mm). The pulling force was given by an Amsler testing machine.

2.3. Materials characteristics

2.3.1. Concretes
In a first stage, we tried to compare the bond strengths of high strength concrete and ordinary concrete. The characteristics of these two concretes are the following :

(a) High strength concrete
It is a concrete made of cement CPA 55R with the addition of CSF from Marignac and of Sika superplasticizer. The limestone aggregates (maximum diameter 10 mm) come from Saint-Beat. The grading used is given in Figure 2. The mix proportions and the mechanical properties of the material are given in Tables 1a, 1b and 2.

(b) Reference ordinary concrete
It is an ordinary concrete with the same aggregates and grading, but without silica fume and superplasticizer. The cement content is 500 kg/m³. Its workability is the same as the high strength concrete. We can notice that this ordinary concrete is a very good one because its compressive strength after 28 days reaches 64 MPa.

2.3.2. Reinforcement
The reinforcing bars used for the test specimens were prestressing wires (TBR, C3, ND1) of 8 mm diameter, with elastic limit of 1500 MPa in order to avoid systematic failure of the bar.

3 Results
We will successively present two series of results. The first series deals with the comparison of bond strength between ordinary and high strength concrete. The second series shows the influence of several constitutive parameters on bond strength.

3.1. Behaviour and strength of bond: comparison between high strength and ordinary concretes
The behaviour of bond can be studied through the variations of the displacement of the embedded bar as a function of the pulling force.
    The embedded length is 12 cm; the tests have been performed on concretes (high strength as well as ordinary) at the age of three days, as soon as the maturity of the concrete can be considered as satisfactory regarding the material microstructure (5). The results are given in Figure 3 and Figure 4, in rough and adimensional values respectively. It can be noticed (Figure 3) that

Figure 1 : Test specimen

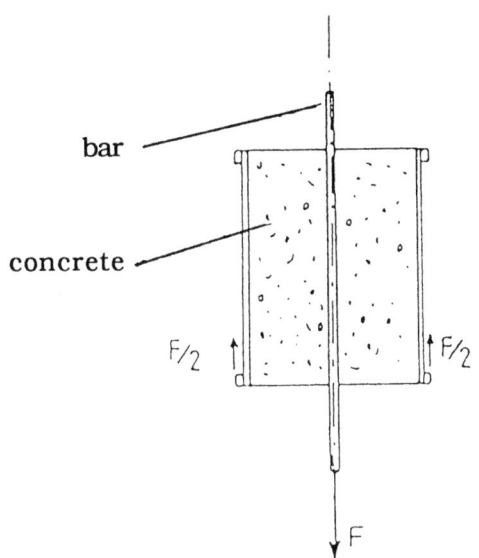

Figure 2 : Grading of aggregates

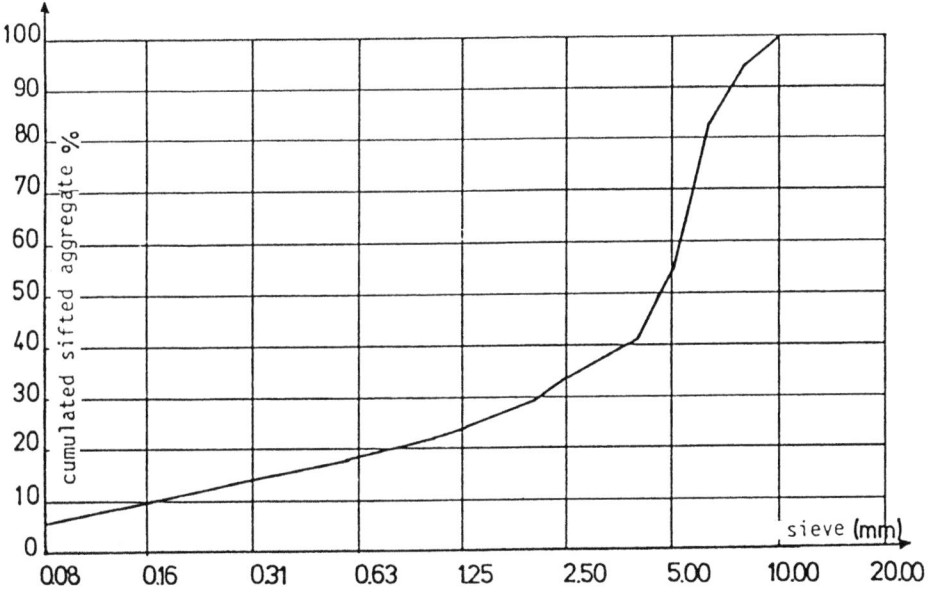

Table 1.a : Characteristics of Condensed Silica Fume

Colour	% SiO$_2$	Density	App. vol. gravity (kg/m³)	ϕ (10^{-3} mm) mean	Specific surface m²/kg
Black	81	2.2	600	0.3	18.2

Table 1.b : Mix proportions of the high strength concrete

Material	Batching (kg/m³ of concrete)
Cement CPA 55R	450
CSF	50, 10 % (C + CSF)
Superplasticizers	12
Water	155
Aggregate	1 870

Table 2 : Mechanical characteristics of high strength and ordinary concretes

Age (days) / Strength (MPa)	1	3	7	14	28
f$_c$ HSC	53.7	72.0	86.0	92.6	106.0
f$_c$ Reference C	33.2	46.5	55.8	61.0	64.0
f$_t$ HSC	5.1	6.0	7.0	7.5	7.9
f$_t$ Reference C	3.2	4.4	4.8	5.5	5.8

130

Figure 3 : Displacement of the bar free end as function of the pulling force

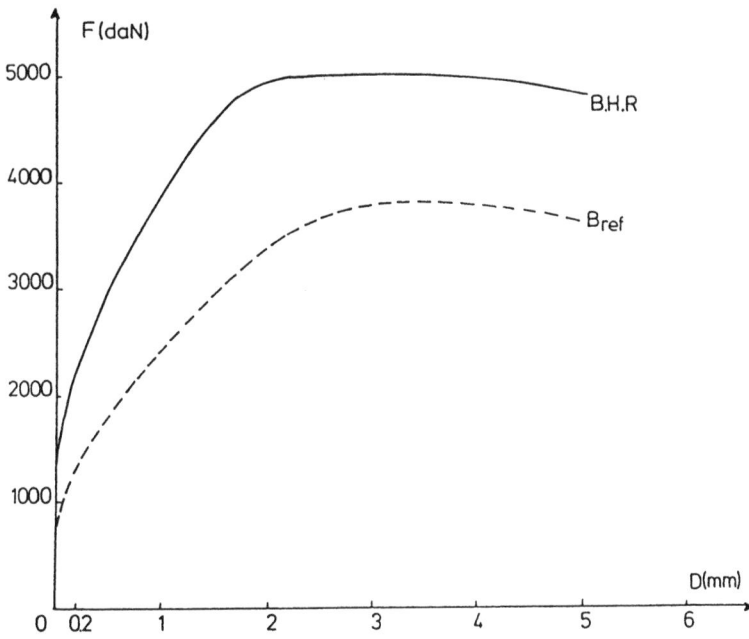

Figure 4 : Displacement of the bar free end as function of the pulling force/concrete tensible stress ratio

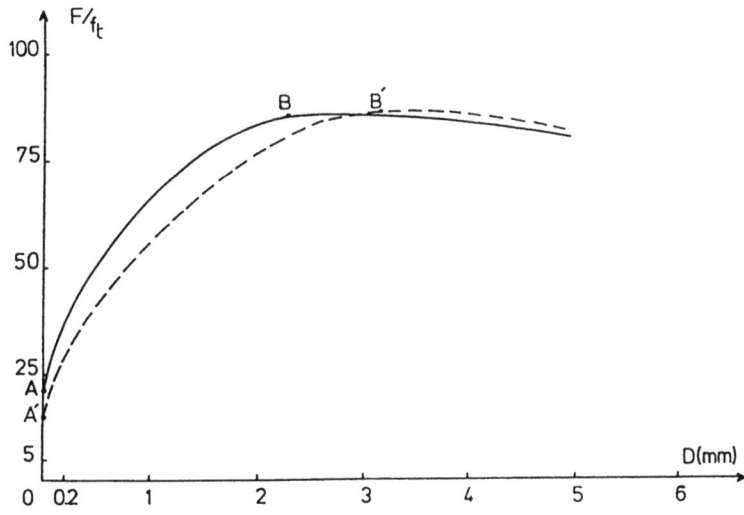

the bond in high strength concrete is better than in ordinary concrete, as far as the following are concerned:

- the intensity of the pulling force corresponding to the first damage,
- the stiffness of the bond, during the stage of controlled slip (ascending branch),
- the maximum value of the pulling force (peak of the curve), giving the value of the bond resistance.

The curves of Figure 4 also show that the improvement of the binding resistance cannot be only explained by the increasing mechanical strength of embedding concrete. After dividing the F values by the tensible strength f_t of the concrete, the D-F/f_t curves are not superimposed, and we can observe:

- a later appearance of the first displacement (A above A'),
- a larger stiffness of the bond (slope AB > slope A'B') in the case of high strength concrete.

The improvement of the density of the microstructure of high strength concrete due to the addition of CSF may be considered as a suitable explanation of these positive effects: a better density of the embedding concrete delays the appearance and propagation of damage in the bond.

This first series of experiments demonstrates that the micro-structure of high strength concrete plays an essential role over the behaviour of bond and we will discuss this aspect later in studying the first irreversible disorders of the bond.

3.2. Influence of the bond constitutive parameters on the damage appearance and propagation

We are here interested in the point A and in the branch AB of the D-F curve. In this view, our experimental procedure analyzes the reversibility capability of the bond, according to a specific experimental procedure. The load F is applied according to successive cycles of continuously increasing magnitude with a return to zero. We measure the relative displacement of the bar in relation to the embedding concrete at the maximum load, so that we can calculate the reversibility rate R such as:

$$R = \frac{d_t - d_r}{d_t}$$

with : d_t = total displacement with maximal load/cycle
d_r = residual displacement.

We notice the appearance of the first damage when R decreases from 100 %. The envelope curve D-F gives the evolution of the damage along the bonding.

We varied the age of the embedding concrete, CSF and cement content, the water/binder ratio, the embedding length and the roughness of the surface of the reinforcement.

3.2.1. Influence of the age of the embedding concrete

The variations of the displacement of the bar in relation to the concrete, as a function of the pulling force, are shown in Figure 5. The reversibility rate/pulling force curves for the same ages are given in Fig. 6. The different ages we considered are: 1, 3, 7, 14, 28 and 90 days, and 6 months.

It can be noticed that the appearance of the first damage is all the latter since the embedding concrete is older. Besides, the stiffness of the binding increases with the age of the concrete, which corresponds to a slowing down of the damage propagation with age.

Figure 5 : Displacement of the bar free end as a function of the pulling force for different ages

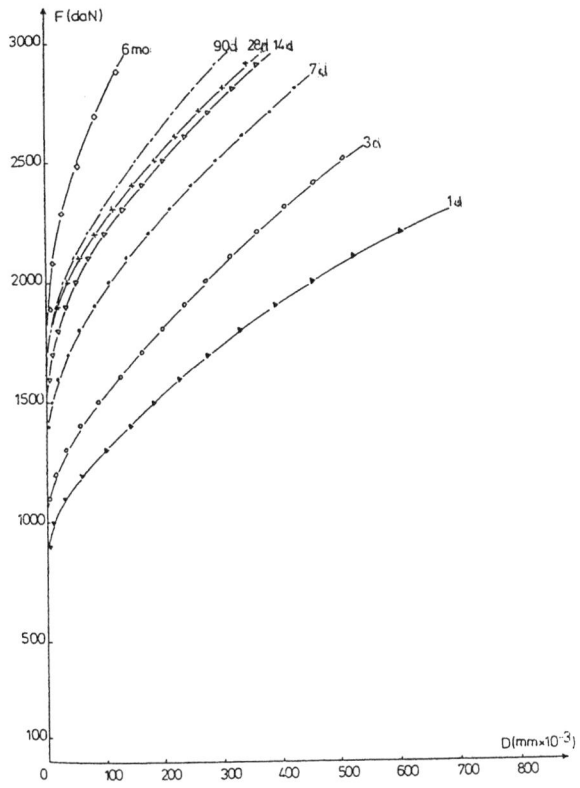

Figure 6 : Reversibility ratio of the bar free end for different ages

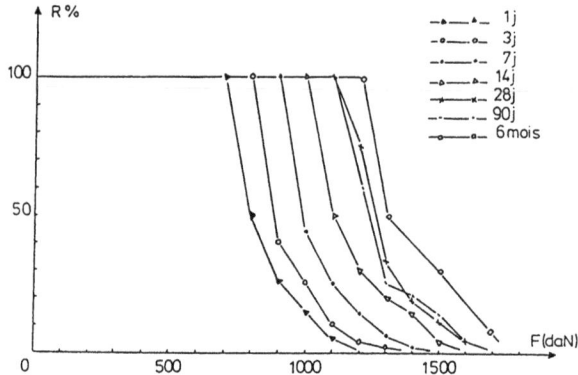

3.2.2. Influence of the percentage of CSF

The percentage by weight of CSF ranged from 5 % to 30 % of the weight of the binding agent, the other parameters being kept constant, except the age of the concrete equal to 3 and 28 days in order to remove any obstacle of the maturity lack of the concrete.

Figs. 7 and 8 show respectively the variations of the total displacement of the reinforcement in relation to the embedding concrete, and of the reversibility rate as functions of the pulling force at 3 days. Figs. 9 and 10 give the corresponding information at 28 days. In each case, the relative curves have been put together with the CSF percentage of 5, 10, 15, 20 and 30 %.

The appearance and the propagation of the damage seem not to vary regularly in relation to the CSF percentage. The classification which we could establish according to the appearance of the damage doesn't stand for the classification of the propagation, and everything is called into question when the age of the concrete goes from 3 days to 28 days.

3.2.3. Influence of the cement content

The batching of the binding agent varied from 400 to 600 kg/m³ of concrete, other parameters being kept constant except the age of the concrete (3 days and 28 days).

The variations of the total displacement of the bar, in relation to the embedding concrete, and of the reversibility rate according to the pulling force are respectively given in Figs. 11 and 12 at 3 days, and in Figs. 13 and 14 at 28 days. In each case, the relative curves have been put together with the cement content 400 and 600 Kg/m³.

The appearance and propagation of the bond damage increases with the cement content. The cement content of 500 kg/m³ can be considered as an optimum, as the concrete is 3 or 28 days old.

Figure 7 : Displacement of the bar free end as a function of the
pulling force for different percentage 3 days

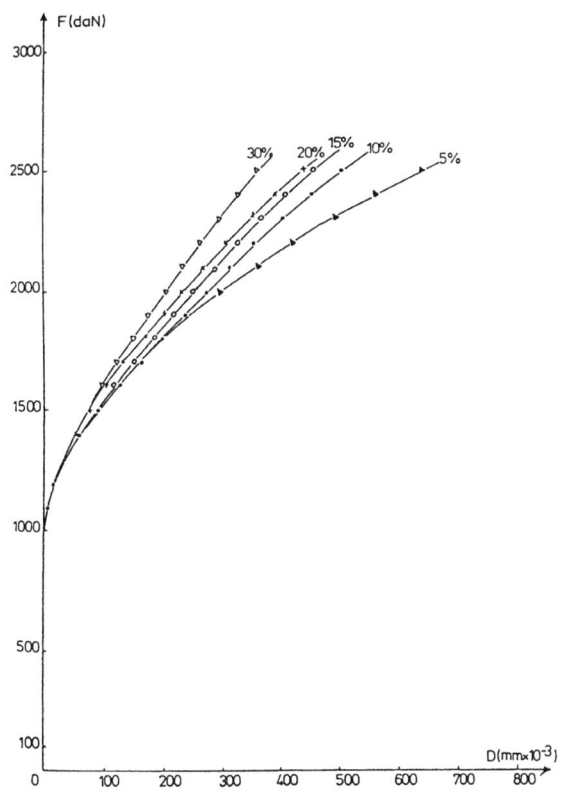

Figure 8 : Reversibility ratio of the bar free end for different
percentage 3 days

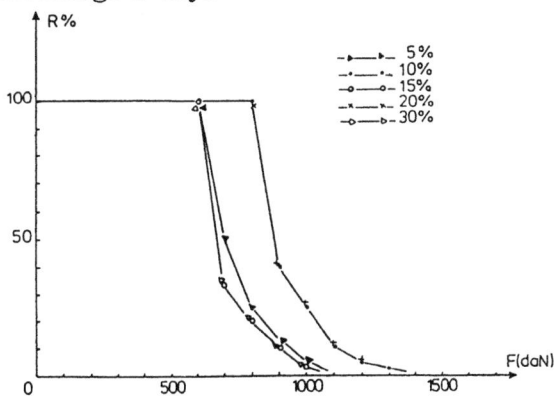

Figure 9 : Displacement of the bar free end as a function of the pulling force for different percentage 28 days

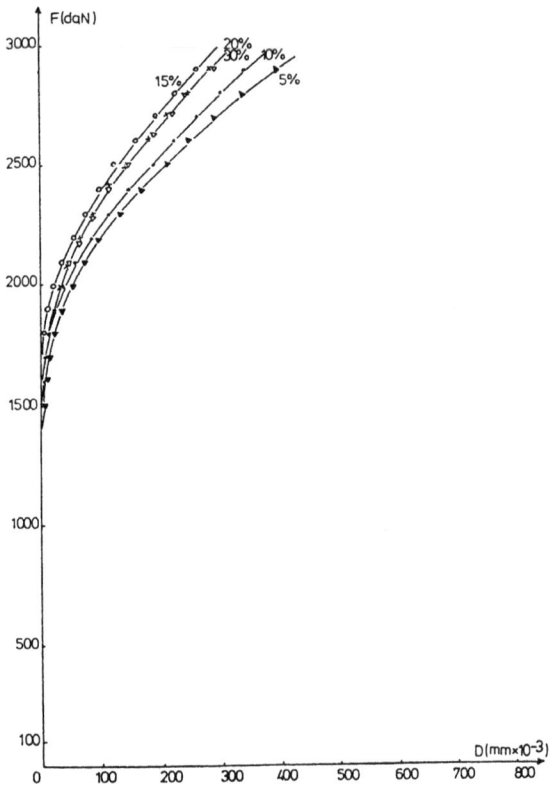

Figure 10 : Reversibility ratio of the bar free end for different percentage 28 days

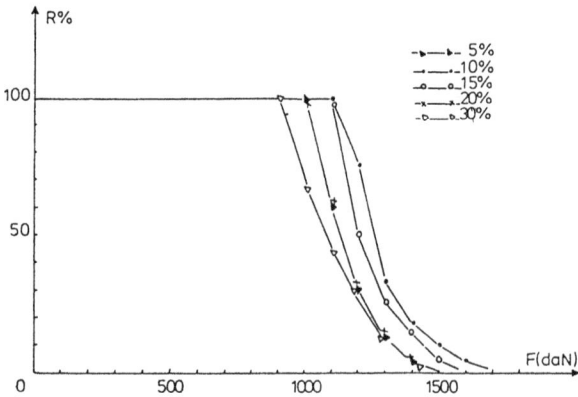

Figure 11 : Displacement of the bar free end as a function of the
pulling force for different cement contents 3 days

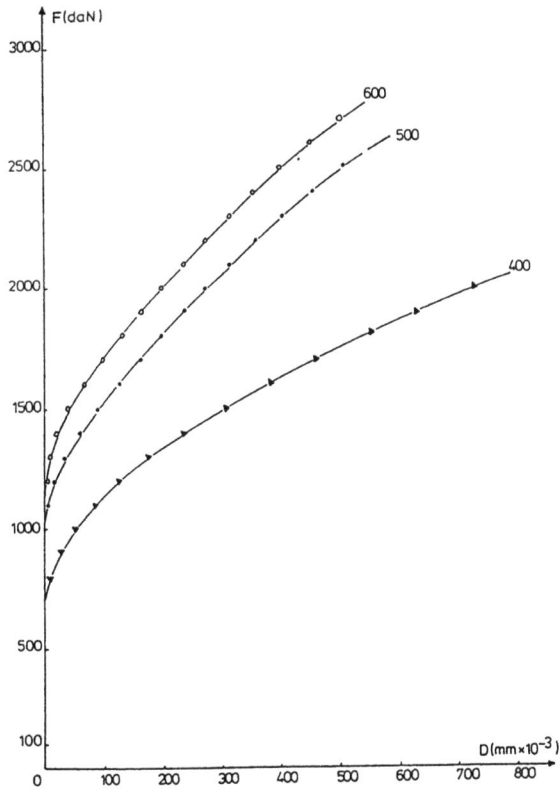

Figure 12 : Reversibility ratio of the bar free end for different
cement contents 3 days

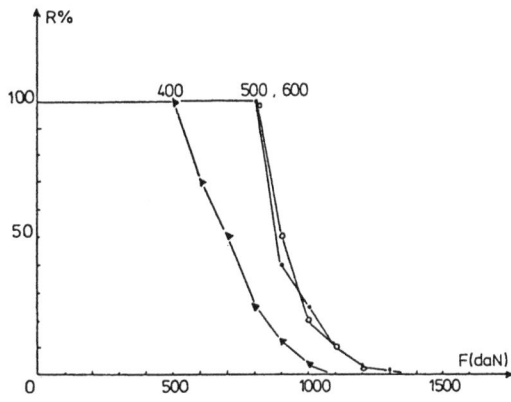

Figure 13 : Displacement of the bar free end as a function of the
pulling force for different cement contents 28 days

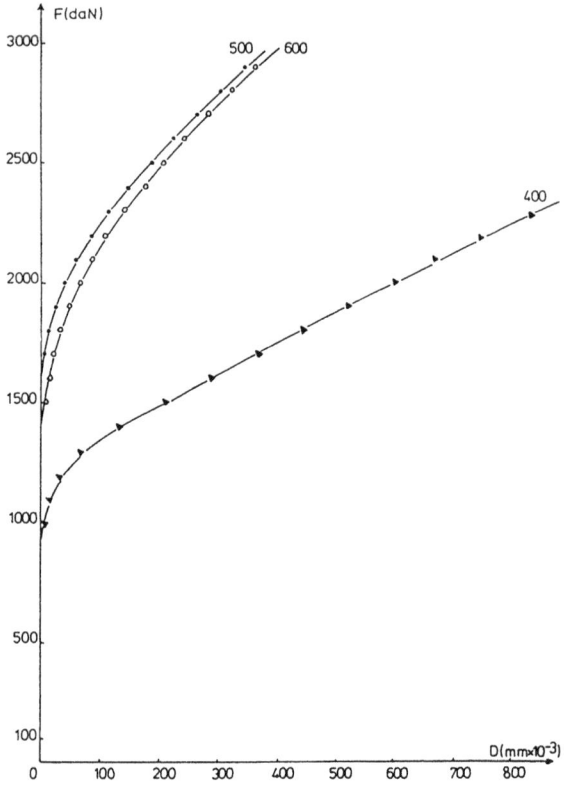

Figure 14 : Reversibility ratio of the bar free end for different
cement contents 28 days

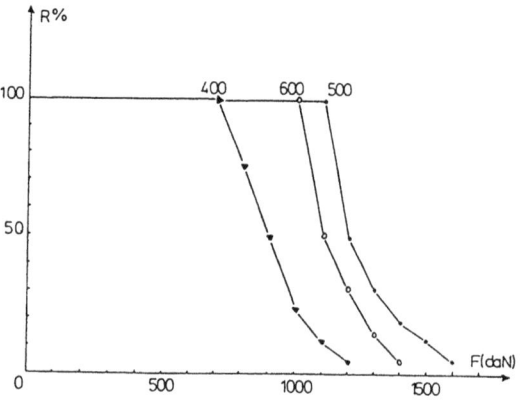

3.2.4. Influence of the water/binder ratio

We varied the water/binder ratio from 0.31 to 0.38, the other parameters being maintained constant except the age of the concrete (3 and 28 days old).

Figs. 15 and 16 show respectively the variations of the total displacement of the bar and of the reversibility rate as functions of the pulling force at 3 days; Figs. 17 and 18 give the same information at 28 days. In each case, the relative curves have been put together with the water/binder ratio 0.31, 0.35, 0.38.

It can be observed that the appearance and the propagation of the bond damage decreases with decreasing water/binder ratio. However, there is small difference between the results for ratios equal to 0.35 and 0.31 at 3 and 28 days.

Figure 15 : Displacement of the bar free end as a function of the pulling force for different water/binder ratio. 3 days

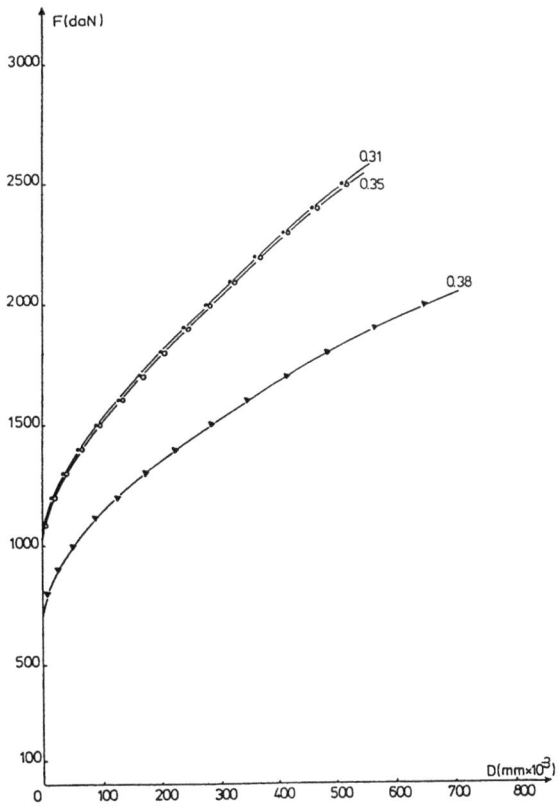

Figure 16 : Reversibility ratio of the bar free end for different water/binder ratio. 3 days

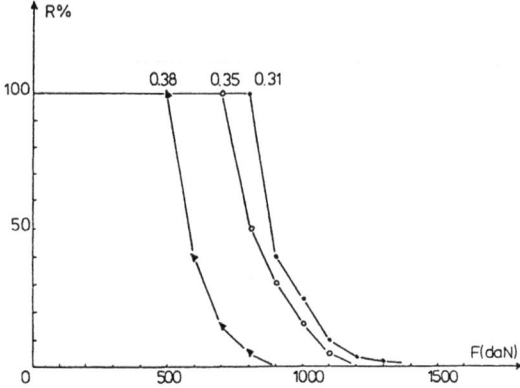

Figure 17 : Displacement of the bar free end as a function of the pulling force for different water/binder ratio. 28 days

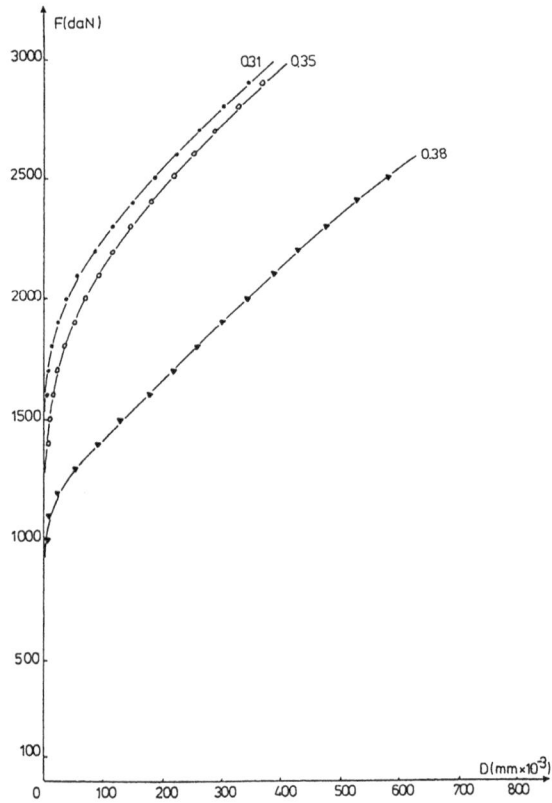

140

Figure 18 : Reversibility ratio of the bar free end for different water/binder ratio. 28 days

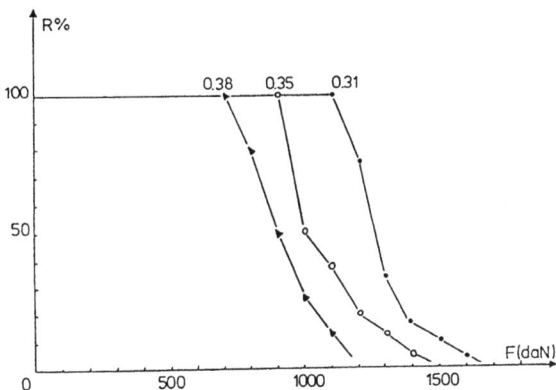

3.2.5. Influence of the embedded lenght of the bar
The embedded lenght of the bar varied from 4 to 25 cm, the other parameters being maintained constant and the concrete being 3 days old.
  The total displacement-pulling force and the reversibility rate-pulling force curves are given in Figs. 19 and 20, in which the relative curves have been put together with the embedded length 4, 8, 12, 16, 20 and 25 cm.
  The appearance and the propagation of the bond damage decreases with increasing embedded length.

3.2.6. Influence of the reinforcement roughness
Two kinds of reinforcement have been studied, deformed and plain wires. The results are presented in Figure 21, giving the variations of the reinforcement displacement at 3 and 28 days as a function of the pulling force.
  It can be observed that the presence of the ribs increases considerably the bond at any moment of its evolution.

141

Figure 19 : Displacement of the bar free end as a function of the
pulling force for different embedding length. 3 days

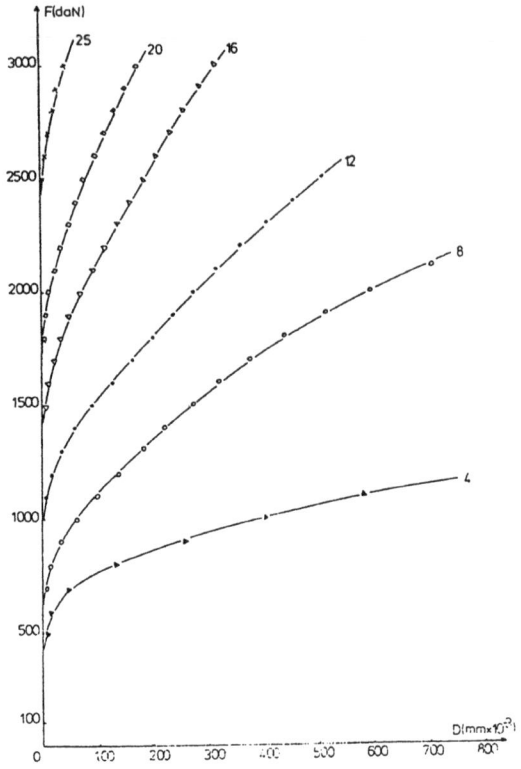

Figure 20 : Reversibility ratio of the bar free end for different
embedding lengths. 3 days

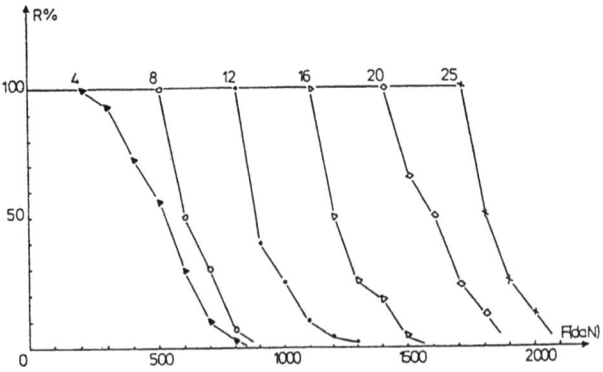

Figure 21 : Displacement of the bar free end as a function of the
pulling force for - reinforcement roughness

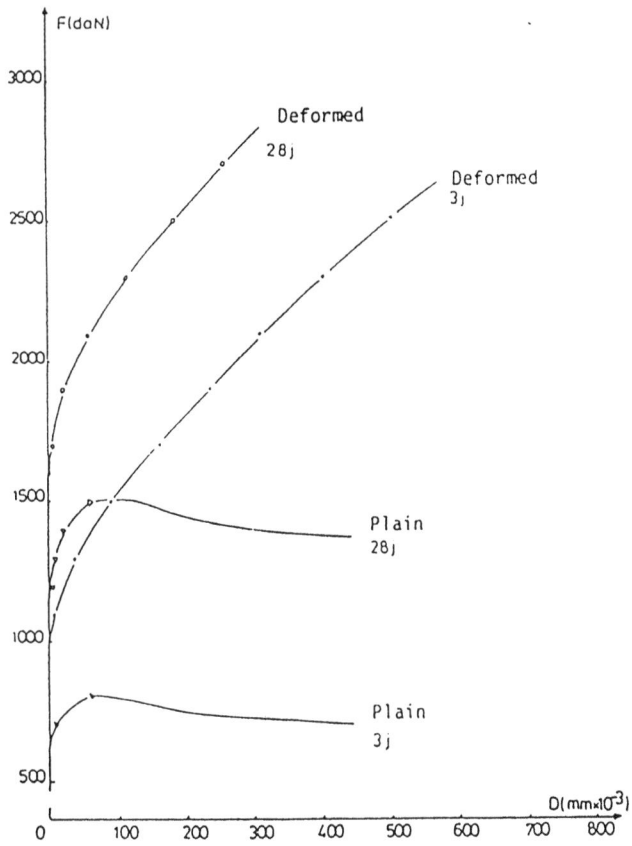

4 Conclusions

The following conclusions can be drawn from our investigations:

- the ultimate strength of bond is directly proportional to the
mechanical strength of the embedding concrete,
- the appearance and propagation of the bond damage not only
results from the mechanical strength of the concrete, but also
from its microstructure,
- increasing the percentage of the binder, the water/binder ratio,
the embedded length of reinforcement andits surface roughness
have all been beneficial whereas the influence of the CSF
percentage in concrete is not yet clear.

143

References

1. H. Khelafi "Construction à l'étude de l'association armature-béton de haute performance". Thèse de Doctorat en Génie Civil - I.N.S.A. de Toulouse, 8 juillet 1988.
2. M. Lorrain and H. Khelafi "Sur la résistance de la liaison armature-béton de haute résistance". Annales de l'I.T.B.T.P., n° 470, décembre 1988.
3. M. Lorrain and H. Khelafi "Construction à l'étude de l'endommagement de la liaison armature-béton de haute performance". Annales de l'I.T.B.T.P., n° 472, février 1989.
4. M. Lorrain and Ph. Bravi "Essai d'arrachement direct axisymétrique pour l'adhérence acier-béton". Conférence Interantionale GAMAC-RILEM "Mesures et Essais en Génie Civil", Lyon, 13-16 septembre 1988.
5. M.N. Oudjit "Réactivité des FSC en présence de chaux ou de CPA". Thèse de Doctorat d'Ingénieur - I.N.S.A. de Toulouse, 13 janvier 1986.

9 MONITORING OF THE CHEMICAL AND MECHANICAL CHANGES IN HIGH PERFORMANCE CONCRETES DURING THE FIRST DAYS

C. VERNET, G. CADORET
Technodes SA Groupe Ciments Français, Paris, France

1 Introduction

Improvements in performance levels of concrete relate not only to strength and durability, but also felxibility of adaptation to the restraints imposed on the construction of buildings and industrial concrete structure components.

We have, with this aim, been pursuing research into better understanding and control of the dynamic properties of cements for several years. Setting time and hardening rate are dependent on many hydration reactions, and on the establishment of chemical links between particles.

High Performance Concrete: From material to structure. Edited by Yves Malier. © 1992 Taylor & Francis.
Published by Taylor & Francis, 2 Park Square, Milton Park, Abingdon, Oxon, OX14 4RN. ISBN 0 419 17600 4.

The need to improve rheological characteristics, and to reduce the quantity of mixing water to reduce porosity, has led to the use of plasticizers or of high range water reducers (HRWR):

For a better control of the duration of the action of these additives and of their secondary effects, we have to study their interactions with hydration processes.

The importance of studying the kinetics of hydration and techniques for continuous measurement of chemical, thermal and mechanical parameters, in the case of High Performance Concretes (HPCs), can thus be appreciated.

Given the interactions which exist in our "chemical gearbox", it is important to carry out these measurements in a simultaneous way. In the past we had developed these techniques on stirred suspensions, with water/cement ratios of around 4. This is a situation in which hydration processes are close to the case of concretes: this has enabled us to better analyse the reaction mechanisms, and to interpret the recorded signals more reliably. We then automatized the simplest measurements, such as the electrical conductivity, and we applyied them to optimisation of cement-additive couples.

The use of this method has allowed us to select the most suitable additives, given the selected binders, to reach the goals set for concrete (rheology, control of short-term and/or 28 day strenth). In particular, it is thus possible to check, if need be, for the existence of strongly non-linear behaviours which leads to hazards on building sites and in factories, and sometimes to major construction incidents, through slight alterations in the quantity of admixtures.

However, it seemed to us that this method could not be applied directly to the formulation of concrete. Slurries, as a result of their dilution, which after the concentration of the additive, do not have exactly the same hydration kinetics. In addition, the absence of sand and, in some cases, of additions which could be used in concrete, removes an important element, the rheological behaviour. We have thus directed our research towards designing an equipment enabling us to work on the mortar constituent of concrete.

We here present the latest development of this kinetic approach, which consists in **relating the chemical processes to the mechanical consequences**. The parameters continuously recorded are as follow:

- electrical conductivity, which gives us information on the variations in composition of the aqueous phase in the capillary pores. In the case of mortars, the conductivity level also depends on the actual water/solid ratio, and enables the consumption of free water to be monitored,
- thermal flow, measured by isothermal calorimetry, is the linear combination of the speed and enthalpies of chemical reactions. By integration we obtain from it the hydration heat, which is an

expression of the overall reaction rate. The Le Chatelier contraction value, which is one of the shrinkage components, can be deduced from this,
- dissipation of ultrasonic waves enables us to follow the progressive change from the viscoplastic to the solid state,
- the speed of sound enables us to find the dynamic module of elasticity. Initially, we measured the resistance under compressive strength directly, at very close intervals, in such a way as to check on the strength/dynamic module relationship.

The examples below give information on the setting and hardening processes, and enable us to consider the relationship between reaction rates and mechanical development. We shall attempt to draw the conclusions of this for HPC optimisation.

2 Setting and hardening

In Appendix 1, we have illustrated in diagrammatic form the relationship between the development of the microstructure and the kinetics of hydration of mortars.

The graph curves in Figure 1 below show the physico-chemical development of a mortar with an initially high resistance, over a 5-period sequence.

2.1 Mixing mortar period
This is a period during which we see the ions from the cement components (aluminates, silicates, calcium, OH-, sulphates) dissolving. The initial dissolution is very rapid and exothermic. We can thus see a rapid increase in electrical conductivity and a thermal flow peak.

In the first minutes two rapid nucleation hydrates are formed:

- hydrated calcium silicate (CSH), formed by the combination of Ca^{2+}, $H_2SiO_4^{2-}$ and OH-ions from calcium silicates in the clinker,
- the ettringite, which is a hydrated calcium tri-sulpho-aluminate formed by the combination of the Ca^{2+}, AlO_4^{0-} and OH-ions, from the aluminate of the clinker and the setting regulator (gypsum, semi-hydrate, anhydrite).

2.2 Dormant period
The rapid rise of pH and of the calcium content of the mixing mortar water slows down the dissolution of the clinker constituents. The thermal flow decreases considerably without, however, disappearing completely. The formation of ettringite and of CSH continues at a slow pace, and the aqueous phase becomes supersaturated with respect to limestone (slow increase in conductivity). The absence of formation of portlandite, $Ca(OH)_2$,

147

Figure 1 : Non-destructive measurements of the mortar of a high initial resistance concrete

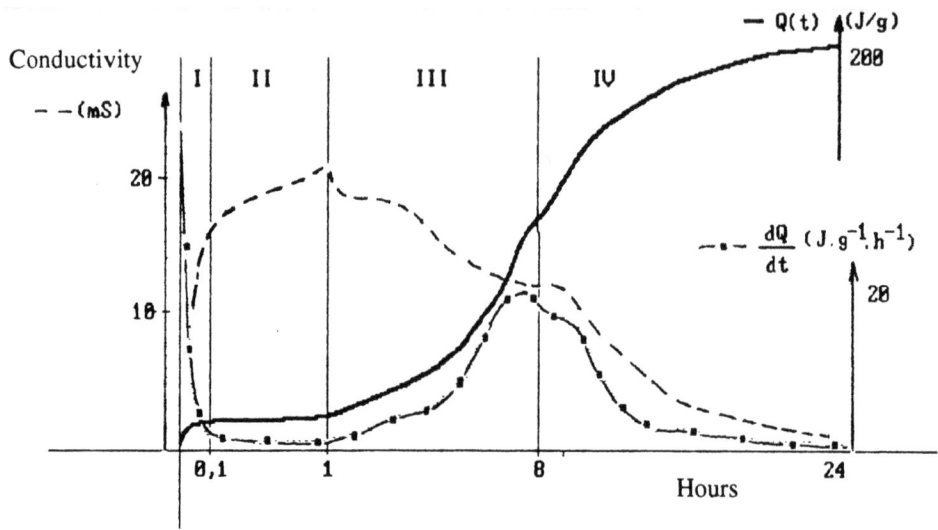

Mechanical aspect Ultrasonic

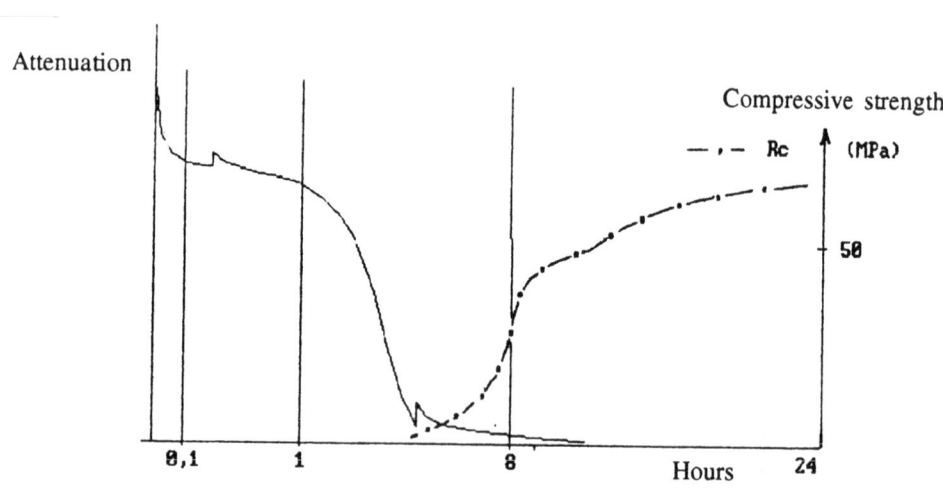

CPA HPR 400 kg/m3   Sand 0-1 450 kg/m3 Melamine 3.6%
E/C 0.35   Sand 1-4 225 kg/m3 Non chlorinated accelerator 3%

could be explained by its slow nucleation speed, given the competing formation of nuclei of CSH which, for their part, form very rapidly.

The ultrasonic signal declines (attenuation increases) with the agglomeration of particles by flocculation (in the case of non-fluidified concretes), and with their interconnection by the first hydrates (in all cases). Structural swelling due to crystals growth can lead to transitory lifts in the sensor, which manifest themselves as peaks in the attenuation curve.

2.3 Setting or acceleration period

Setting reactions are set off by a "chemical trigger", which is the precipitation of portlandite. This takes place when the level of concentration of silicates of the aqueous phase has dropped below a threshold of around a few micromoles per litre. The sudden consumption of calcium and OH^- ions through the formation of $Ca(OH)_2$ accelerates the dissolution of all the constituents of the clinker. This is reflected in a dropping off of the conductivity curve, followed by a pseudo-stationary state. An increase in thermal flow is observed, first quite slow since the precipitation of portlandite is endothermic.

Since the formation of hydrates is more rapid, the ultrasonic signal grows weaker, indicating the rapid increase in the "sticking points" between the grains. The beginning of setting, measured using a Vicat sensor, is generally situated throughout this decline, in the absence of premature stiffening of the mortar through the ettringite needles.

The end of the setting period is characterised by a very low ultrasonic signal, and an acceleration in hydration.

The first compressive strength measurements can now be taken. Mineral additions are likely to accelerate the triggering of setting by catalysing the nucleation of portlandite due to their very high surface area. Silica fumes have a very variable level of reactivity, and have generally only a minor effect, although it is detectable in this sense.

2.4 Hardening period

In most portland cements, the molar proportion of gypsum is lower than that of aluminate. The formation of ettringite, which is very rapid during the setting period, leads ultimately to the consumption of all the gypsum, generally after between 9 and 15 hours. The ettringite then becomes the source of sulphate, to form mono-sulpho-aluminate with the excess aluminate.

This reaction consumes 2 moles of aluminate for each mole of ettringite, and we see a very rapid reaction taking place with the aluminate in the clinker. This is reflected in a very marked thermal peak for cements rich in C3A.

The heat thus produced may accelerate hydration of the silicates (thermal coupling of reactions). This is how cements rich in C3A are considered more "nervous", and generally preferred in prefabrication.

In Figure 1, we see a conductivity peak related to the solution of alkalis incorporated into the C3A lattice.

The rapide rise of hydration heat will also be noted, and thus the rate of the overall reaction. It is surprising that this development is not accompained by a concomitant increase in the compressive strength. In the case of a non-accelerated mortar (Fig. 2), the level of mechanical resistance is extended to twenty hours, whereas the heat curve shows continuous rise.

We shall return to this point in the following chapter.

2.5 Period of slowing

The grains of clinker are covered in a thicker and thicker layer of hydrates, which slows down the diffusion of water towards the reactive interface. After fifteen or so hours, we come to the diffusion slow-down period. The chemical reactions nevertheless continue, as long as the hydric state of the concrete allows it. The pozzolanic action of the additions such as silica fumes only becomes noticeable after a few days. Endogenous shrinkage becomes one of the main factors in variation of volume, in concretes with low water and high concrete content. Since concentrations in the aqueous phase are then close to their stationary value, electrical conductivity then becomes an indicator of the water content of the mortar, and tends to become lower and lower.

3 Links between reaction rates and mechanical change

The most striking experimental fact obtained using these continuous measurements is the existence of significant differences between the shape of hydration heat curves and that of compressive strength curves during the hardening period. We now propose to find an interpretation for this. To this end, we shall classify the phenomena, at the microscopic scale, into two groups:

- those contributing to the manufacture of the concrete "glue", i.e. the formation of hydrates and mainly CSH which, by virtue of its highly divided structure, is liable to form many sticking points per unit of mass (auto-healing),
- those contributing to the creation of internal strains or of microcracks beyond the local elastic limit, and which can lead locally to variations in volume (self-stress).

Heat of hydration gives us information as to the quantity of "glue" formed. But the reactions producing the glue take place with

150

Figure 2 : Concrete with silica fumes

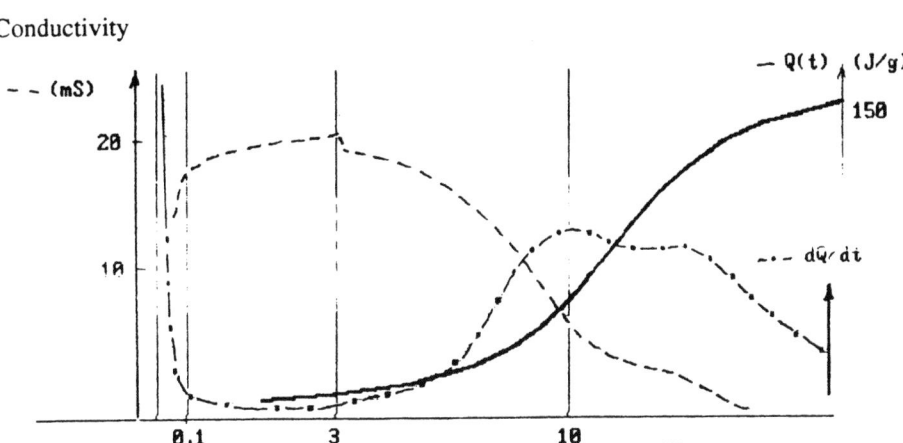

Mechanical aspect

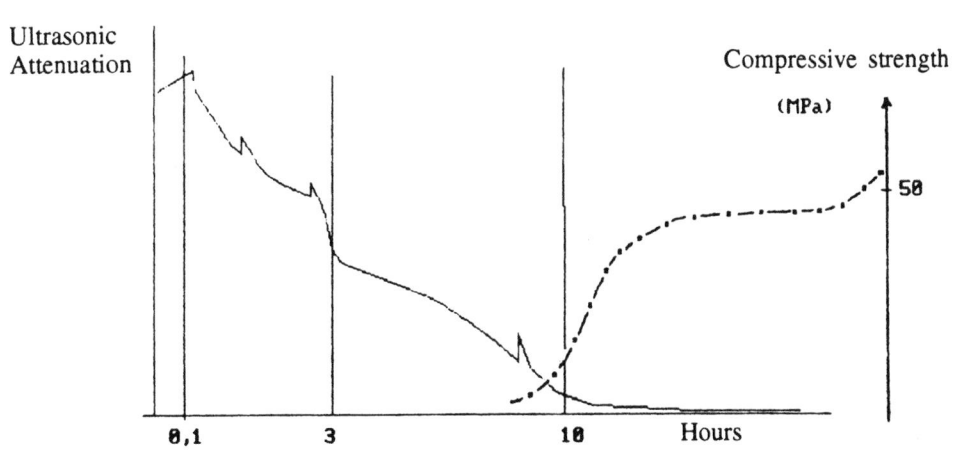

CPA HP 415 kg/m3   F.S. 35 kg/m3   Melamine 2.5%
E/C 0.35   Sand 0-5 755 kg/m3

corresponding diminution of the molar volume: Le Chatelier contraction is proportional to the heat of hydration, and the production of glue leads to the production of internal stresses.

If these processes were acting alone, we could expect to find a simple relationship between the strength and the rate of reaction, and the fact of their common origin would then lead us to consider using a glue with a lower "efficiency".

But in addition some of the chemical reactions produce crystals which, as they grow, exert a crystallisation pressure (Riecke's law) on the surrounding walls. This is related to the super - saturation ratio μ signiifying the deviation from equilibrium.

$$P = R.TLn\mu/v$$

(where v is the molar volume of the species under consideration).

Growing crystals can thus behave like "micro-jacks" and produce a spurious porosity.

This applies to crystals of ettringite and portlandite (Fig. 3), which are used for this reason in certain expansive cements. In these cements, expansion is produced while the molar volume diminishes! The variation in volume is the result of the **increase in porosity** due to crystalline growth and chemical contraction.

Concrete shrinkage results not only from the above processes, but also from self-desiccation due to consumption of water from the hydrates when the capillary pores have started to empty under the effect of hydration reactions. In the case of building constructions, thermal shrinkage must also be taken into account, and varies according to the thermal loss in the thickness of the structure components and the intensity of the thermal flow.

Consequently, at the microscopic scale, strains are liable to occur at all the interfaces, those of the grains of sand, and also those of the large portlandite crystals. To some extent, CSH will be able to compensate for the variations in local volumes. A proportion of the "glue" will be used for healing of the micro-cracks due to crystals growth.

Mechanical resistance will be produced as a result of all these phenomena, and it seems that during the setting period, the time when the crystals growth is at its fastest, the speed of healing is high enough for no effect on the rise in resistance level to be noticeable (Fig. 4).

From the results of figures 1 and 2, the hardening period seems to be the one during which the crystallisation pressure remains high, whereas the speed of formation of CSH has begun to diminish ; this would be a good explanation for the change in the slope of the curve of increase in resistance.

152

Figure 3 : Increase in porosity through pressure of big crystals
during growth

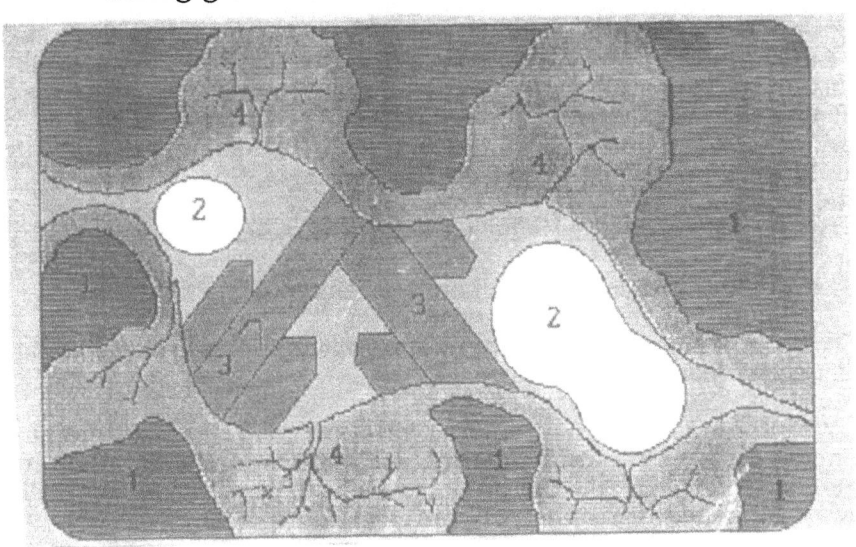

1 Grains undergoing hydration   3 Big portlandite
2 New pores   4 Capillary pores of micro-cracks

Figure 4 : Origin of variations in volume during hydration

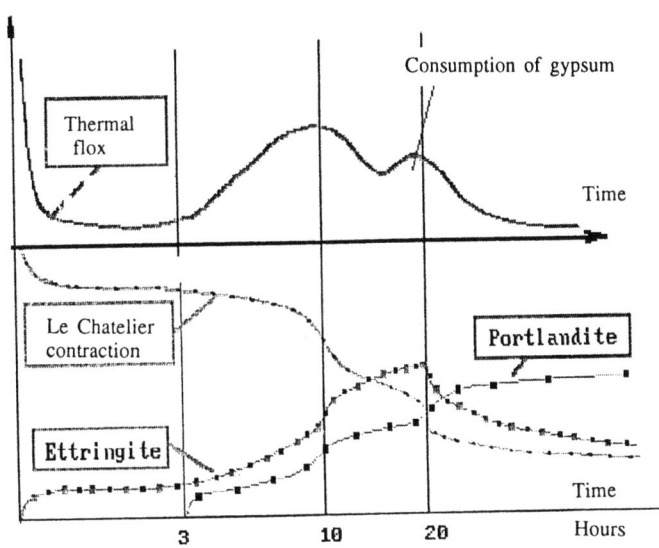

4 Consequences for optimisation of high performance concretes

To increase the performance of concretes further, it would be possible to reduce the secondary effects leading to internal stresses.

But this would imply diminishing the reaction yield, with equal resistance. This implies, for example, that concretes must be made where a low proportion of the cement can be hydrated, i.e. only surface of the grains is made to react. Only a small quantity of water would then be needed, far below the corresponding quantity for total hydration.

Of course, this can only be achieved if two conditions are met:

- overcoming rheological difficulties related to substantial reduction in the water/cement ratio,
- accomplishing quite a careful granular filling to avoid creating useless pores.

We should observe that current formulations of high performance concretes, resulting from empirical techniques, are a first step in this direction: in HPCs the proportion of mixing water is generally lower than the stoichiometric quantity, granular filling is paid particular attention, and the reaction rate of the clinker can be seen to be lower than in traditional concretes. The use of pozzolanic siliceous additions will also help increase the proporition of CSH (glue) relative to the crystalline types.

5 Conclusion

Monitoring of chemical and mechnical development of HPCs during the first days enables the properties already found in building applications to be explained.

Understanding of the mechanism used, combined with understanding of the changes in micro-structures, is opening up new lines of development. The latter could concern both the nature of additives and the composition of binders or manufacturing procedures.

Finally, it should be stressed that the development of HPCs implies carrying out, therefore designing structures which will turn this new material to the best account.

Appendix 1 : Explanation using images of microstructure of concrete over time.

We felt that using pictures to illustrate the changing characteristics of hydration of cement paste would enable the reader to have a better idea of what is happening at the microstructural scale.

Initially (1h), the clinker grains (in grey) start to hydrate, and hydrated calcium silicates (CSH) form (in yellow). The sulphate ions from the dissolution of gypsum (in green) diffuse into the mixing water (in blue).

Microstructure of cement paste (t = 1h)

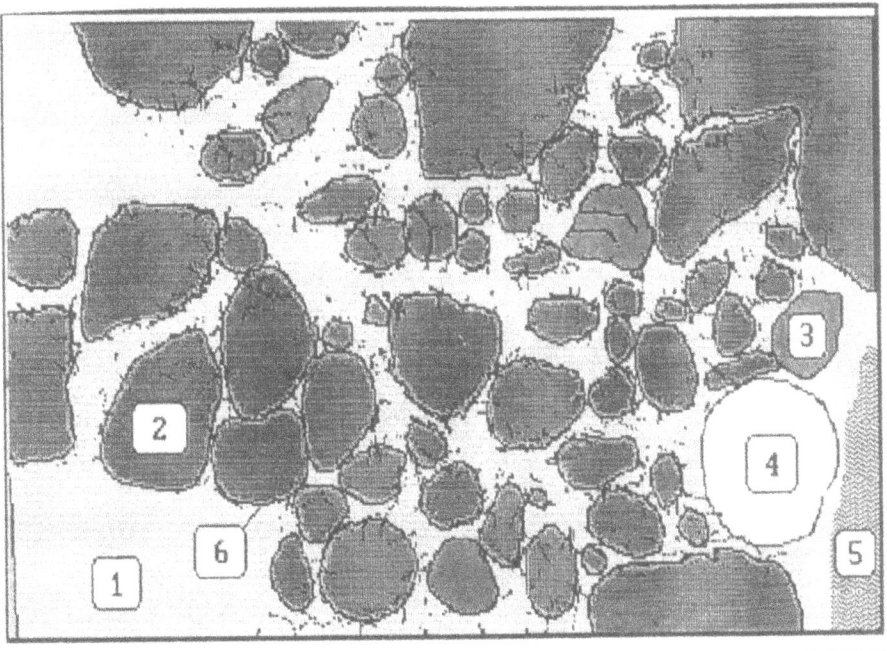

5 microns

1 - WATER	3 - GYPSUM	5 - SAND
2 - CLINKER	4 - BUBBLE	6 - HYDRATES

Situation of the diagram in the kinetics curves.

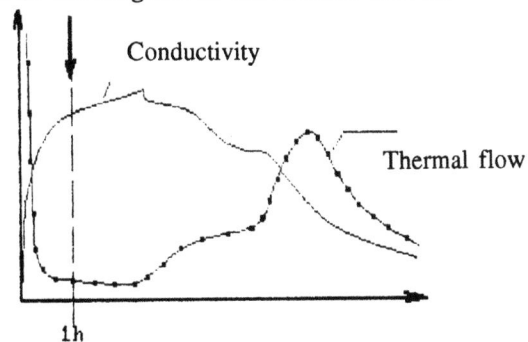

A little later, ettringite (hydrated calcium trisulpho-aluminate, shown here in dark green) is formed at the edge of the clinker grains: two hours after mixing there is already a non-negligible quantity of this.

Microstructure of cement paste (t = 2h)

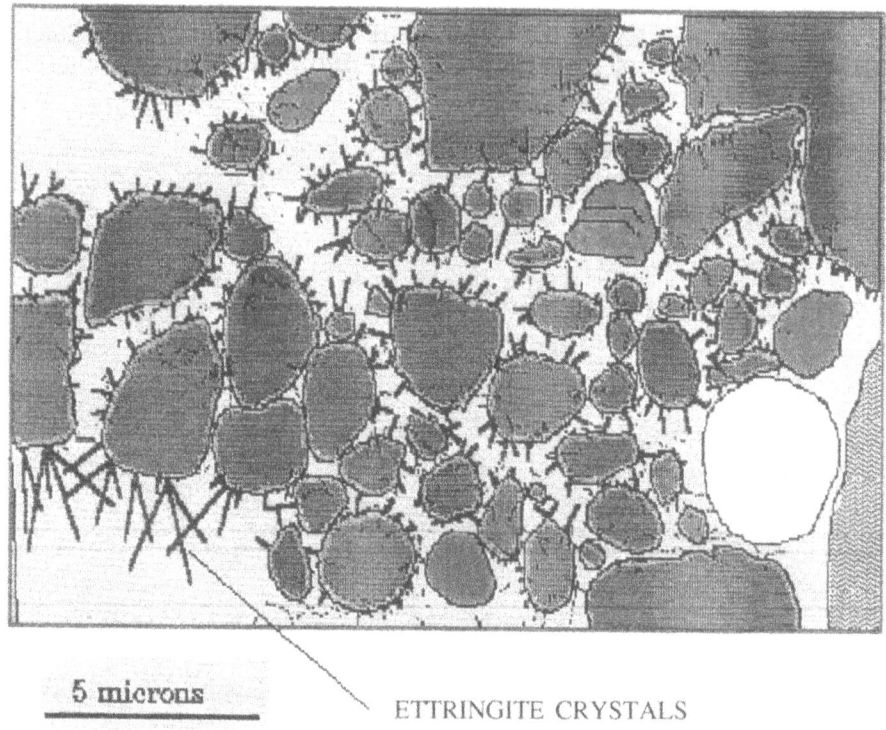

5 microns

ETTRINGITE CRYSTALS

Situation of the diagram in the kinetics curves.

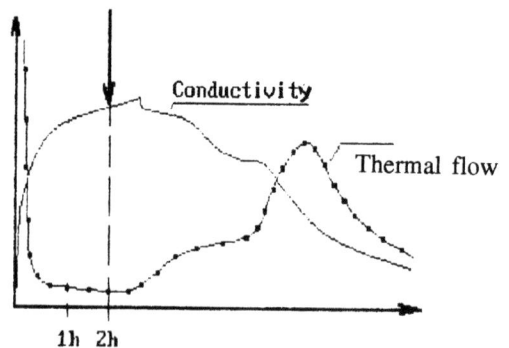

Conductivity

Thermal flow

1h 2h

Hydration goes on, the portlandite Ca (OH)$_2$ precipitates after a few hours. The CSH layers on the surface of the clinker grains, ettringite and portlandite crystals, start to inter-penetrate. The material becomes stiffer: this is the start of setting.

Microstructure of cement paste (t = 4h)

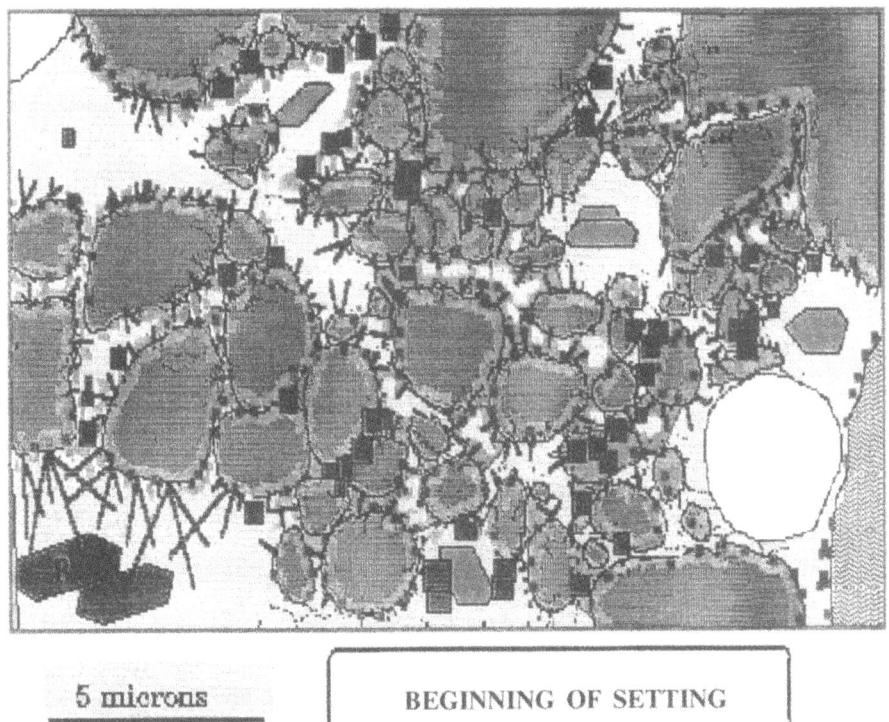

5 microns

BEGINNING OF SETTING

P : PORTLANDITE

Situation of the diagram in the kinetics curves.

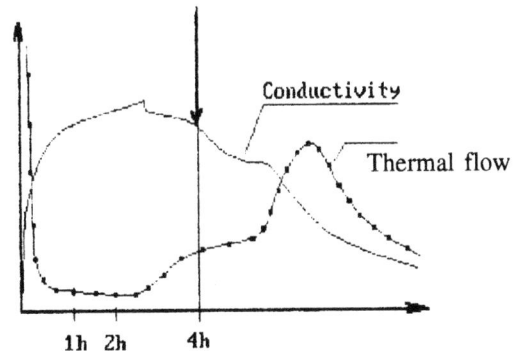

Conductivity

Thermal flow

1h 2h   4h

When gypsum has been used up, ettringite begins to dissolve and the reactions begin to accelerate again. The CSH develops rapidly and the last still separate layers start to join up. The crystals do not have enough room to grow, there is high crystal pressure: this is structural swelling. With the large amount of hydrates formed, the Le Chatelier contraction becomes significant. This is more or less how the cement paste is 9 hours after hydration.

The cement paste has now reached its slow hardening phase.

Microstructure of cement paste (t = 9h)

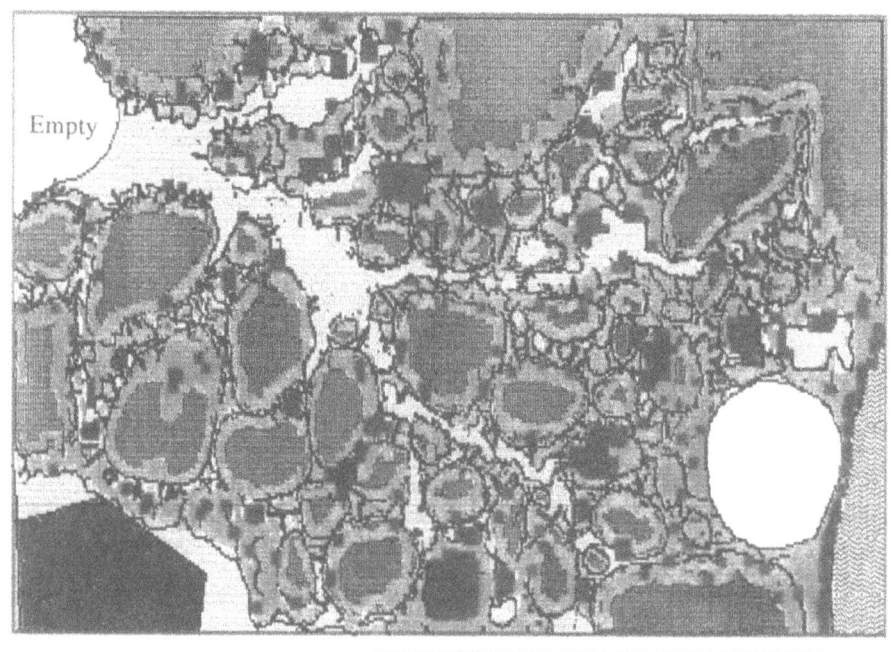

5 microns

AFTER GYPSUM HAS BEEN USED UP, ETTRINGITE DISSOLVES

Situation of the diagram in the kinetics curves.

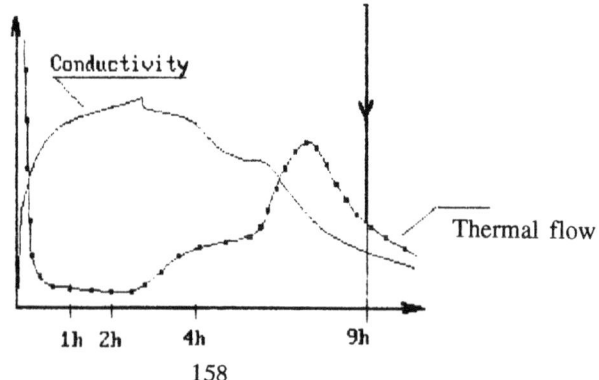

28 days later the material has become more dense. The CSHs truly form the matrix of the concrete. Some portlandite crystals have become larger than the initial grains of clinker. The residual grains of clinker have greater affinity for water than the CSHs. In certain impermeable zones hydration can only take place through consumption of a part of the hydrates'water: this is the self-desiccation process.

Microstructure of cement paste (1 month)

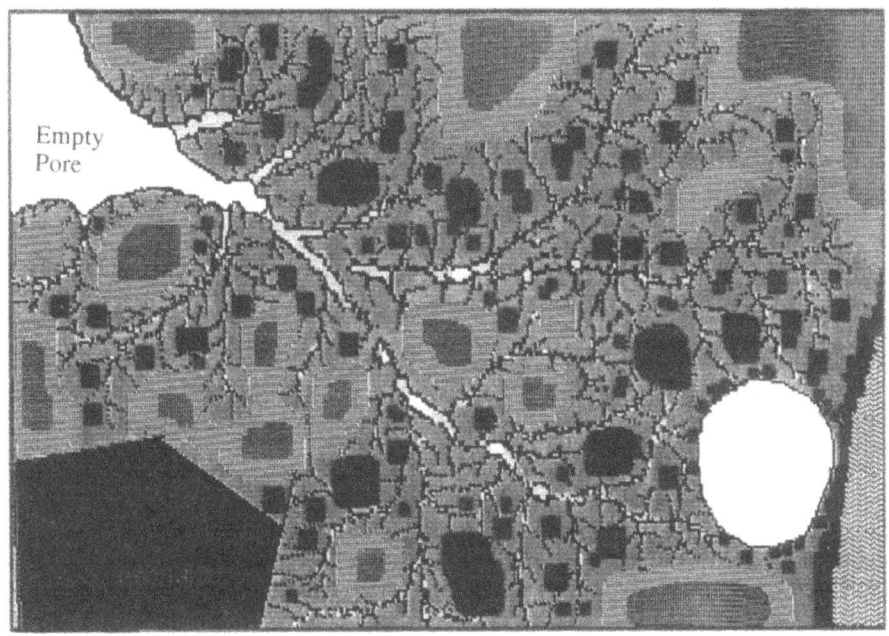

Empty Pore

Hydration leads to self-desiccation.
The capillary pores are becoming more and more empty.

10 LARGE REDUCTION OF DEFLECTIONS DUE TO HPC

R. FAVRE, H. CHARIF, J. -P. JACCOUD
IBAP, E P F|L, Lausanne, Switzerland

1 Introduction

The Institute of Reinforced and Prestressed Concrete (IBAP) at the Ecole Polytechnique Fédérale in Lausanne has undertaken during the past two decades several research topics concerning the serviceability of concrete structures, particularly in the realm of deformations. In fact, the long-term deformation of concrete structures has become increasingly important in the last few years. This has resulted from advances in the scientific and technological fields permitting the construction of structures which have longer and more slender spans than in the past.

There are several ways to reduce the long-term deformations of a concrete structure. For example, we can mention the use of prestressing or an increase of the reinforcement. Today, a new approach in concrete material technology is rapidly developing. This approach seemed to us an interesting proposal for the reduction of deformations. In fact, the use of this "new concrete", called high performance concrete (HPC), is no longer limited to certain parts of structures highly subjected to compression (columns of a skyscraper, arch of a bridge), but tends towards the construction of whole and not only special structures (offshore).

This paper summarises the principal findings of a thesis work (2) which has been done in continuation of previous researches. The objective of this research was to show that it is possible to greatly reduce, almost by a factor of two, the long-term deflections of RC structures, in particular RC slabs, through the use of HPC.

Reduction of slab deflections is obtained by exploiting four improved properties offered by these concretes, three of a mechanical nature and one of a rheological nature:

1. a better steel-concrete bond, which increases the tension stiffening effect and favourably influences the deflection and the crack widths;

High Performance Concrete: From material to structure. Edited by Yves Malier. © 1992 Taylor & Francis.
Published by Taylor & Francis, 2 Park Square, Milton Park, Abingdon, Oxon, OX14 4RN. ISBN 0 419 17600 4.

2. a better tensile strength, which results in an increase of the cracking moment and a limited and delayed appearance of the cracks;
3. a better elastic modulus, which decreases the elastic deformations;
4. less creep, which decreases the time-dependent deformations.

Thus, this research work was divided into four parts :

1. Study of materials which are used in the mix design of HPC.
2. Experimental study. This consisted of tests on concrete cylinders and RC slabs (3).
3. Development of finite element non-linear programs, to calculate beam or slab deformations taking into account time-dependent effects (creep, shrinkage), reinforcement and cracking.
4. Parametrical studies which determine the influence of the principal parameters and demonstrate the aptitude of HPC to reduce deflections.

In the present paper we are going to concentrate on the last three points (2, 3 and 4), i.e. the experimental study, the theoretical model and the calculation method of deflections, and finally the parametrical study which was carried out by the use of the finite element programs.

2 Experimental study

2.1 Materials
Ordinary portland cement was used with a specific surface of 0.3 m^2/g and a compressive strength (ISO) of 40 MPa. Only one type of superplasticizer was used : Sikament 320. The dry extract of this superplasticizer is 40%. The silica fume used was imported form Germany (trade mark : V.A.W. RW-Füller). It is composed of 97% of SiO$_2$. Its specific surface varies from 20 to 22 m^2/g. The sand and the coarse aggregates used in this study were siliceous limestone moraine materials. These were graded as follows :

0/3 mm rounded sand
3/8 mm rounded sand
8/16 mm rounded or angular coarse aggregates
16/32 mm rounded or angular coarse aggregates

2.2 Concrete composition
A large number of concrete cylinders (160 x 320 mm) were tested to determine the following properties: compressive strength, tensile strength, elastic modulus, creep and shrinkage. The results

are described in detail in references (3) and (4). For the slab tests, four different concretes were used. Two reference concretes called A1 and A4 (A1 is the standard concrete typically used in building construction in Switzerland) and two HPC called CS4 (without silica fume) and CF4 (with silica fume). Table 1 shows the composition of these concretes.

2.3 Mechanical properties

Table 2 gives the principal results from the concrete cylinder tests at 3, 28, 182 and 364 days.

2.4 Rheological properties (shrinkage, creep)

Figure 1 shows the shrinkage deformations ε_{cs} of the cylinders with carton protection for concretes A1, A4, CS4 and CF4 (RH=50 ± 5%. T = 20 ± 1°C). Measurement of the shrinkage started after 6 days of curing (identical cure for the slabs and the cylinders). It was noticed that the ε_{cs} of the HPC CS4 and CF4 were higher than the ε_{cs} of concrete A1 at an early age. However, after one year the ε_{cs} of concrete CF4 was less than the ε_{cs} of concrete A1.

Figure 2 shows the creep deformations ε_{cc} for concretes A1, A4, CS4 and CF4. Loading (N = 150 kN, $\sigma_c \cong 7.5$ MPa) was applied after 28 days on cylinders with carton protection. After one year of loading, the ε_{cc} of concretes CS4 and CF4 were approximately reduced by 30 and 60% respectively, with regard to ε_{cc} of concrete A1.

2.5 Series E : one-way slab tests

The series E test is described in detail in reference (3). Only a brief description of the test and the results concerning deflections will be given here. The other results (curvatures, cracking, etc.) can be found in references (3) and (4).

Series E consisted of 16 reinforced concrete slabs working in the same manner as a simply supported beam (see figure 3). The reinforcement in each slab was identical and principally consisted of high bond steel bars S500 (see figure 4). The principal tensile reinforcement consisted of 5 bars Ø12mm at intervals of 150 mm which represents a ratio ρ of 0.58%. The transverse reinforcement consisted of Ø 8 mm stirrups at intervals of 200 mm in the central zone where the measurements were concentrated.

Beside the type of concrete the slabs differed by the level and sign of loading as follows (see figure 5) :

Table 1 : Mix proportions of test concretes

Concrete name	Cement (kg/m^3)	Superplas -ticizer (kg/m^3)	Silica fume (kg/m^3)	Water (l/m^3)	W/C (-)	Slump (cm)
A1	300	0	0	180	0.6	6.5
A4	375	0	0	179	0.48	5.5
CS4*	425	8.5	0	159	0.38	11
CF4*	388	7.8	39	156	0.4	6

* with angular coarse aggregates

Table 2 : Mechanical properties

	Concrete age :	3 days	28 days	6 months	1 year
A1	Splitting tensile (MPa)	1.87	3.85	3.85	4.62
	Compressive (MPa)	13.6	28.0	29.3	40.3
	Direct tensile (MPa)	---	2.09	2.40	3.06
	Modulus (GPa)	---	29.0	31.2	33.1
A4	Splitting tensile (MPa)	2.74	4.04	4.48	4.97
	Compressive (MPa)	23.0	36.4	44.1	49.5
	Direct tensile (MPa)	---	2.47	2.89	3.61
	Modulus (GPa)	---	31.0	33.6	34.5
CS4	Splitting tensile (MPa)	3.48	4.56	5.17	5.35
	Compressive (MPa)	32.6	56.7	67.1	70.1
	Direct tensile (MPa)	---	3.19	3.99	3.97
	Modulus (GPa)	---	33.8	35.6	37.7
CF4	Splitting tensile (MPa)	3.63	4.56	5.18	5.78
	Compressive (MPa)	39.0	71.4	71.7	78.2
	Direct tensile (MPa)	---	3.60	4.05	4.52
	Modulus (GPa)	---	34.6	40.3	37.9

--- No tests at 3 days

Figure 1: Shrinkage deformations of concretes A1, A4, CS4 and CF4 (t_0 = 6 days)

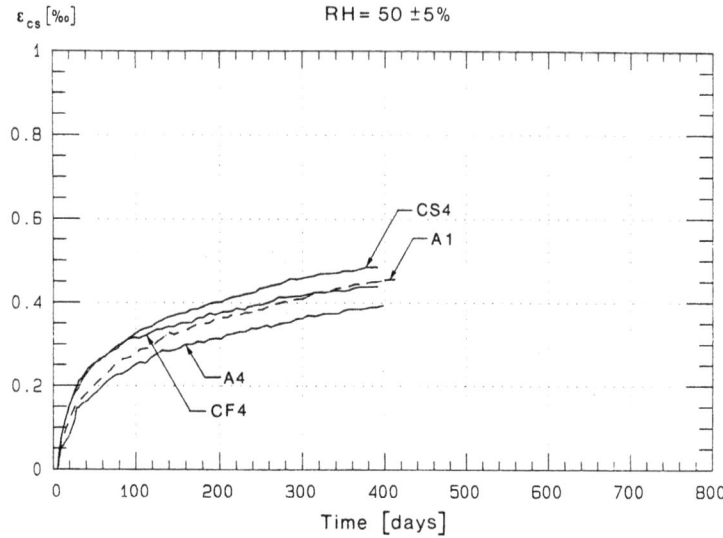

Figure 2: Creep deformations of concretes A1, A4, CS4 and CF4 (t_0 = 28 days)

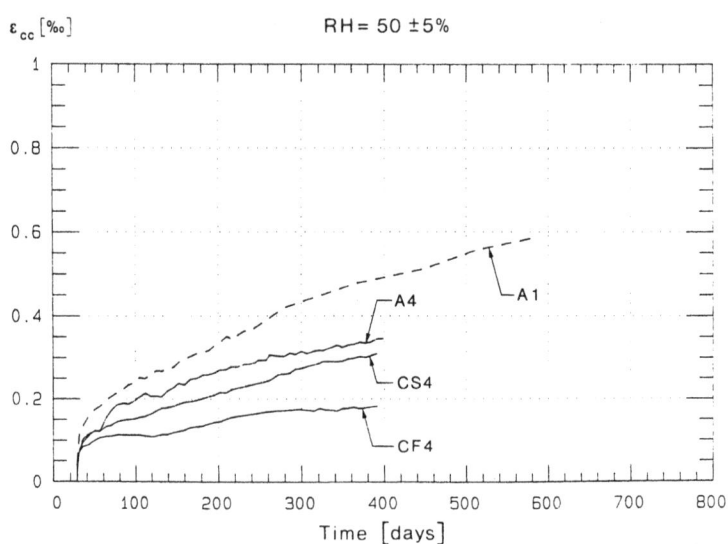

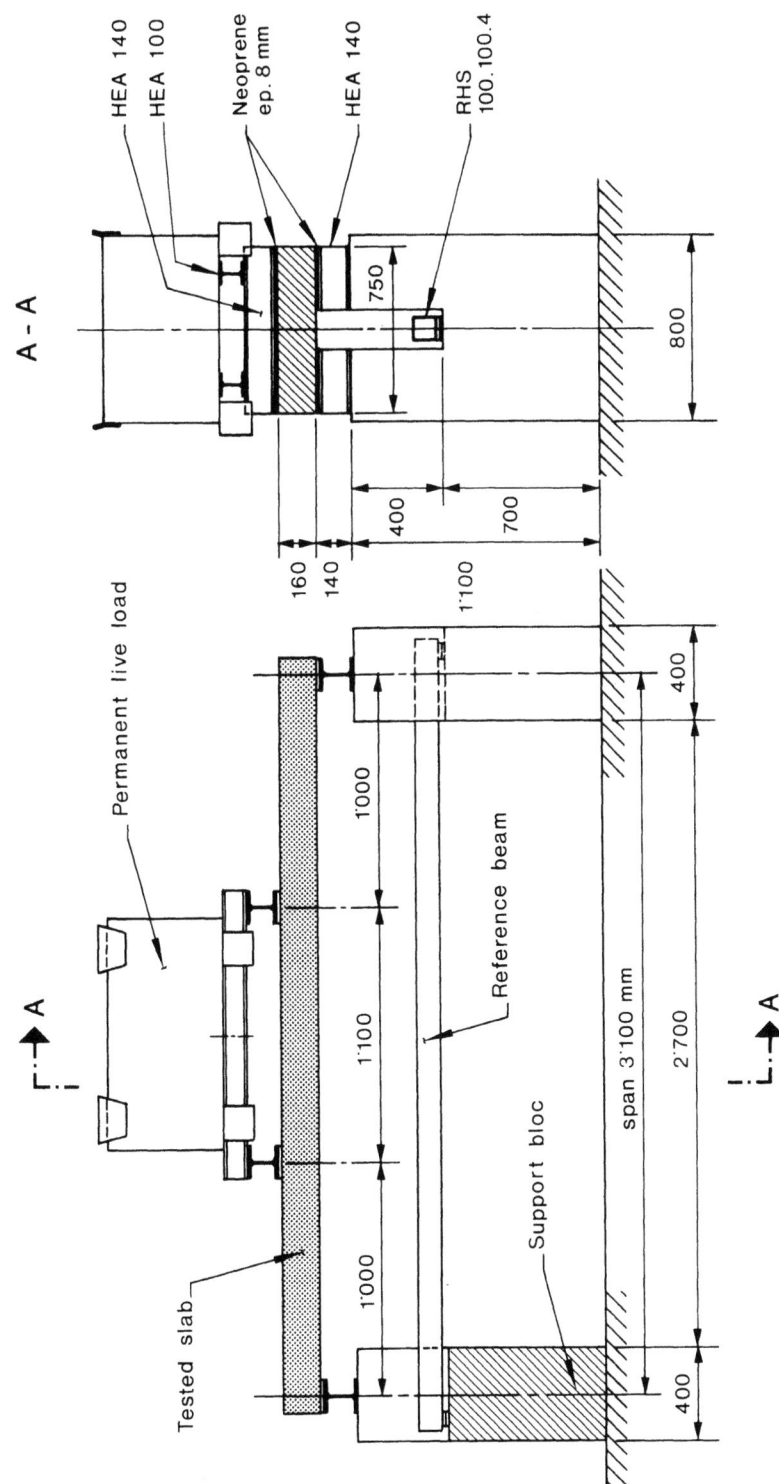

Figure 3: Test arrangement for series E slabs

Figure 4: Formwork and reinforcement plan of series E slabs

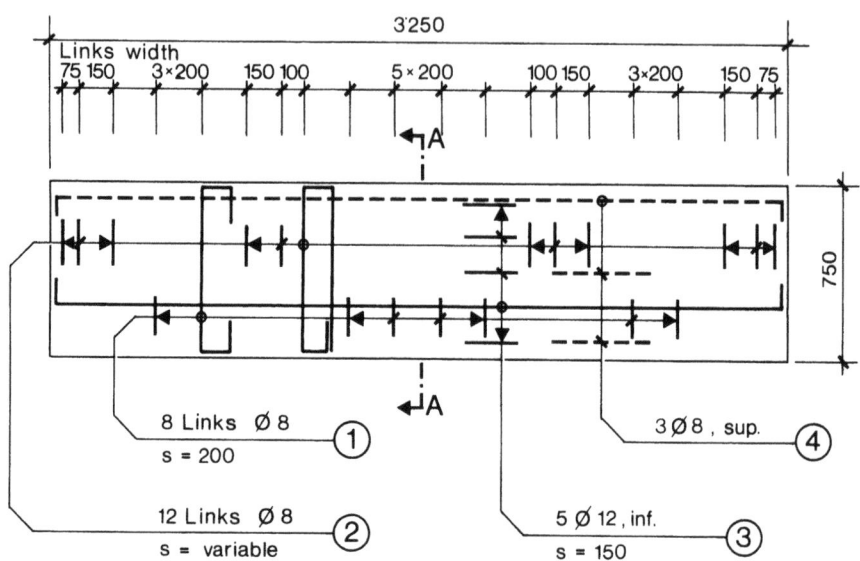

3'250

Links width
75 150   3×200   150 100   5×200   100 150   3×200   150 75

A

750

8 Links Ø 8   ①
s = 200

3 Ø 8 , sup.   ④

12 Links Ø 8   ②
s = variable

5 Ø 12 , inf.   ③
s = 150

SECTION A-A

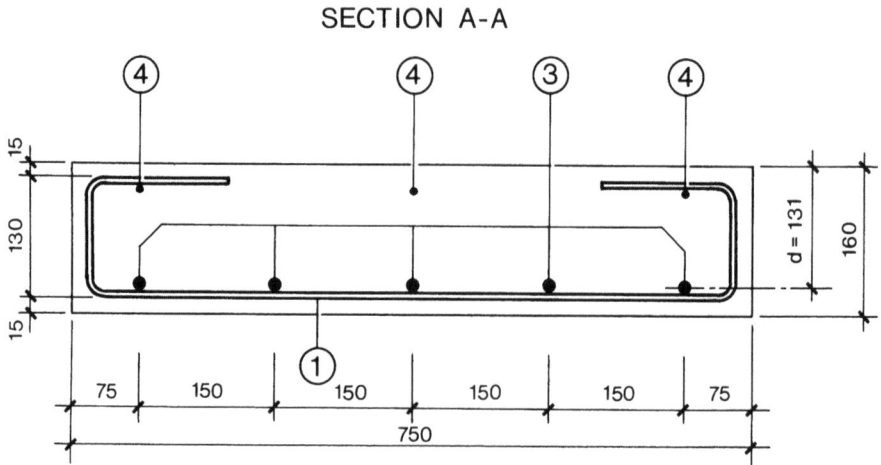

④   ④   ③   ④

15   130   15

d = 131   160

①

75   150   150   150   150   75

750

166

Figure 5: Load levels and signs

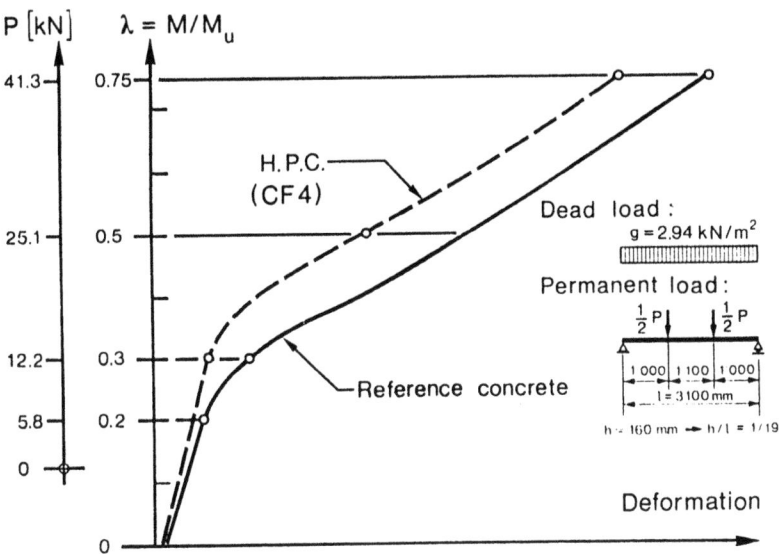

Type of concrete	Load level	Load signs	
	$\lambda = M/M_u$	M+ Test N°	M− Test N°
A1 (Reference)	0.2 0.3 0.75	E 101 E 102 E 104	E 111 − E 114
A4 (Reference)	0.3 0.75	E 202 E 204	− −
CS4 H.P.C.	0.3 0.5 0.75	E 302 E 303 E 304	− − E 314
CF4 H.P.C.	0.3 0.5 0.75	E 402 E 403 E 404	E 412 − E 414

1. The bending moment due to the permanent load varied between 20 and 75% of the conventional ultimate bending moment M_u (M/M_u = 0.2, 0.3, 0.5 and 0.75). M_u = 32.3 kNm calculated with f_y = 460 MPa.
2. 11 slabs were made to act under a positive moment, i.e. the tensile reinforcement was placed near the lower surface during concreting.
3. 5 slabs were made to act under a simulated negative moment; in order to achieve this, the slabs were concreted with the tensile reinforcement placed near the upper surface, they were then turned upside down to carry out the experiment. This is to simulate the behaviour of a continuous RC member over a support.

All the slabs were subjected to an identical curing regime. They were cast inside the laboratory and maintained humid for 6 days by covering them with plastic sheeting. The formwork was struck after 6 days and the slabs were stored in the air conditioned laboratory (RH = 50 ± 5%, T = 20 ± 1°C).

The slabs were loaded at t_0 = 28 days. The loading was carried out quasi-instantaneously by means of a metal chassis or a special metal frame, previously loaded with lead ingots or concrete blocks and placed in position by an overhead travelling crane.

The central zone of each slab was subjected to a constant bending moment. In this zone we measured :

1. the deflection,
2. the surface strain of the concrete, on the top and bottom surface of the slab, from which it was possible to determine the curvatures and the average crack width.

Figure 6, which is a typical example of the results obtained, shows the evolution of deflections for the first year of the slabs made with concretes A1, A4, CS4 and CF4; for the loading level M/M_u = 0.3. Note that after 1 year of loading the reduction of the deflection with respect to that of the reference A1 slab was about 20% for slabs made with concretes A4 and CS4 and about 40% for the slab made with concrete CF4.

3 Numerical study

3.1 Calculation model : approach by curvatures

There is an abundance of technical literature in Europe concerning the modelling of concrete structures and even more in North America. From this bibliography, three possible theoretical models can be extracted: approach by layers, approach by additional stiffness to steel and approach by curvatures. Only the third

Figure 6: Evolution of measured deflections at mid-span of slabs made with concretes A1, A4, CS4 and CF4 (t_0 = 28 days)

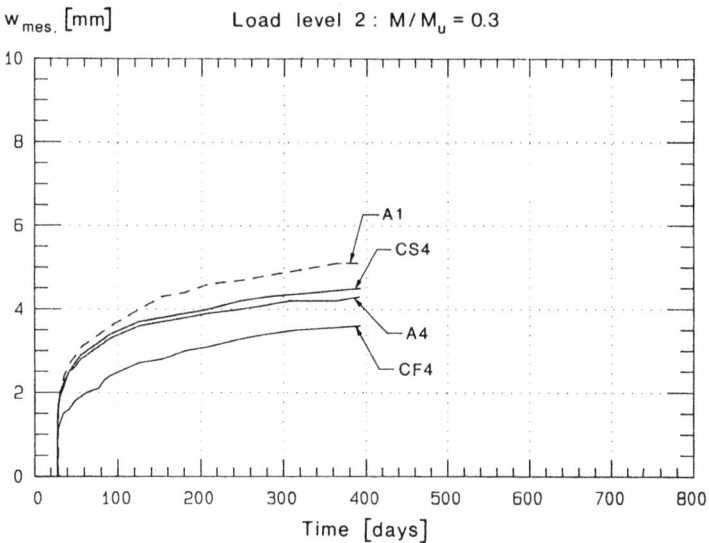

approach is presented here which has been retained and implemented thanks to the finite element method.

The approach by curvatures is a more direct method than the approach by layers to model concrete structures. In fact, instead of working with the stress-strain ($\sigma-\varepsilon$) laws and integrating these laws over the depth of the structure to obtain the curvatures and the moments, we work directly with non-linear moment-curvature laws. The principal difficulty is to find moment-curvature laws which properly represent the behaviour of a concrete structure taking into account the cracking, the time-dependent effects and the reinforcement. The moment-curvature relationship of the CEB (6) appears to us to satisfy all these criteria. It has been developed and verified experimentally in the case of ordinary concrete. Our experimental study (3) on one-way slabs has shown that this relationship is still valid in the case of HPC.

3.1.1 CEB moment-curvature relationship

The CEB moment-curvature relationship defines the curvature (called mean curvature) at each point of the structure. This mean curvature is a combination of the curvature in state 1 (uncracked) and the curvature in state 2-naked (fully cracked without participation of tensile concrete). Figure 7 shows the relationship in the case of simple bending.

The mean curvature ψ_m (instantaneous or long-term) can be calculated at any point (section) of a structure as follows:

169

Figure 7: CEB moment-curvature relationship (simple bending)

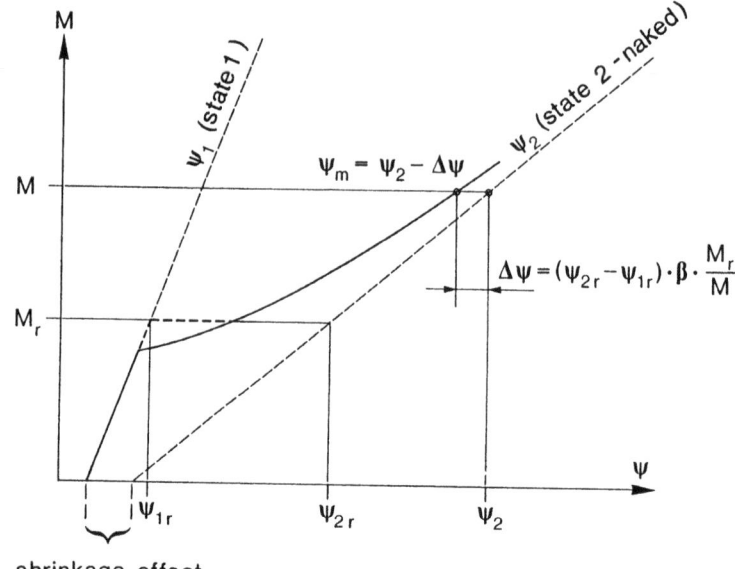

shrinkage effect

$$\psi_m = \psi_1 \qquad \text{for non-cracked state}$$

$$\psi_m = \psi_2 - (\psi_{2r} - \psi_{1r}) * \beta * \frac{M_r}{M} \qquad \text{for cracked state}$$

where :

ψ_1 - curvature in state 1

ψ_2 - curvature in state 2-naked

ψ_{1r} - curvature in state 1 for $M = M_r$

ψ_{2r} - curvature in state 2-naked for $M = M_r$

M - acting bending moment

M_r - cracking moment

$M_r = W_1 . f_{ct} \cong W_c . f_{ct}$

W_1 - section modulus in state 1 (including the reinforcement)

W_c - section modulus of concrete only

$f_{ct} = f_{ctk\ 0.05}$ - if it is a question of avoiding structural damage

$\quad = f_{ctm}$ average tensile strength if it is a question of calculating the deformations

$\beta = \beta_1 . \beta_2$ coefficient characterizing the tension stiffening effect

β_1 coefficient characterizing the bond quality of the reinforcement bars:

$\beta_1 = 1.0$ for high bond bars and 0.5 for smooth bars

β_2 coefficient representing the influence of the duration of application or of the repetition of loading (permanent, quasi-permanent, cyclic):

$\beta_2 = 0.8$ at first loading and 0.5 for long-term loading or for a large number of load cycles.

The curvatures ψ_1 and ψ_2 in states 1 and 2-naked respectively can be calculated as a function to the basic curvature ψ_c :

$$\psi_c = \frac{M}{(EI)_c}$$

where $(EI)_c =$ the basic stiffness of just the concrete section without taking into account reinforcement or time-dependent effects.

The curvatures in state 1 and 2-naked are determined as follows :

$$\psi_1 = \kappa_{s1} \cdot (1 + \kappa_{\varphi 1} \cdot \varphi) \psi_c + \kappa_{cs1} \cdot \frac{|\varepsilon_{cs}|}{d}$$

$$\psi_2 = \kappa_{s2} \cdot (1 + \kappa_{\varphi 2} \cdot \varphi) \psi_c + \kappa_{cs2} \cdot \frac{|\varepsilon_{cs}|}{d}$$

The calculation of the correction coefficients κ_{s1}, κ_{s2}, $\kappa_{\varphi 1}$, $\kappa_{\varphi 2}$ κ_{cs1} and κ_{cs2} is given in reference (7). d is the effective depth, ε_{cs} is the shrinkage strain and φ is the creep coefficient.

3.1.2 Comparison with experimental results (series E slabs)
The curvatures for the slabs made with concretes A1, A4, CS4 and CF4 were calculated. The composition of these concretes is given in Table 1. These curvatures have been calculated by taking the values of the mechanical and rheological properties of the cylinder tests as indicated in Table 3.

Figure 8 shows, as an example, the mean curvatures calculated and measured after 1 year of loading for slabs made with the HPC CF4. Note the good agreement between measured and calculated values.

Table 3 : Mechanical and rheological properties of concretes considered for the calculation of the moment-curvature relationship

Concrete :		A1	A4	CS4	CF4
f_{cm}	MPa	28.0	36.4	56.7	71.4
E_{co}	GPa	26.6	27.3	30.2	37.5
f_{ctm}	MPa	2.52	3.0	3.72	4.1
$f_{ct,sp}$	MPa	3.6	4.04	4.56	4.97
$\varphi(t - t_0 = 1 \text{ year})$	(-)	1.47	1.16	1.41	0.89
$\varepsilon_{cs} (t - t_0 = 1 \text{ year})$	(‰)	0.455	0.390	0.485	0.439

Figure 8: Measured and calculated moment-curvature relationship of concrete CF4 after one year of loading

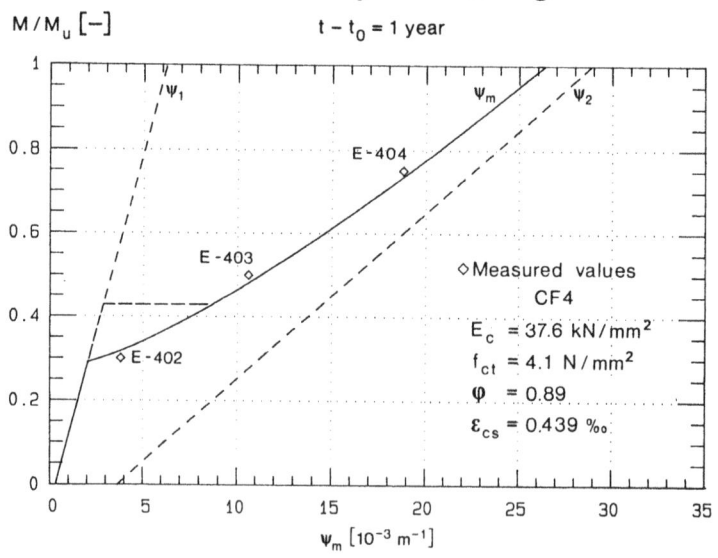

3.2 Calculation method : finite element

Today, the technical literature concerning the finite element method is copious and very detailed. For this reason, it seemed to us unnecessary to go into detail on the theory of this method. However, we can simply stated that globally there are two formulations: an integral formulation based on the principle of virtual work and a variational formulation based on the minimisation of the total potential energy of a structure. Actually, these two formulations give the same matrical equation:

$$[K (U)]. \{U\} = \{F\}$$

The finite element method seems to us vastly superior to other calculation methods. This is especially due to the rapid progress in computers science. In the case of slabs, the finite element method allows the coming together of two theories: a linear and a non-linear geometrical theory. As non-linear geometrical theory we can mention Van Karman's theory developed in 1910 which calculates the deformation tensor with quadratical terms. In the practical case of RC slabs, there is no real problem of geometrical non-linearity and in the majority of cases we can use a linear geometrical theory.

3.3 Definition of the slab finite element

A simple rectangular finite element based on the linear geometrical theory of bending thin plates (Kirchhoff theory) was chosen. It is a classical finite element which was developed in several publications and which has given good results in linear elastic calculations.

In Kirchhoff's theory, the displacement field at any point of the slab can be defined by the following equations :

$$u = - z \frac{\partial w}{\partial x} \qquad v = - z \frac{\partial w}{\partial y} \qquad w = w (x, y)$$

Thus, at a geometric node of this finite element, we have to define three nodal variables (3 degrees of freedom): the deflection w and the two rotations $\frac{\partial w}{\partial x}$ and $\frac{\partial w}{\partial y}$. The finite element has in total 12 degrees of freedom.

This finite element is non-conform, but its convergence has been checked through the use of numerical techniques like for example the "Patch test".

Figure 9 shows this finite element with its positive sign convention for deflections w_i and rotations $\theta_{xi} = \left(\frac{\partial w}{\partial y}\right)_i$ and $\theta_{yi} = \left(\frac{\partial w}{\partial x}\right)_i$ at each node i.

173

3.4 Iterative procedure of non-linear calculation

We incorporated the CEB non-linear moment-curvature relationship into the aforementioned finite element. This was made possible thanks to an iterative procedure which uses a combination of the substitution method (secant matrix method) and the incremental method (step by step method). The different calculation assumptions have been described in detail in reference (2). Two main problems were encountered during this calculation. The first problem was the determination of the torsional stiffness D_{xy} of the RC slab. To avoid this problem we pass through the two principal directions of a slab where the torsional moment is zero. Nevertheless, we needed an assumption on the D_{xy} value to calculate the initial solution. For this we adopted the assumption of Huber and Timoshenko:

$$D_{xy} = \frac{1-v}{2} \sqrt{D_x\, D_y}$$

$$\text{where } D_x = D_y = \frac{E_c h^3}{12\,(1 - v^2)}$$

The second problem concerned the reinforcement area A_s which is required to calculate the correction coefficients κ_{s1}, κ_{s2}, $\kappa_{\varphi 1}$, $\kappa_{\varphi 2}$, κ_{cs1} and κ_{cs2}. To resolve this problem we chose the following equation :

$$A_s = A_{s\theta 1} \cos^2 \gamma_1 + A_{s\theta 2} \cos^2 \gamma_2$$

Figure 10 shows any point in the slab (infinitesimal element) where the principal direction I is defined by the angle α_1 with respect to the x axis. The reinforcement is defined by the areas $A_{s\theta 1}$ and $A_{s\theta 2}$ and by their directions (angles θ_1 and θ_2 with respect to the x axis). Angles γ_1 and γ_2 are easily calculated as a function of the known angles θ_1, θ_2 and α_1. The previous equation represents the plastification criterion of Johansen (5).

Figure 11 shows the flowchart of the iterative calculation procedure for the slab finite element. The matrix (D), called $(D)_m$ (m for "mean") is modified at each iteration and is calculated by the following equations :

$$D_{xym} = \frac{M_{xy}}{2\psi_{xym}} \qquad\qquad D_{xm} = \frac{(M_x - D_{1m}\,\psi_{ym})}{\psi_{xm}}$$

174

Figure 9: Slab finite element. Sign convention

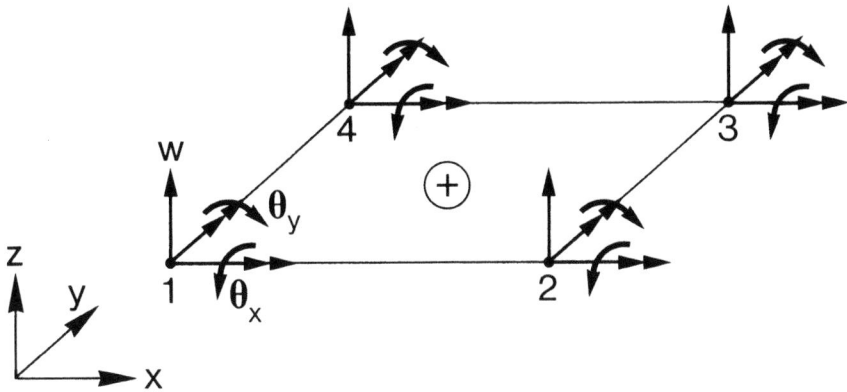

Figure 10: Calculation of the reinforcement sections in the principal directions

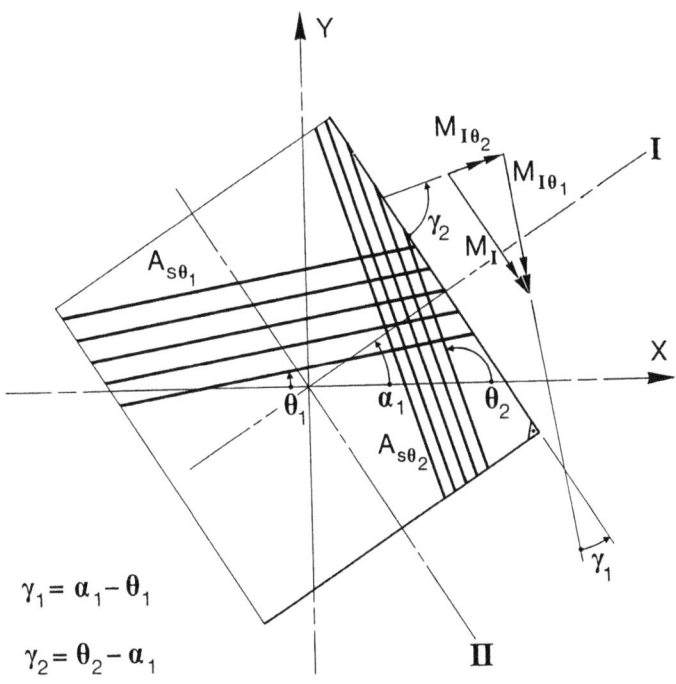

$$\gamma_1 = \alpha_1 - \theta_1$$

$$\gamma_2 = \theta_2 - \alpha_1$$

Figure 11: Calculation procedure for matrix [D] incorporated into slab finite element routines

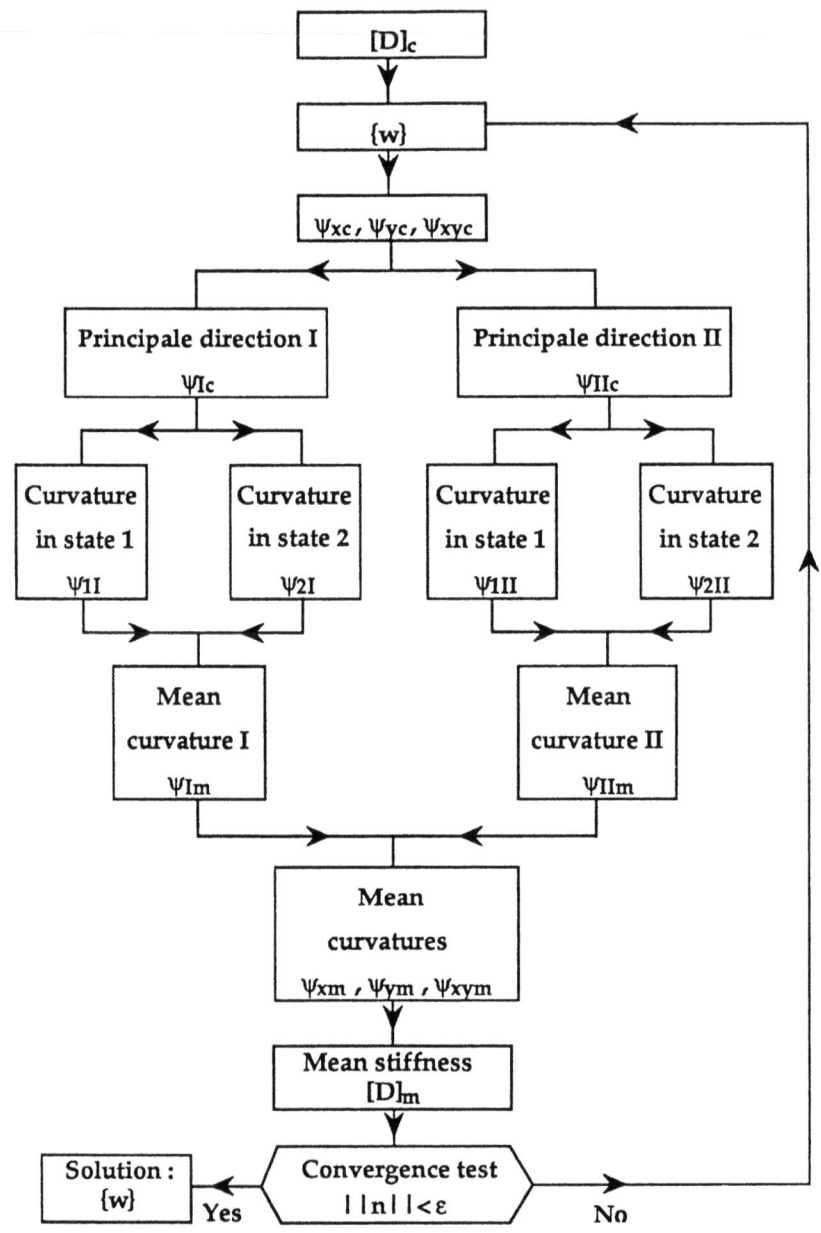

$$D_{1m} = \frac{2\nu \, D_{xym}}{(1-\nu)} \qquad\qquad D_{ym} = \frac{(M_y - D_{1m}\psi_{xm})}{\psi_{ym}}$$

3.5 Parametrical study

The objective of this parametrical study is to show the influence of some mechanical and rheological properties of concrete on the long-term deflection of a RC slab. The chosen example is an industrial building flat slab. Figure 12 shows the plan of this flat slab which has a span of 8.30 m.

Although the determinant field for deflection is the angle field, we show here the deflection calculation at the edge field B-C/1-2. Figure 13 shows the finite element mesh.

The principal characteristics of this flat slab are :

1. Geometrical data :
    depth : h = 300 mm
    effective depth in the span : $d^+ \cong 260$ mm
    effective depth over the support : $d^- \cong 400$ mm
    span : l = 8.30 m
    slenderness ratio : l/h = 28

Figure 12: Plan of the flat-slab

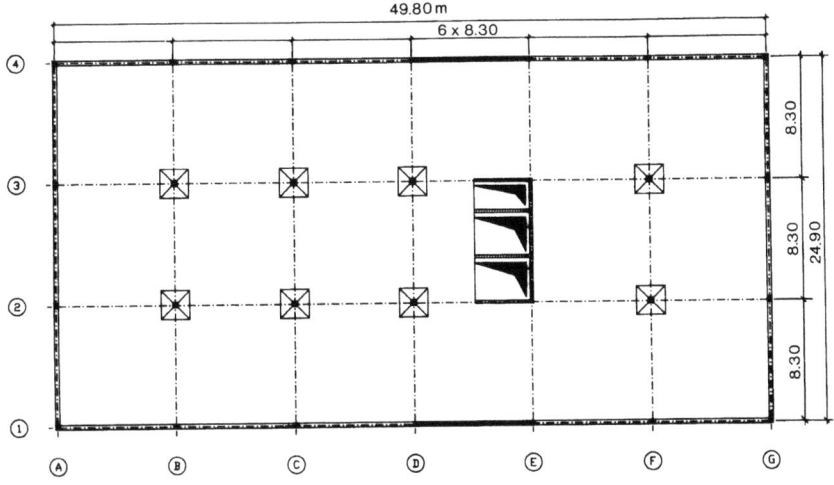

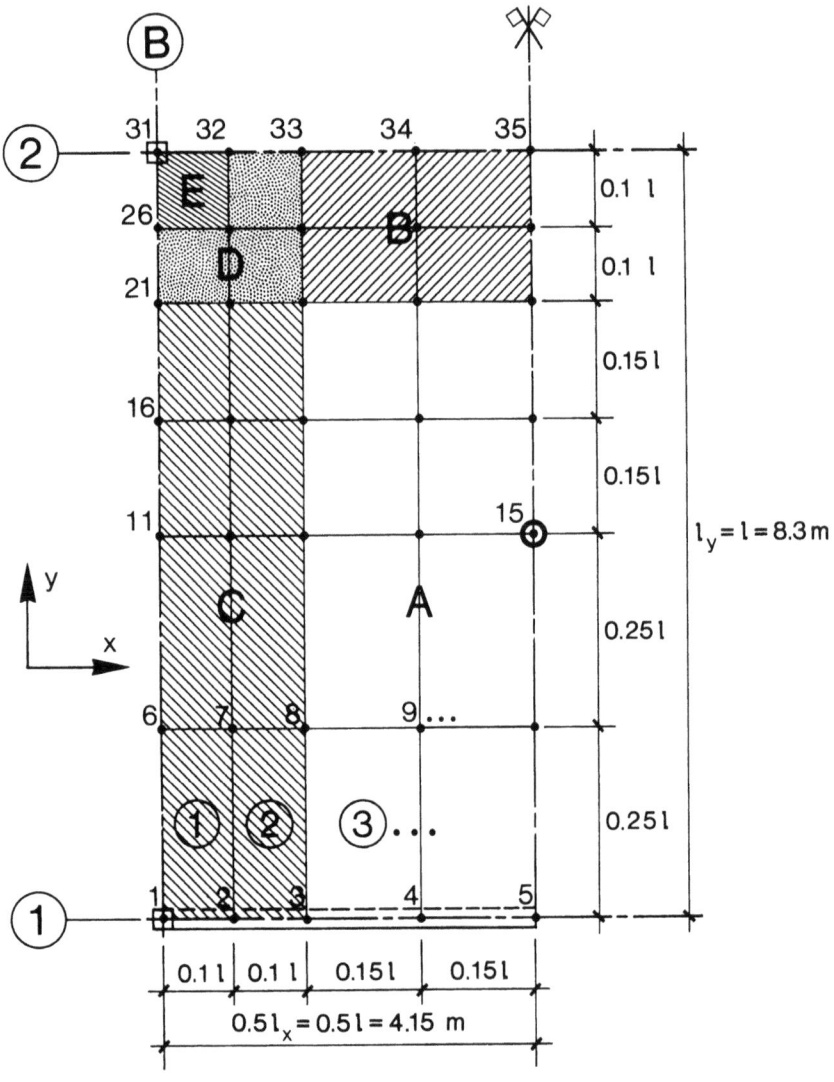

Figure 13: Finite element mesh of the flat slab

2. Materials
ordinary concrete (C20 according to the CEB)

E_c = 35 GPa

f_{ct} = 2.5 MPa

φ = 2.8

ε_{cs} = 0.47 ‰

β = 0.5

HPC :

E_c = 45.5 GPa	(+ 30%)
f_{ct} = 3.75 MPa	(+ 50%)
φ = 1.68	(- 40%)
ε_{cs} = 0.47%	(without change)
β = 0.5	(without change)

steel S500 (passive reinforcement)

f_y = 460 MPa

The quantity of reinforcement has been chosen on the basis of a plastic design calculation according to the Kinetic yield-line theory. The reinforcement plan is shown in figure 14.

3. Loads

dead load :	11 kPa
short-term live load :	8 kPa
long-term live load :	3 kPa
ultimate load (q_u) :	31.6 kPa
cracking load (q_r) :	6.5 kPa

3.5.1 Influence of the tensile strength of concrete f_{ct}

We have calculated the deflection of this flat slab by varying the tensile strength from 1.5 to 5 MPa. In fact we have considered the following values :

f_{ct} = 1.5   2.0   2.5   3.0   4.0   5.0   MPa

and this for four loading levels:

q/q_u = 0.2   0.3   0.45   0.6

All other parameters remained constant. Figure 15 shows the results obtained. The deflections calculated with a tensile strength of 2.5 MPa have been taken as reference values.

This figure clearly shows that the reduction of the slab deflection due to an increase of the tensile strength of concrete depends, above all, on whether the slab is cracked or not. When the slab is cracked we can obtain important reductions of about 50%. As soon as the tensile strength is sufficiently high, so much so that the slab

179

Figure 14: Reinforcement plan of the flat slab

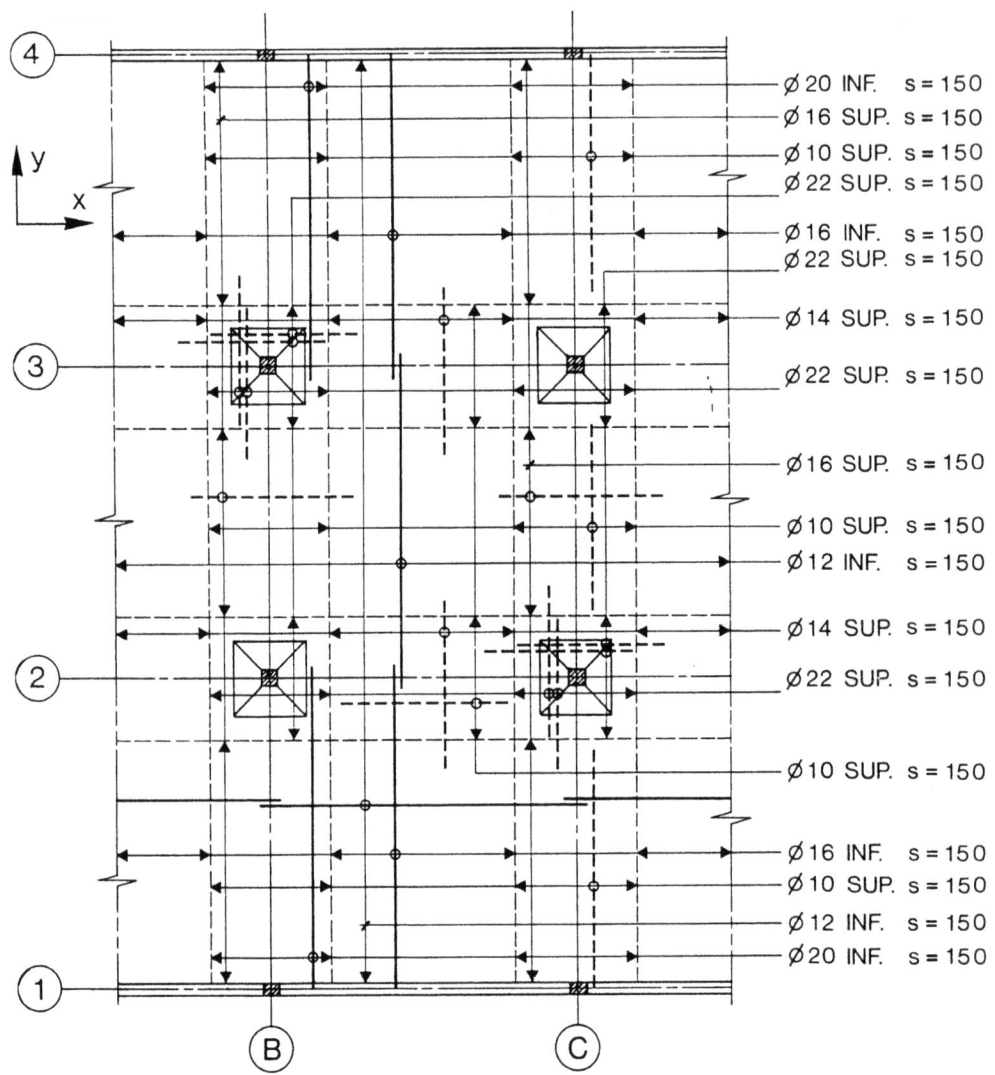

Figure 15: Influence of the tensile strength f_{ctm} on the long-term deflection of the flat slab

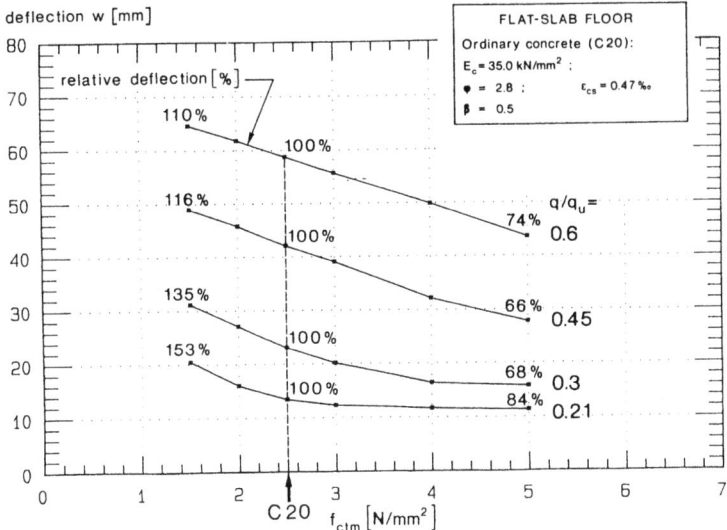

is no longer cracked, all supplementary increases of tensile strength have practically no influence on the deflection.

3.5.2 Influence of creep coefficient φ

We calculated the flat slab deflection at the four previous loading levels while varying the creep coefficient as follows:

$$\phi = 0.5 \quad 1.0 \quad 1.5 \quad 2.0 \quad 2.5 \quad 2.8 \quad 3.0 \quad 4.0$$

All other parameters were kept constant.

Figure 16 shows the results obtained. The deflections calculated with a creep coefficient of 2.8 have been taken as reference values. Note that when the creep coefficient decreases, the deflection decreases in a proportionally constant fashion. Also, the reductions are of a greater importance when the loading level is low.

3.5.3 Influence of shrinkage strain ε_{cs}

We also determined the influence of shrinkage strain ε_{cs} on the deflection of this flat slab. The calculations were carried out with the following values of ε_{cs}:

$$\varepsilon_{cs} = \quad 0.15 \quad 0.25 \quad 0.35 \quad 0.47 \quad 0.55 \quad 0.65\%o$$

Figure 16: Influence of the creep coefficient φ on the long-term deflection of the flat slab

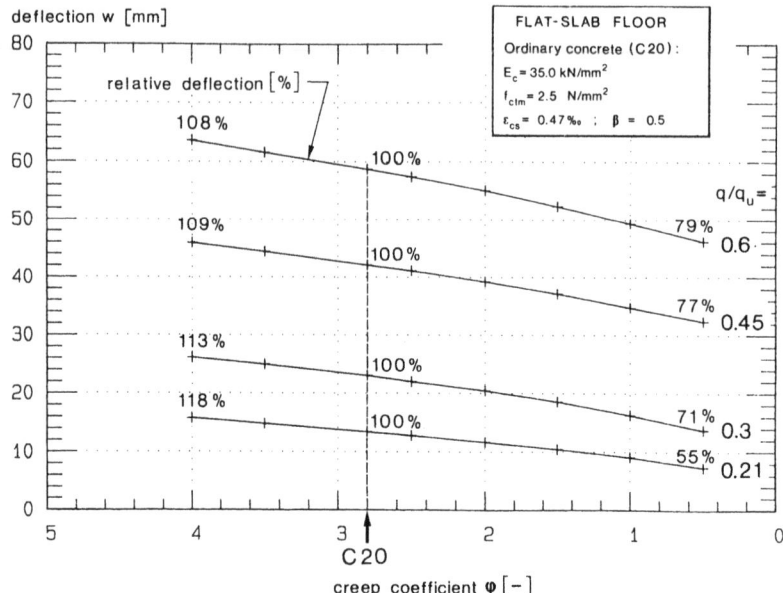

creep coefficient φ [−]

11

Figure 17 shows the results obtained. The deflections calculated with ε_{cs} = 0.47‰ have been taken as reference values.

Here we also notice that the deflection decreases in a proportionally constant fashion with the reduction of ε_{cs}.

3.6 Synthesis : possible reduction of slab deflection thanks to HPC.

We finish this chapter by examining two load-deflection curves at mid-span of the edge (see figure 13, point 15) of the flat slab. These two curves were calculated using the iterative non-linear procedure. They were calculated using the properties of an ordinary concrete and a HPC respectively. The mechanical and rheological properties of these two concretes were given in paragraph 3.5. Figure 18 shows the two calculated load-deflection curves.

From these two curves we have determined the relative reductions of deflection in function of the loading level through the use of a HPC. Figure 19 shows the reductions obtained. We can see that these reductions depend on the loading level. The reductions curve is characterized by a "peak" whose magnitude and location depend upon the static system of the slab and the HPC used. Here, the maximum reduction is about 50%. It extends from a loading level $q/q_u \cong 0.2$ until about 0.4.

Figure 17: Influence of the shrinkage strain ε_{cs} on the long-term deflection of the flat slab

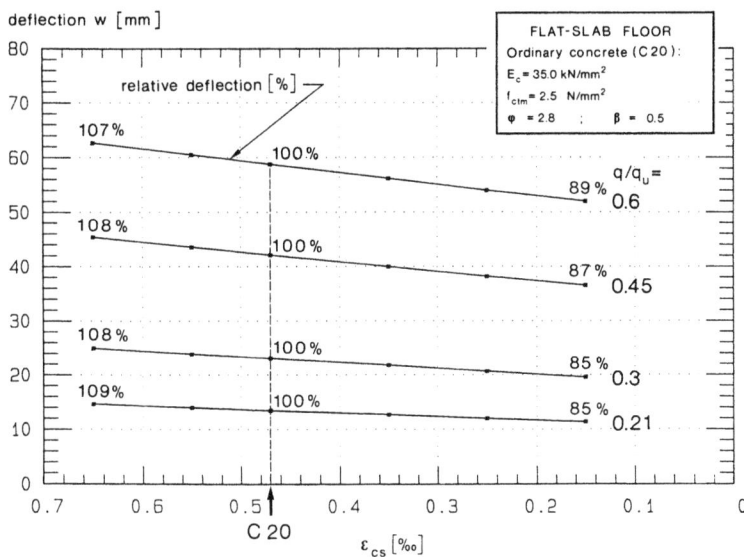

Figure 18: Load-deflection curves at the mid-span for flat slab made with ordinary concrete and HPC respectively

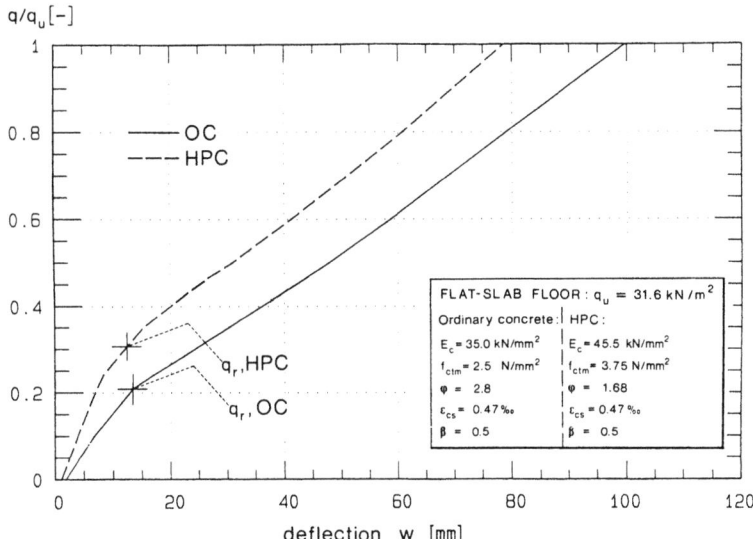

Figure 19: Relative reduction of the long-term flat-slab deflection versus load level

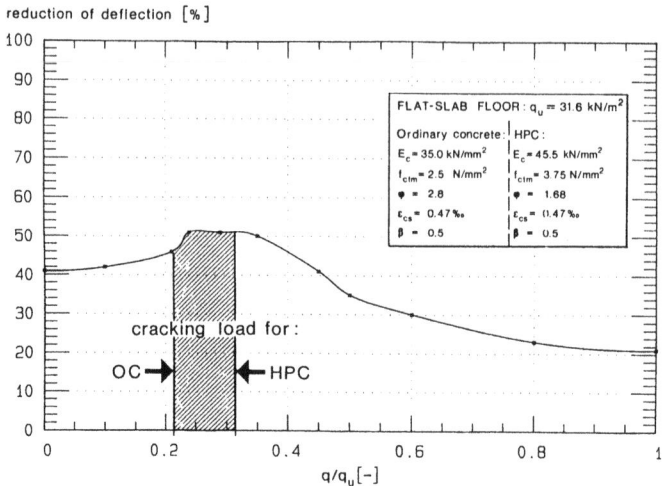

4 Conclusions

The utilization of HPC improves not only the strength and the durability of concrete structures but also their serviceability. In fact, the use of HPC provides an interesting solution to the problem of their excessive deformations.

The CEB moment-curvature relationship which has been established for ordinary concretes is still valid for HPC. This relationship which takes into account the reinforcement, the cracking and the time-dependent effects (shrinkage and creep) has been incorporated into a finite element program which calculates the long-term deflections of RC slabs. An iterative procedure based on the secant matrix method and the step by step method has been used to do the non-linear calculation.

The parametrical study has shown the influence of some mechanical and rheological properties of concrete on the deflection of a slab. Thus, we can say that an increase of concrete tensile strength would greatly reduce the deflection of a slab if this slab is cracked. Also, the reduction of the time-dependent effects (shrinkage and creep) reduce in a practically linear way the deflection of a slab.

This parametrical study has shown also the important reductions of deflection that are possible thanks to the utilization of HPC. The reduction obtained with respect to an ordinary concrete is a function of the loading level. It is characterized by a curve with a "peak" whose magnitude can attain 50% reduction or even more in function of the mechanical and rheological properties of the HPC which could be used. The extension of this "peak" is set into the range between 20 and 40% of the ultimate load of the slab. It corresponds to the quasi-permanent loading level to which structures are usually subjected.

5 Acknowledgments

The authors acknowledge all the institutions and personnel who have helped with this research. In particular we wish to thank the Computer Centre of the Civil Engineering Department at the Ecole Polytechnique Fédérale of Lausanne, for the exceptional means that were made available. We wish also to thank the Laboratory of Buildings Materials at the EPFL for its large contribution to the experimental study. Finally, we wish to thank the Swiss Cement Industry (VSZKGF) and the firm Sika for their financial support.

References

1. Favre R., Charif H., Jaccoud J.P. : Improved serviceability of reinforced concrete slabs with the use of high-strength concrete; Second International Symposium On The Utilization of High-Strength Concrete, ACI, SP-121, Berkeley, California, U.S.A., May, 1990
2. Charif H. : Réduction des déformations des structures en béton grâce à l'utilisation de béton à hautes performances; étude numérique et expérimentale. Doctoral thesis No 844, EPFL, Lausanne 1990.
3. Charif H., Jaccoud J.P. : Réduction des déformations et améliorations de la qualité des dalles en béton. Rapports d'essais, 2 volumes, EPFL, Lausanne; 1 rapport intermédiaire sur les études expérimentales, publication IBAP No 126, mai 1988.
4. Charif H., Jaccoud J.P., Alou F. : Reduction of deformations with the use of concrete admixtures. International Symposium on admixtures for concrete, Rilem, Barcelona-Spain, May 1990, Chapman & Hall pp. 402-428.
5. Favre R., Jaccoud J.P., Koprna M., Radojicic A. : Dimensionnement des structures en béton. Traité de génie civil de l'EPFL, volume 8, Edition PPR, Lausanne 1990;
6. CEB-FIP Model Code 1990, bulletin d'information No 195 et 196, draft March 1990.
7. CEB Manuel : Cracking and deformations, Task group : Favre R., (Reporter), Beeby A.W., Falkner H., Koprna M., Schiessl P. Jaccoud J.P., EPFL, Lausanne, 1985.

11 HIGH PERFORMANCE CONCRETES – ENGINEERING PROPERTIES AND CODE ASPECTS

L. PLISKIN
Engineer, L.P. Paris, France

1 Introduction

High Performance Concretes (HPC) open new opportunities in the utilization of the material concrete. Their use implies significant changes in the conceptual approach and in the construction methods for concrete structures. Civil Engineers have yet neither mastered nor measured the full potentialities of these new materials.

The exceptional properties of HPC proceed essentially from their reduced porosity. As a matter of fact porosity and permeability are the governing parameters which account for their performance.

Contrary to widespread opinion, the main characteristic of HPC is therefore not their high compressive strength which is only one of their many facets; it is their reduced porosity which makes them new materials with multiple advantages.

It is possible to categorize the main advantages of HPC as follows.

2 Durability

The increased durability of HPC constitutes the chief and most important contribution to construction. It is this property on which their use will be based since durability is an owner requirement.

3 Ease of placing

The use of superplasticizers follows the manufacture of high slump flowing (very workable) concretes. The placement of these concretes is much easier and less costly.

4 Early age strength

HPC make it possible to meet the requirements of contractors who ask for high early strength in order to achieve a quicker re-use of

High Performance Concrete: From material to structure. Edited by Yves Malier. © 1992 Taylor & Francis.
Published by Taylor & Francis, 2 Park Square, Milton Park, Abingdon, Oxon, OX14 4RN. ISBN 0 419 17600 4.

formwork or to accelerate production rates.

5 Productivity increase

The last two properties lead to higher productivity and therefore to a greater competiveness of concrete.

6 Mechanical, physical and chemical properties

All HPC properties proceed from their reduced porosity. They may differ largely from those of ordinary concretes.

6.1 High compressive strength
The high compressive strength of HPC is one of their most striking features. When ordinary concretes exceed with difficulty strengths of 35 to 40 MPa (5000 to 6000 psi), HPC reach currently 60 to 90 MPa (9000 to 13000 psi). Tomorrow the 100 MPa (15000 psi) barrier with be passed.

6.2 Impermeability
The impermeability of concretes to liquids and gases is directly dependent upon their W/C ratio. HPC being low W/C concretes, they offer a smaller permeability and hence a better durability.

6.3 Freeze-thaw resistance
HPC show an enhanced freeze-thaw resistance. The need for air-entraining agents is still a controversial point.

6.4 Abrasion resistance
HPC have a much higher abrasion resistance. This is of particular interest in road and pipe construction.

7 Quality improvement

The use of HPC will pull the construction industry up to higher levels of quality.

8 Costs

The increased cost of HPC is largely balanced by concrete quantity savings and by productivity improvements. Indirect benefits are also gained through durability improvements leading to maintenance savings.

Although HPC are made with the same basic components as ordinary concretes, their much higher qualitative and quantitative performances make them new materials. According to their use, they offer different advantages such as enhanced durability, impermeability or higher strength. This is why we have termed what are usually named high strength concretes High Performance Concretes.

a. Concrete classes

The generic expression HPC covers a wide range of concretes with different properties. It seems advisable to propose a general classification.

This classification can be based upon a structural engineers' approach, that is to say on the consideration of the 28 day characteristic strength of the concrete, f_{c28}, although this characterization does not describe actually the different properties of HPC. It must be emphasized that whenever speaking of concrete compressive strength, the test specimen type and dimensions should always be indicated in order to preclude inaccurate comparisons.

According to the French Code, this strength is to be measured on cylinders with 16 cm diameter (200 cm2 sectional area) and 32 cm height. When the maximum aggregate size is smaller than 16 mm, the cylinder dimensions should be reduced to 11 cm in diameter and 22 cm in height.

The classification could be as follows:

		f_{c28}
OC	Ordinary Concretes	20 to 50 MPa
HPC	High Performance Concretes	50 to 100 MPa
VHPC	Very High Performance Concretes	100 to 150 MPa
EC	Exceptional Concretes	> 150 MPa

High Performance Concrete class should be divided into two subclasses according to whether or not they contain ultrafines. They need a content of 400 to 450 kg/m^3 of high strength Portland cement. High range water reducing admixtures must be added in order to maintain the W/C ratio between 0.35 and 0.40 without impairing the concrete workability. No special aggregates are required.

Very High Performance Concretes are obtained by reducing the W/C ratio to between 0.20 and 0.35. They need high dosages of HRWR admixtures. All the concrete components should be high quality components. It is necessary to use ultrafines such as silica fume. They generally need retarding admixtures. The aggregates must have a high strength and a Young's modulus not too different from that of the hardened cement paste in order to minimize differential deformations between the aggregates and the matrix. Some authors advocate the use of aggregates with a maximum size of 10 to 12 mm.

Exceptional Concretes are still laboratory concretes. 300 MPa concretes, with W/C ratios around 0.16, are currently obtained in laboratories.

As a matter of fact this classification does not lead to a simple and unequivocal modelling of the mechanical properties of concretes. Like any classification it builds discontinuities which do not exist actually.

In France, 50 to 90 MPa cast in-situ concretes have already been used in bridge construction. The Joigny bridge, designed and built in 1989 as an experimental bridge used a 60 MPa concrete.

b. Modelling of concrete properties

The structural engineer is particularly interested in a limited number of parameters which describe the physical and mechanical properties of concrete playing a direct part in the design of a structure.

A material is therefore characterized by a reduced number of parameters allowing a correct modelling of the behavior of the structure. It should nevertheless be emphasized that this modelling is only an oversimplified representation of the actual behavior of the structure.

c. Design parameters

The main design parameters are :

- the constitutive law of the concrete establishing a relationship between stress and strain
- the 28-day compressive characteristic strength
- in some cases, the 28-day tensile characteristic strength f_{t28}
- the instantaneous Young's modulus E_i
- the creep coefficient ϕ
- the value of the partial safety coefficients γ_b

In France a National Project of Research and Development on "New developments of concrete" was launched several years ago. One of the main topics of this Project is HPC. Numerous tests and experimentations have been or are being carried out in order to get a better knowledge of HPC and VHPC properties. The same kind of researchs were made in Norway. This probably explains why the French and the Norwegian Concrete Codes are the only ones to take these new concretes into account.

It is necessary to know and to set the particular values of the parameters applicable to HPC. As a matter of fact these values vary with the HPC classes and within each class. Considering the present state of knowledge it begins to be possible to lay down design properties for concrete grades in the C50.. C100 MPa range.

The French Codes BAEL (Béton Armé aux Etats Limites: limit state design of reinforced concrete) and BPEL (Béton Précontraint aux Etats Limites: limit state design of prestressed concrete) have been updated in order to incorporate C 60 concretes, that is to say concretes with a 60 MPa 28-day characteristic strength measured on 16 x 32 cm cylinders.

The recent Norwegian Code NS 3473 for the Design of Concrete Structures extends its specifications to concretes with a 85 MPa characteristic strength. Since these strengths are measured on 10 x 10 cm cubes this upper limit is comparable to the French limitations.

c.1 Behavior diagram of HPC

The behavior diagrams of HPC differ according to the maximum concrete strength. It is to be noted that

- the stress-strain curve remains linear up to the highest value of the stress/ultimate stress ratio as the latter increases;
- the strain at the stress peak increases with the strength;
- the post-peak branch becomes steeper as the strength increases;
- the ultimate compressive strain decreases as the strength increases.

The behavior of ordinary concretes and of HPC are different:

- in ordinary concretes microcracking begins as soon as the compressive stress reaches about 30 to 40% of the ultimate compressive stress. At 80-90% of this latter stress cracks become interconnected.
- in HPC little microcracking appears until the compressive stress reaches 65 to 75% of the ultimate stress. At 80-90% of this latter stress cracks occur mainly in the transition zone and are not yet fully interconnected.

The fracture modes are also different:

- the typical fracture surface of an ordinary concrete is rough. The fracture occurs along the aggregate-matrix interface; the aggregates are not broken.

- conversely the typical fracture surface of an HPC is smooth. The cracks pass without any discontinuity through both the matrix and the aggregates. This transgranular fracture is more sudden.

The structural engineer needs simplified schematic diagrams defining the constitutive law of concrete. Various simplified diagrams have been proposed for the HPC such as triangular or trapezoid diagrams. After comprehensive studies the French Codes kept the classical parabola-rectangular diagram for HPC up to C 60. These studies showed that the shape of the stress-strain curve does not influence significantly the value of the ultimate state resistance of elements with different cross-sections.

c.2 Modulus of elasticity

The general relationship linking the value of the modulus of elasticity of concrete to its strength remain valid for HPC. Several formulations can be found. Because of the dispersion of the experimental results they all give correct values. When comparing them the type of specimens tested should be kept in mind.

Here are several expressions giving the value of the instantaneous Young's modulus :

France:
BAEL,BPEL $E_{id} = 11,000 \, f_{cd}^{1/3}$ (MPa) [1]

Norway:
NS 3473 $E_c = k_E \, f_c^{0.30}$ (MPa) [2]

USA:
ACI 318 $E_{cd} = 5,000 \, f_{cd}^{1/2}$ (MPa) [3]

Notes:

[1] - up to C60 concrete
[2] - up to C85 concrete
[3] - for ordinary concrete

The Poisson's ratio keeps its value of 0.20 to 0.25.

c.3 Allowable compressive stress f_{c28}

In most Codes the allowable compressive stress is given by :

$$f_c = \frac{0.85 \, f_{ck}}{\gamma_c}$$

where f_{ck} = compressive characteristic strength at the considered date

       0.85 = a coefficient taking into account the effect of sustained loads

       γ_c = the partial safety coefficient applicable to concrete, generally equal to 1.50

It is worth discussing these values.

Age of reference

Historical reasons base the choice of the age of 28 days for characterizing the concrete strength. The use of HPC may justify a change of this age of reference. Some HPC, like some ordinary concretes, carry on hardening after 28 days and reach a 90-day or a 180-day strength much higher than the 28-day strength. When the live loads are applied at older ages it would be worthwhile to take this increased strength into account.

Behavior under sustained loads

The behavior of concretes under sustained loads depends upon their mode of microcracking. Laboratory tests show three steps of behavior following the increasing value of the compressive stress :

- linear creep step: creep strains are proportional to the stress. the creep coefficient is constant.
- non-linear creep: creep strains become much more important. the creep coefficient increases quickly.
- rupture step: concrete failure occurs under a stress smaller than the intantaneous ultimate compressive stress. The value of the failure stress depends on the duration of the loading.

Tests on the effects of sustained loads which have been carried out in the US show the differences between ordinary concrete and HPC.

	Stress/Ult. instant. stress	
	Ord. Concrete	HPC
Linear creep	45%	65%
Non linear creep	75%	85%
Rupture	> 75%	< 85%

The strength of HPC under sustained loads appears therefore to be higher than that of ordinary concrete. These results seem questionable to some French researchers who found similar behaviors for ordinary concrete and HPC under long-lasting loadings. It would be of great economical interest to clarify this point since this could allow a larger coefficient than 0.85 in assessing the allowable compressive stress, should the American findings be confirmed.

It should be also emphasized that the consideration of this coefficient has a meaning only when dealing with loads applied during an infinite time. This is generally not the case for live loads.

The new French Codes take this reduced effect into account by defining the allowable compressive stress as

$$f_c = \frac{0.85 \ f_{ck}}{\theta \ \gamma_c}$$

where θ is a coefficient which depends on the probable loading duration

$\theta = 1$	if the loading lasts more than 24 h
$\theta = 0.90$	if it lasts within 1 h and 24 h
$\theta = 0.85$	if it lasts less than 1 h

Partial safety coefficient γ_c

Up to now most Codes consider $\gamma_c = 1.50$ for cast in-situ concretes. This is the case of the French Codes. It is worth wondering whether this safety coefficient could not be reduced for HPC since they imply a higher level of quality in fabrication and in placement. This is still an open question in France.

It is worth stressing that the Norwegnia Code specifies $\gamma_c = 1.40$. This latter value may even be reduced to 1.25 when deviations in dimensions do not exceed the specified tolerances.

Maximum design stress σ_c

The French Codes BAEL and BPEL prescribe a maximum compressive design stress in serviceability state equal to

0.60 f_{cd} during construction and under frequent loads
0.50 f_{cd} under quasi-permanent loads

f_{cd} being the allowable stress at the age of d days.

This limitation is questionable since it has been shown that HPC offer an elastic and linear field up to 0.70 or 0.80 f_{cd}. It would therefore be advisable to check whether values of 0.60 f_{cd} could not be accepted for HPC under quasi-permanent loads.

c.4 Allowable tensile stress f_{t28}

The value of the conventional allowable tensile stress for HPC can be assessed by the following expression which is also valid for ordinary concretes:

$$f_{td} = 0.60 + 0.06\ f_{cd}\ \text{(MPa)}$$

c.5 Bond

The French Rules consider that the bond stress between concrete and rebars τ_u is proportional to the tensile stress f_{td}. It is therefore a function of the compressive stength of the concrete. This means that HPC allow a reduced anchoring length of the reinforcements. Tests have been carried out in France which confirm this statement.

This fact may have an important economic aspect as it can lead to large savings in reinforcement quantities by using higher strength steel rebars such as steel with a 700 to 1000 MPa elastic limit. Given their reduced anchoring length in HPC, the crack widths and spacings will keep their accepted usual values. The durability of such reinforced HPC will therefore not be impaired.

c.6 Allowable shear stress

Tests confirm the validity of the French approach concerning the allowable shear stress, at least for HPC with a strength smaller than 60 MPa.

At ultimate limit state, the French Code BPEL considers a participation of the concrete in the shear resistance. This participation is taken equal to 1/3 f_{td}. Since fracture surfaces of HPC are smoother than those of ordinary concretes it will be necessary to check the validity of this assumption which leads to a reduction of the shear reinforcements.

c.7 Shrinkage

The shrinkage of HPC can be considered to be the same as the shrinkage of ordinary concrete.

c.8 Creep

The creep of HPC is smaller then that of ordinary concretes. The creep coefficient decreases while the strength increases that is to say while the W/C ratio decreases. For ordinary concretes, the value of the creep coefficient φ is generally taken equal to 2.00 when loading is applied at 28 days. It seems that this coefficient may be as low as 1.00 for some C 60 concretes and 0.50 for some C 100 concretes. The Norwegian Code links the value of the creep coefficient to the strength as follows.

$$\varphi = \varphi_0 \times 8.3 / \qquad\qquad (= f_{ck}^{1/2})$$

where φ_0 is the creep coefficient at 28 days and $f_{ck}^{1/2} = 3$.

An important research program is being carried out in France to get a better knowledge of the creep properties of HPC, particularly at early ages that is to say between 1 and 28 days. For the time being, it is accepted to consider a value of $\varphi = 1.50$ for C 60 concretes in preliminary designs. For definitive calculations the actual value of the creep coefficient is to be measured on the actual HPC used in the construction.

Attention should be given to the fact that the rate of creep is much greater in HPC than in ordinary concretes.

It must be emphasized that the total deformation, instantaneous and deferred, has the same order of magnitude as with ordinary concretes since higher stresses are applied to HPC.

c.9 Ductility
Compression tests show that the stronger the concrete the more brittle it is. This could be of concern since modern design methods take into account the plasticity of materials. As a matter of fact the question of the brittleness of HPC is a false problem since structural engineers deal with reinforced and/or prestressed concretes the ductility of which can be adjusted according to any particular needs.

Flexural tests run on reinforced HPC beams show that their ductility is similar to that of beams made with ordinary concrete.

Care should nevertheless be exercised concerning HPC elements such as columns submitted to axial loads. Experiments are being carried out in France in order to set the rules concerning minimum longitudinal and transversal reinforcements in such pieces.

c.10 Fatigue
Microcracking is less severe in HPC than in ordinary concretes. Fatigue built up damages should therefore be smaller. This seems confirmed by Norwegian tests which found an endurance limit for HPC. Practically, the Miner cumulative damage method can be used to assess the design life of reinforced or prestressed HPC structures.

12 TESTING HIGH PERFORMANCE CONCRETE

M. LESSARD and P-C. AITCIN
Université de Sherbrooke, Québec, Canada

1 Introduction

The recent development of high performance concrete (HPC) is having a tremendous impact on the course of the construction technology. The intensive search for higher compressive strength has led to a new area of over 120 MPa HPC. Research has been so polarized by the improvement of the mechanical properties that adjustments to equipment as well as testing techniques have so far been ignored or taken for granted.

Testing high performance concrete is a critical issue on which no consensus has yet been reached [1]. Among the many questions that often arise and that are mostly under passionate discussions, the following are of paramount importance:

- At what age must the compressive strength of HPC be tested ? 28,56 or 91 days ?
- What size should the specimens be? 52x104,75x150,100x200 or 150x300 mm ?
- Should the faces of cylindrical specimens be capped with a high strength capping compound or be ground on a lathe ?
- What are the qualitative and quantitative influence of the diameter of ball bearing as well as the influence of the eccentricity of the loading and that of the rigidity of the press ?

It is the intention of the authors to discuss each of these questions separately based on the many years of laboratory research in the field of high performance concrete testing at the Université de Sherbrooke, Québec, Canada.

High Performance Concrete: From material to structure. Edited by Yves Malier. © 1992 Taylor & Francis.
Published by Taylor & Francis, 2 Park Square, Milton Park, Abingdon, Oxon, OX14 4RN. ISBN 0 419 17600 4.

2 The questions one should ask

2.1 Age of testing

It is generally agreed that the compressive strength of HPC must be tested at 91 days because most of these concretes contain cementitious materials (silica fume, ground granulated slag, fly ash, etc) which hydrate at a slower rate than Portland cement. Moreover, the 28 day deadline is not critical as far as design is concerned since most structures made of HPC are not fully loaded at either 28 or 91 days although precast elements may be an exception to this pratical rule.

2.2 Shape of the specimens

Concrete technicians in America are familiar with cylindrical specimens cast into plastic molds that are very easy to strip with compressed air and clean, and are reusable hundred of times. However, even with careful striking of the top of the cylinder, the cylinders must be capped before testing.

On the contrary cubes don't need any capping, but cube molds are heavy (most of them are made of steel), they take time to demold, time to clean and are thus not very popular in North America. In Europe, this does not seem to be the case and cubes are used more often. Whatever shape is selected, it should be noted that the cube strength is approximately 1.25 times the cylinder strength [2].

2.3 Size of specimens

The choice of size is important because it relates to the capacity of the testing machine. In Table 1, the necessary loads to fail 150 x 300,100 x 200 and 75 x 150 mm cylinders of 100 and 150 MPa concretes are presented. The corresponding suggested maximum load capacity that a press should be able to handle safely, such as a very high strength concrete on a long term basis, is also indicated. In North America this load capacity is customarily taken as 150% of the specimen load to failure.

It is seen from Table 1 that the capacity required for testing 150 mm diameter specimens is quite high limiting therby the number of laboratories able to test it to failure.

It is a well known fact that the diameter of a normal concrete specimen has an influence on the resulting mechanical characteristics. In high performance concrete, however, no systematic and exhaustive investigations have been made so far. The only attempts have resulted in scattered results.

Gonnerman [3] observed that, generally speaking, for normal concrete (~ 33 MPa) compressive strenght decreased with increasing specimen diameter. The variation was however very small for cylinders having a diameter less than 305 mm. For instance a 102 x 203 mm specimen failed at 101% of the load at which a

Table 1: Maximum press capacity required to test 100 and 150 MPa concrete specimens of different diameters

	Load at rupture		Press capacity required	
Dimension of specimens	100 MPa	150 MPa	100 MPa	150 MPa
150 x 300 mm	1.76 MN	2.65 MN	2.65 MN	4.0 MN
100 x 200 mm	0.785 MN	1.18 MN	1.2 MN	1.75 MN
75 x 150 mm	0.44 MN	0.66 MN	0.66 MN	1.0 MN

152 x 305 mm specimen failed. RILEM [4] found that the compressive strength ratio of 150 x 300 mm over that of 100 x 200 mm specimen is equal to 97%.

Malhotra [5], investigating 8 to 46 MPa concretes, found that this ratio ranged from 84% to 132%. He noted also that the standard deviation increased with decreasing diameter. This increase was so important that the author recommended the number of 102 x 203 mm specimens be twice as much as that of 152 x 305 mm specimens in order to achieve a comparable degree of accuracy.

Carrasquillo et al. [6], investigating 23 to 80 MPa concretes, found that the compressive strength ratio of 152 x 305 mm over 102 x 203 mm specimens was 90%. Forstie and Schnormeier [7] pointed out that the 102 x 203 mm specimens had a significantly higher strength than 152 x 305 mm ones for strength above 34 MPa. For instance, for 48 MPa concrete, they reported a compressive strength difference of 6.9 MPa between 102 x 203 and 152 x 305 mm speicmens. Strength is comparable in both cases at about 20.7 MPa. In contrast, under 19.8 MPa, 102 x 203 cylinders tested slightly lower than their 152 x 305 mm conterparts. These findings support Malhotra's [5] contention, that, at lower strength levels, larger specimens will test higher.

Nasser and Kenyon [8] compared the results of tests performed on 76 x 152 mm and conventional 152 x 305 mm specimens, and found an average compressive strength ratio of 86%. At 28 days however this average ratio was only 79%. This rather low ratio was attributed by the authors to a premature setting of the concrete due to the admixtures as well as the time of placing which was longer than required.

Date and Schnormeier [9], investigating 24 to 36 MPa concretes, reached the same conclusion, that is, the compressive strength of 102 x 203 mm specimens is generally higher than that of 152 x 305 mm specimens and the strength difference is not significant for the concretes considered. The same tendency has been observed by Nasser and Al-Manaseer [10] who compared the compressive strength of 75 x 100 mm to 150 x 300 mm specimens.

The results obtained by Carrasquillo and Carasquillo [11] for concretes having a compressive strength ranging from 48 to 80 MPa were opposed to those presented above for normal concretes. They found the 102 x 203 mm specimens is compressive strength to be approximately 93% of that of 152 x 305 mm specimens.

Intensive investigation has been made at the Université de Sherbrooke, Québec, Canada, where 11 concretes were considered. The compressive strength ranged from 35 to 122 MPa as can be seen in Table 2. All compressive strengths are based on an average of 3 cylinders unless otherwise specified. The standard deviation(S) as well as the coefficient of variation (S/fc) are also displayed in the table.

It is reasonable to assume that concrete samples made of the same batch are homogeneous. Therefore the variations in the measured compressive strength of concrete cylinders made from the same mix are mainly due to those related to fabrication, curing and testing. In laboratory conditions, the curing factor can reasonably be eliminated and the variations can be assumed to be mainly due to fabrication and testing parameters.

The results obtained at the Université de Sherbrooke presented in Table 2 show that the overall average value of the coefficient of variation obtained on the 152 x 305 mm specimens (2.3%) and 100 x 200 mm specimens (2.2%) is almost identical. This result tends to confirm those of Moreno [12] but are in contradiction with others indicating that the coefficient of variation is larger when smaller specimens are used [5,13].

It is interesting to study in more detail the results obtained on concrete N° 11 (Table 2). Note that the 100 x 200 mm cylinder mean strength is based on 3 values while that of the 152 x 305 mm cylinder is based on 7 values. The 100 x 200 mm cylinder coefficient of variation (2.4%) is approximately equal to that of 152 x 305 mm cylinder (2.7%). This suggests that the same number of specimens of either size can be used to obtain approximately the same degree of accuracy.

Table 3 shows, in addition to the average compressive shown in Table 2, the correlations between the 100 x 200 mm cylinder compressive strengths and those of 152 x 305 mm cylinders. Thus, for 28 as well as 91 days, the 100 x 200 mm specimen compressive strength is 106% that of 152 x 305 mm specimens, on average, for concretes ranging from 35 to 126 MPa. This reflects the general tendency already pointed out by other researchers (3,4,5,6,7and 9) who obtained 101 to 110% for 23 to 80 MPa concretes.

Table 2: Compressive strengths of 100x200 mm and 152x305 mm cylinders (average of 3 specimens)

Concrete identification		1	2	3	4	5	6	7	8	9	10	11
28 days	100 x 200 mm											
	$\overline{f}_c$ (MPa)	37.5	75.2	93.4	96.7	101	99.3	110	105	90.0	114	112
	S (MPa)	0.6	1.5	1.7	1.0	1.2	2.5	1.0	1.0	0.8	1.5	2.7
	S / $\overline{f}_c$ (%)	1.6	2.0	1.8	1.0	1.2	2.5	0.9	1.0	0.9	1.3	2.4
	152 x 305 mm											
	$\overline{f}_c$ (MPa)	35.0	71.7	91.4	88.3	95.4	94.5	105	104	85.6	108	108*
	S (MPa)	0.5	1.7	2.1	0.7	3.4	0.2	1.0	1.7	3.6	1.5	2.9
	S / $\overline{f}_c$ (%)	1.4	2.4	2.3	0.8	3.6	0.2	1.0	1.6	4.2	1.4	2.7
91 days	100 x 200 mm											
	$\overline{f}_c$ (MPa)	44.4	—	111	104	122	108	117	121	111	126	—
	S (MPa)	0.7	—	2.9	6.7	2.7	3.1	2.0	4.0	5.5	4.5	—
	S / $\overline{f}_c$ (%)	1.6	—	2.6	6.4	2.2	2.9	1.7	3.3	5.0	3.6	—
	152 x 305 mm											
	$\overline{f}_c$ (MPa)	38.2	—	105	96.5	114	106	117	116	103	115	—
	S (MPa)	0.6	—	3.0	3.6	0.6	2.7	5.3	1.2	2.7	4.0	—
	S / $\overline{f}_c$ (%)	1.6	—	2.9	3.7	0.5	2.5	4.5	1.0	2.6	3.5	—

* Average of 7 specimens

Table 3: 100x200 mm compressive strengths in percentage of 152x305 mm compressive strengths

Identification of concrete	Average Compressive Strengths at 28 days (MPa)			Average Compressive Strengths at 91 days (MPa)		
	100 x 200 mm cylinders	152 x 305 mm cylinders	$\frac{f'_c 100}{f'_c 152}$ (%)	100 x 200 mm cylinders	152 x 305 mm cylinders	$\frac{f'_c 100}{f'_c 152}$ (%)
1	37.5	35.0	107	44.4	38.2	116
2	75.2	71.7	105	—	—	—
3	93.4	91.4	102	111	105	106
4	96.7	88.3	110	104	96.5	108
5	101	95.4	106	122	114	107
6	99.3	94.5	105	108	106	102
7	110	105	105	117	117	100
8	105	104	101	121	116	104
9	90.0	85.6	105	111	103	108
10	114	108	106	126	115	110
11	112	108	104	—	—	—
Maximum value of strength ratios	116 %					
Minimum value of strength ratios	100 %					
Average value of strength ratios	106 %					
Standard deviations of ratios	3.6 %					

However, these results conflict with those presented by Carrasquillo and Carrasquillo [11]. We believe this is attributable to the difference in experimental procedures, conditions and set ups.

The strength of high performance concrete is seen to depend on specimen size. This suggests that any related results should include the diameter of the specimens considered.

2.4 Bearing block dimensions

In order to evaluate the influence of the bearing block dimension on the 100 x 200 and 152 x 305 mm cylinder compressive strength, two different bearing blocks were used at the Université de Sherbrooke. Both blocks comply with ASTM standard C39-83b [14]. The first block is 102 mm diameter and is appropriate for testing 100 x 200 mm samples while the second is 152 mm diameter and is appropriate for 152 x 305 mm cylinders. The bearing blocks are displayed in Figures 1 and 2.

Table 4 gathers the 28 day compressive strength of the two concretes. The compressive strength of the first concrete was

Figure 1 : 102 mm diameter normalised bearing block for 100 x 200 specimens

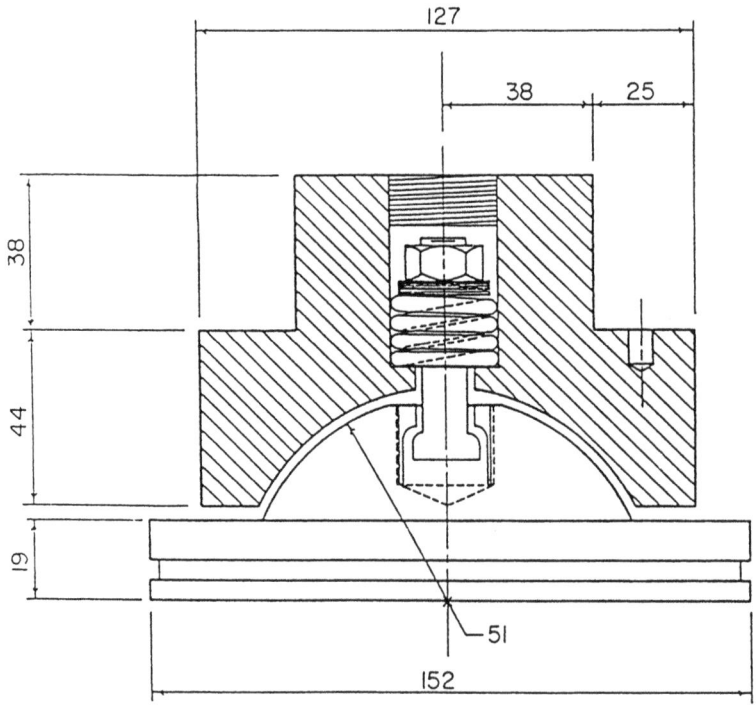

Figure 2 : 152 diameter normalised bearing block for 152 x 305 mm specimens

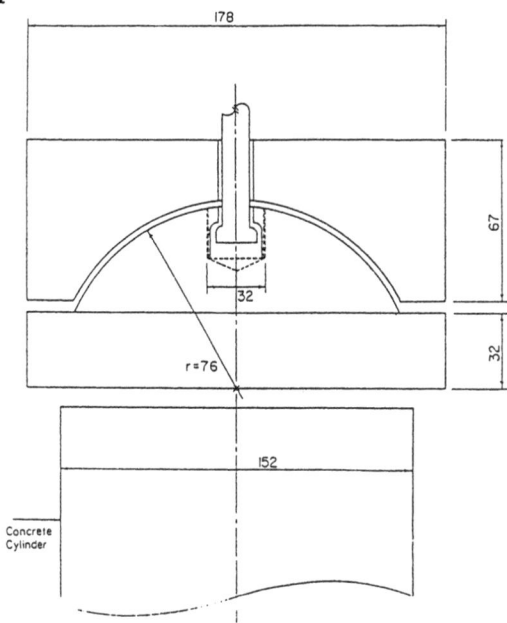

obtained on 12 capped as well as ground specimens. The second concrete however, is based on the average of three 100 x 200 mm and seven 152 x 305 mm all ground specimens. The average compressive strength of 100 x 200 mm cylinders is also expressed in terms of that of 152 x 305 mm cylinders and is given in the last column of Table 4.

In analyzing the results shown in Table 4, the following observations can be made.

In using the 102 mm diameter bearing block for both testing 100 x 200 and 152 x 305 mm specimens, then, the 100 x 200 mm compressive strength is seen to be as great as 125 to 127% of that of 152 x 305 mm sample.

In complying with ASTM standards, that is by using the 102 mm bearing block to test 100 x 200 mm cylinders and the 152 mm diameter block to test 152 x 305 samples, this percentage drops to 104% which is approximately equal to the percentage found earlier.

If one examines the 152 x 305 mm compressive strengths obtained on the 102 mm diameter block on one hand and on the 152 mm diameter block on the other, one can notice that the compressive strength, when using the 152 mm diameter block is 121% greater. The rupture type is also seen to be different and dependant upon the bearing diameter. Indeed, and as illustrated in Figure 3, conical type of rupture has been observed on specimens

Table 4: Average compressive strengths for different bearing diameters

Identification of concrete	Type of bearing	Average compressive strengths (MPa)					
		Capped			Ground		
		100 x 200 mm cylinders	152 x 305 mm cylinders	$\overline{\varepsilon'c}\,100 / \overline{f'_c}152$ (%)	100 x 200 mm cylinders	152 x 305 mm cylinders	$\overline{\varepsilon'c}\,100 / \overline{f'_c}152$ (%)
1	Ø 102 mm	118*	93.0*	127	119*	93.5*	127
2	Ø 102 mm	—	—	—	112**	89.3***	125
	Ø 152 mm	—	—	—	—	108***	104****

*	Average of 12 specimens	***	Average of 7 specimens
**	Average of 3 specimens	****	112/108

Figure 3 : Failure types governed by bearing block diameter

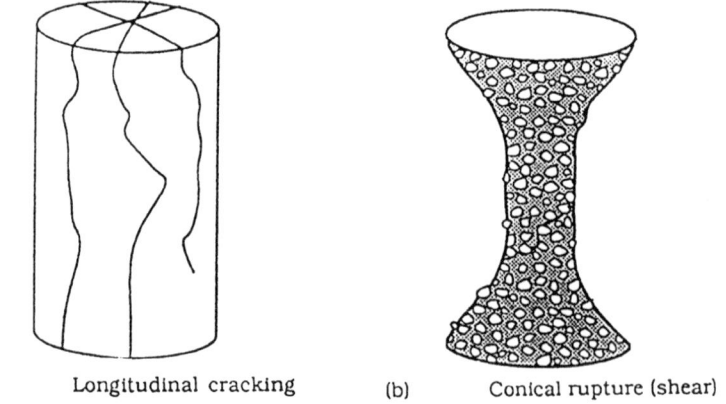

(a)   Longitudinal cracking   (b)   Conical rupture (shear)

tested on the 152 mm block and splitting type of rupture on specimens tested on the 102 mm block.

According to ASTM standard C39-83b [14], in performing a classical compressive test, a conical type of rupture is to be expected because the lateral constraint is not prevented. This is explained as follows : when a solid is subjected to compression in one direction, it tends to expand in all other perpendicular directions [15]. However, there exists a frictional force between the press platens and the cylinder [16]. This frictional force, in reacting to lateral displacement, gives rise to a lateral compressive force which is responsible for the formation of a cone at rupture [15]. When the lateral constraint is eliminated, the lateral compressive force disappears and a splitting type of rupture is obtained. Since this force significantly increases the apparent strength concrete, it becomes normal that the strength drops drastically as the lateral constraint is eliminated.

According to Neville [17] this strength drop would be of the order of 80% for concrete having a compressive strength around 45 MPa. Our results on 152 x 305 mm cylinders show that the rupture type conforms to ASTM standard when testing with a 152 mm diameter bearing. However, when using a 102 mm diameter bearing, the rupture seems to characterize the case where the lateral constraint has been eliminated, that is a splitting type with a strength drop of 83%. This would explain the results obtained on 100 x 200 and 152 x 305 mm specimens tested on 102 mm diameter bearing, that is a strength decrease of 20% between 100 x 200 and 152 x 305 mm cylinders, a conical type of rupture for 100 x 200 specimens and a splitting type of rupture for 152 x 305 mm specimens.

Nevertheless, it is important to note that the lateral constraint cannot be eliminated when a 102 mm diameter bearing is used. Therefore, other phenomena, such as those related to the diameter of the bearing or that of bottom platen (152 mm in our case), would be responsible for the observed behaviours.

This study shows that particular attention should be paid to the dimensions of the bearing block in order to obtain comparable results.

2.5 Preparation of specimen ends
Concrete cylindrical specimens are normally capped before testing to be sure that the two faces on which the uniaxial load is applied are parallel. Nobody questioned this practice until the testing of very high performance concrete started to cause some problems. The following questions then arose : what exactly is the strength of commercial capping compounds ? Does its compressive strength measured on 50 x 50 x 50 mm (2x2x2 in.) cubes represent the actual strength of the material when it is used as a very thin cap squeezed between the plates of the testing machine and the two end surfaces of the specimen ?

In such a situation, most of the capping compound is confined and a biaxial compressive test should be more appropriate to measure the actual compressive strength of this compound. However, in such a case, what is the lateral stress that needs to be applied on the capping compound specimen ?

At the Université de Sherbrooke, we have been using different so-called high performance capping compounds sold in North America. These compounds have a typical 50 mm cube compressive strength of 60 to 70 MPa and can be used to cap 100 to 130 MPa HPC cylinders provided the capping is as thin and hence as confined as possible. A thickness ranging from 1.5 to 3 mm has been suggested in the litterature [14,17]. At the Université de Sherbrooke a thickness ranging from 1 to 2 mm was found optimum and is therefore currently used. Beyond 130 MPa, capping compound crushes at final rupture making it thereby difficult to know whether or not this phenomenon influences the ultimate compressive strength of the concrete specimen.

Table 5 represents some of the results obtained at the Université de Sherbrooke, the average compressive strengths (fc) are based on 10 specimens unless otherwise specified. The standard deviation (S) and the coefficient of variation [(S/fc) x 100] for each value are also given in the table. It may be of interest to note that the capping compound used in this study is called hi-cap from Forney and its specified 50 mm cube compressive strength ranges from 55 to 62 MPa. The producer specifies also its compressive strength, when used a cylindrical slice of 152 mm diameter and 6.4 mm thick, as being 110 MPa.

A statistical study on the compressive strengths of different concretes revealed no significant average strength difference between ground and capped cylinders. A 5% statistical error was assumed in this study.

This suggests that grinding is only necessary if the accuracy is to be excellent (coefficient of variation below 3%) in the case of concretes below 130 MPa or if the concrete has a compressive strength above 130 MPa. It is worth noting that the above results are based on high quality capping compounds.

Concrete specimens can be prepared in advance and placed in a curing room. During testing, the cylinder ends are put directly into contact with the testing machine platens. In doing so, nicely shaped conical ruptures are obtained in most cases.

Despite research, we do not really know the behavioral difference between capped and ground specimens whose strengths is beyond 130 MPa. However, tests on six ground cylinders yielded an average compressive strength of 149 MPa. (Standard deviation = 2.8 MPa and coefficient of variation = 1.9%). The specimens were water cured for 4 to 9 months before testing. Also a seventh cylinder was capped and tested. The compressive strength attained was129 MPa, that is 87% of the average compressive strength of the ground cylinders. However, we cannot draw a general tendency based on this result alone and more research and testings are to be performed on concrete above 130 MPa in order to be more conclusive.

Table 5: Compressive strength of capped and faced 100x200 and 152x305 mm specimens (average of 10 specimens)

			Identification of concrete				
			1	2	3	4	5
100 x 200 mm cylinders	Capped	$\overline{f_c}$ (MPa)	115*	118*	120	121	129*
		S (MPa)	4.8	5.7	3.1	3.8	4.0
		S / $\overline{f_c}$(%)	4.2	4.8	2.6	3.1	3.1
	Ground	$\overline{f_c}$ (MPa)	117*	119*	121	123**	132**
		S (MPa)	2.5	2.7	2.3	1.9	2.8
		S / $\overline{f_c}$(%)	2.1	2.3	1.9	1.5	2.1
152 x 305 mm cylinders	Capped	$\overline{f_c}$ (MPa)	—	93.0	115	114**	119**
		S (MPa)	—	2.8	3.1	4.9	4.5
		S / $\overline{f_c}$(%)	—	3.0	2.7	4.3	3.8
	Ground	$\overline{f_c}$ (MPa)	—	93.5*	114	117***	122**
		S (MPa)	—	1.6	2.9	3.6	3.5
		S / $\overline{f_c}$(%)	—	1.7	2.5	3.1	2.9

*   Average of 12 specimens
* *   Average of 9 specimens
* * *   Average of 7 specimens

2.6 Effect of eccentricity on concrete compressive strength

Two concretes have been made in order to evaluate the eccentricity effect on concrete compressive strength. The first is a 115 MPa high performance concrete while the second is a 30 MPa normal strength concrete. The cylindrical specimens were tested for three different eccentricities : 12.5, 6 and 4 mm. The maximum eccentricity, that is 12.5 mm,was chosen on the basis of the principle of central kernel limits (18, 19, 20). This principle does not rigorously apply to this case as the loading is not centered. However, it was only starting point we could think of at that stage.

The specimens have been subjected to compressive tests and the results are presented in Table 6 where the compressive strength is displayed for each type of concrete and at different eccentricities. Moreover, the type of rupture observed after failure is also given. Figure 4 illustrates the three basic types of rupture encountered in practice.

A statistical student "t" test analysis of the data given in Table 6 yielded the following: for normal strength concrete, there is no significant average compressive strength difference between specimens with no eccentricity and those with an axial eccentricity

Table 6: Compressive strengths for different excentricities (average of 5 specimens)

	Compressive strengths (MPa)			
Ecentricity (mm)	Concrete 1	Type of rupture	Concrete 2	Type of rupture
0	29.3	conical	115**	conical
4	30.0	conical	115	conical
6	28.7	conical	108	diagonal
12.5	26.5	diagonal	95.4	splitting

*   The type of rupture is described in figure 4
** Average of 9 specimens

Figure 4 : Failure types encountered during eccentricity testing

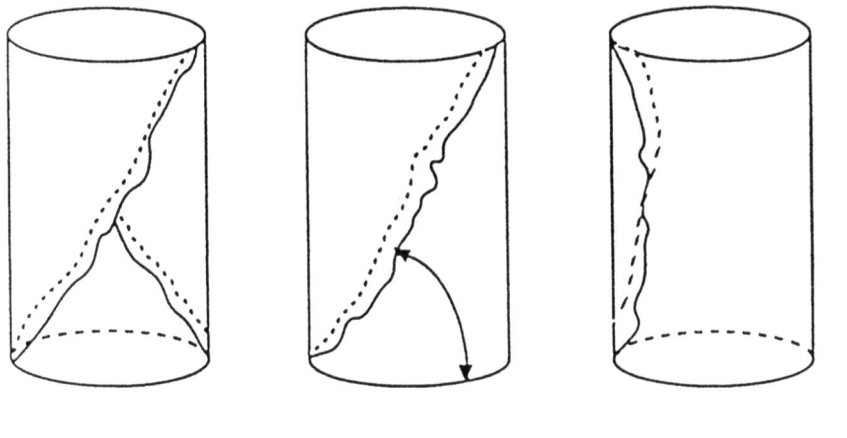

CONICAL RUPTURE   DIAGONAL RUPTURE   SPLITTING RUPTURE

of 4 to 6 mm. However, this difference becomes more important when comparing specimen with no eccentricity to those with an eccentricity of 12.5 mm. For high strength concrete an eccentricity of as low as 6 mm gave rise to a significant difference in compressive strength compared to specimens with no eccentricity, 5% error allowance has been admitted for the above analysis.

It can be concluded that, within 5% error, the measured compressive strength is not affected provided the eccentricity with respect to the central axis of the testing machine bearing is kept below 6 mm for normal strength concrete and below 4 mm for high performance concrete.

2.7 Testing machine capacity and rigidity

Table 7 gives the maximum diameter of two different HPCs (100 MPa and 150 MPa) and the corresponding testing machine capacity that can be used safely. Thus the maximum load that has to be carried by the testing machine for 100 and 150 MPa high compressive stength cylinders is:

- 1.0 MN the maximum size of the specimens that can be safely tested is 75 x 150 mm
- 2.0 MN the maximum size of the specimens that can be safely tested is 100 x 200 mm
- 4.0 MN the maximum size of the specimens that can be safely tested is 150 x 300 mm

Table 7: Maximum concrete specimen diameters that can be tested on testing machine of different capacities

	Maximum specimen diameter that can be tested safely			
Capacity of the press	1.0 MN	1.5 Mn	2.0 Mn	4.0 Mn
100 MPa	75 mm	100 mm	100 mm	150 mm
150 MPa	75 mm	75 mm	100 mm	150 mm

The problem of the press rigidity has been mostly pointed out by those researchers interested in post-peak behavior of concrete (21). However, the lack of testing machine rigidity can give rise to some problems at final failure of the specimens. These problems are related to the elastic energy stored in the testing machine column and which is suddenly released at failure. This repeated phenomenon can offset the testing machine calibration (22) and ultimately put it out of order.

The difference between a conventional and a rigid testing machine is illustrated in Figure 5 (21). Figure 5 displays a stress-strain curves of two testing machines. As can be seen the press having the coefficient of rigidity kp 1 is less rigid than that of the one having that coefficient kp2.

The difference between a conventional and a rigid testing machine is illustrated in Figure 4 and can be explained as follows : first the decrease Δx in specimen length at ultimate strength reduces its capacity by a force decrement equal to $\Delta Pe = (dP/dx) \Delta x$, in which dP/dx is the slope of AB and hence the coefficient of rigidity of the concrete specimen related to the displacement Δx. In the case of the conventional testing machine, $\Delta Pp1 = kp1 \times \Delta x$ and since $\Delta Pp1 < \Delta Pe$, then some energy is liberated from the system and that creates instability. Contrariwise, in the case of the rigid testing machine, $\Delta Pp2 = kp2 \times \Delta x$ and since $\Delta Pp2 > \Delta Pe$ the system requires energy and that creates stability.

On the other hand, the press should be mounted on a solid base capable of safely dissipating this energy without generating harmful vibrations.

It may be worth noting that a rigid testing machine would not

Figure 5 : Difference between conventional and rigid testing machine

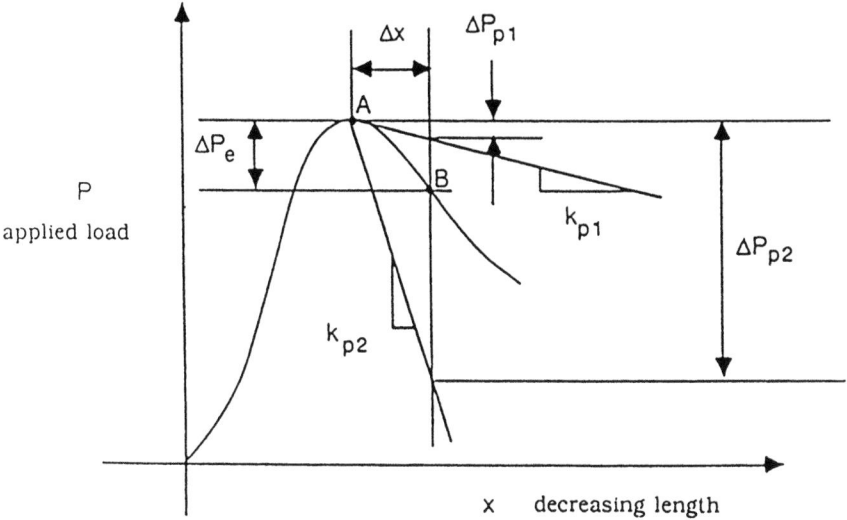

deform as the specimen deforms even at ultimate strength. The rate at which the load is applied decreases and a greater strength is obtained. In the case of a less rigid testing machine, however, the loading stress follows closely enough the specimen stress-strain curve and as cracks start propagating, the energy stored in the testing machine is quickly released. This results in a loading stress at failure which is smaller compared to the loading stress attained in the case of a rigid testing machine. This type of rupture often results on a violent explosion (17).

3 Practical recommendations

The following practical recommendations can be formulated :

- Use reusable plastic cylinders having a thick wall in order to avoid the ovalization of the upper face of the specimens. The concrete can be withdrawn quite easily from the mould by applying air through the orifice of the bottom part of the cylinder.
- Face the top of the concrete specimen when fresh and leave it harden in an upright position. In so doing, the grinding and the polishing efforts, which are done on a turning machine with a diamond wheel, are substantially reduced for above 130 MPa concretes.
- Check that the bearing block is adequate.

- Make sure that the testing machine used has the maximum rigidity. Whenever possible, try to increase this rigidity.
- Make sure the testing machine is well anchored to the floor.
- Equip the testing machine to be able to
  . measure the elastic modulus of concrete
  . trace the stress-strain curve up to failure
  . trace the stress-strain curve of different deformation rates.

These stress-strain curves allow a better insight into the behavior of concrete and the role played by coarse aggregate in HPC : the form of the hysteresis, the relaxation time and the residual deformation are some of the information that can be very useful during interpretation of the results.

4 Conclusion

We do not pretend to have come up with definitive answers to all problems related to the testing of HPC. However, we have tried to share our own experience in the field which, we believe, should be treated as a whole and urgently in order to come up with standard and precise testing procedures. This would facilitate the resolving of interpretation problems that can be quite an exercise when monetary interests are involved.

5 References

1. ACI Committee 363, (November 1987) Research Needs for High-Strength Concrete, ACI Materials Journal, pp. 559-561.
2. Neville, A.M., Brooks, J.J. (1987) Concrete Technology, 1st edition, p. 304-308.
3. Gonnerman, H.F. (1925) Effect of Size and Shape of Test Specimen on Compressive Strength of Concrete, Proceedings ASTM, Vol. 25, Part 2, pp. 237-250.
4. Commission RILEM, (September 1961) Coefficients de correspondance entre les résistances de différents types d'éprouvettes, Bulletin R.I.L.E.M. Vol. 12, pp. 237-250.
5. Malhotra, V.M. (January 1976) Are 4 x 8 Inch Concrete Cylinders as Good as 6 x 12 Inch Cylinders for Quality Control of Concrete?, ACI Journal, Vol. 73, No. 1, pp. 33-36.
6. Carrasquillo, R.L., Nilson, A.H., State, F.O. (May 1981) Properties of High Strength Concrete Subject to Short-term Loads, ACI Journal, Vol. 78, No. 3, pp. 42-45.
8. Nasser, K.W., Kenyon, J.C. (January 1984) Why Not 3 x 6 Inch Cylinders for Testing Concrete Compressive Strength?, ACI Journal, Vol. 81, No. 1, pp. 47-53.

9. Date, C.G., Schnormeier, R.H. (August 1984) Day-to-day Comparison of 4- and 6- in Diameter Concrete Cylinder Strengths, Concrete International, Vol.6, No 8, pp. 24-26.

10. Nasser, K,W., Al-Manaseer, A.A. (May 1987) It's time for a Change from 6 x 12 to 3 x 6 in. Cylinders, ACI Journal , Vol. 84, No. 3, pp. 213-216.

11. Carrasquillo, P.M., Carrasquillo, R.C. (January 1988) Evaluation of the Use of Current Concrete Practice in the Production of High Strength Concrete, ACI Journal, Vol. 85, No 1, pp. 49-54.

12. Moreno, J. (January 1990) 225 W. Wacker Drive, Concrete International, Vol. 12, No. 1, 35-39.

13. De Larrard, F., Torrenti, J.M., Rossi, P. (March 1988) Le flambement à deux échelles dans la rupture du béton en compression, Bulletin de liaison des laboratoires des ponts et chaussées, No. 154, pp. 51-55.

14. American Society for Testing Materials (1984) Standard Test Method of Compressive Strength of Cylindrical Concrete Specimens, ASTM C39-83b, pp. 24-29.

15. Thaulow, S. (December 1962) Apparent Strength of Concrete as Affected by Height of Test Specimen and Friction Between the Loading Surfaces, Bulletin RILEM, Matériaux et Constructions - Recherches et Essais, No. 17, pp. 31-33.

16. Torrenti, J.M (June 1987) Comportement multiaxial du béton : Aspect expérimentaux et modélisation, Ecole Nationale des Ponts et Chaussées, Thèse de doctorat, pp. 104-107.

17. Neville, A.M. (1981) Properties of Concrete, Pitman Publishing Ltd., 3rd edition, pp. 534-539.

18. Nachtergal, C. (1966) Aide-mémoire pratique de résistance des matériaux, Boeck, Bruxelles, 5th edition, pp. 737-737.

19. Dreyfuss, E. (1956) Leçons sur la résistance des Matériaux, Eyrolles, Paris, Vol. 2, pp. 268-283.

20. Massonnet, C. (1962) Résistance des Matériaux, Dunod, Paris, pp. 368-373.

21. Vutukuri, V.S., Lama, R.D., Saluja, S.S. (1984) Handbook on Mechanical Properties or Rocks, trans Tech Publications, Claausthal, Germany, Vol. 1, pp. 253-267.

22. Holland, T.C. (June 1987) Testing high strength concrete, Concrete Construction, Vol. 32, No. 6, pp. 534-536.

13 MIXING AND CASTING HP AND VHP CONCRETE IN THE LABORATORY

G. BERNIER
ENS de Cachan, Civil Engineering Department, France

During a symposium on high performances concrete, two half-days devoted to the making of HP and VHP concrete were organized by the E.N.S. de Cachan. This paper summarizes the objectives and results of these practical experiments.

1 Experiments

Two types of concretes were considered: HP concrete (80 MPa compressive strength at 28 days) with local materials found in the Paris area; and VHP concrete (115 to 120 MPa compressive strength at 28 days and 140 to 150 MPa at 91 days) with materials designed by our Canadian colleagues. The concretes were cured at 20°C and tested at 28 days. The results are shown in Table 1.

2 Formulation and realization

2.1 HP Concrete made with local materials
One of the present preoccupations is to see how it is possible to make an HP concrete with local materials, in order to ensure that the increase of cost of these designed HP concretes stays reasonable, and their use becomes feasible.

2.1.1 Materials
Gravel (5-20mm): Original silico-calcareous rolled aggregate of dry density 2.52 from the Ballois pit. These aggregates are relatively porous and their water absorbtion is 2% by weight after one hour. Their sphericity coefficient is relatively low, only 0.09.

Sand (0-5mm): Coming from the Seine, it has a density of 2.56. It is essentially siliceous. So clean (SE = 95) its fineness modulus is 2.6, the fines content below 160μm is less than from 2 to 3%. Its water absorption is negligible.

High Performance Concrete: From material to structure. Edited by Yves Malier. © 1992 Taylor & Francis.
Published by Taylor & Francis, 2 Park Square, Milton Park, Abingdon, Oxon, OX14 4RN. ISBN 0 419 17600 4.

Cement: The cement was HTS of the Lafarge Company from the Theil plant. This cement was choosen because it has a low C_3A (3%) content, although its performance is more moderate than other cements of type CPA-HP. There is a little formation of primary ettringite so no early stiffening occurs and a retarder is not required.

Silica fume: from Pechiney Company (Laudun) in a densified form. Dry mixing is necessary to reduce agglomerates and split up the "microscopic particles" inside the materials.

Plasticizer: Naphthalene sulfonate plasticizer occurs in two forms: liquid with 40% dry extract, or powder. The commercial brandused, Lomard, is supplied by the Henkel-Nopco Company.

2.1.2 Mix design
The mix proportions are derived from research of the maximum density (Baron and Lesage method) and are based on the local availability of the constituents. For $1m^3$ of concrete (in kg):

Gravel 5-20 mm	1225.2
Sand 0-5 mm (volume ratio G/S = 1.97)	631.5
HTS cement	407.7
Silica fume (10% by weight of cement)	40.7
Lomard D plasticizer (2% by weight of cement)	8.1
Prewetting water (2% of the gravel mass)	24.5
Water	122.3
	_____
	2460 kg

The composition may appear fiendishly exact. In reality, we have considered the real volume mass of fresh concrete to correct the theoretical composition.

2.1.3 Placing and controlling fresh concrete
The procedure is as follows:

- Prewetting of gravel (using 2% of water by mass).
- Dry mixing of granular materials for 1 minute.
- Introduction of water and 1/3 of plasticizer and mixing for 2.5 minutes.
- Introduction of the remaining plasticizer (2/3) in dry extract. Mixing for 2 minutes.
- Measurement of slump.
- Casting test pieces (16x32 cm).
- Measurement of density.

The measurements have given the following results:
Slump ≈ 20cm.
Density 2460kg/m^3.

2.2 Concrete made with good materials

In order to show the possibilities of VHP concrete, the Canadian team Michel Lessard and Pierre-Claude Aitcin from the Faculté des Sciences Appliquées, Université de Sherbrooke, Québec, used a hard river glacial gravel and crushed dolomitic limestone, as well as a different plasticizer. Two mixes have been tested.

2.2.1 Materials

Gravel (2.5-10mm): The gravel used in the first mix was for the most part a granitic gravel with a density (saturated surface dry) of 2.72. It has very good volume coefficient (0.21) and a very rough surface without sharp edges because it does not come from an artificial crushing.

The gravel in the second mix was a hard dolomitic limestone from a crushed massive stone and has a density (saturated surface dry) of 2.80 and a volume coefficient of 0.15.

Sand: Sand of the Seine region of density 2.56.
Cement: Same as HTS.
Silica fume: The material has a density of 2.22 and being from North America is not in a densified form, it is more difficult to transport and has a low apparent density.
Plasticizer: Two types of plasticizer were used. In the first mix, a naphthalene, Disal from the Handy chemical product company was used, (density 1.22). It contains 42% of dry extract. A GT resin from Chryso company was used in the other.

2.2.2 Mix proportions.

The first composition from the Canadian works was as follows:

Granitic gravel	1075.6
Seine Sand	753.9
HTS cement	502.6
Silica fume	50.2
Plasticizer (Disal)	12.1
Water	115.6
	2510 kg

2nd composition:

Dolomitic limestone gravel	1093.6
Seine Sand	772.6
HTS cement	506.6
Silica fume	50.7
Plasticizer	21.1
Water	115.4
	2560 kg

2.2.3 Placing and control of fresh concrete

The procedure is as follows:

- Mixing of the sand for 1 minute to homogenise it.
- Addition of cement and silica fume, 2 minutes.
- Addition of water and plasticizer over a period of approximately 3 minutes.
- Addition of aggregates and mixing for 5 minutes.
- Sampling

The measurements have given the following result:

1st concrete:
   Slump 23cm
   Density 2510kg/m^3

2nd concrete:
   Slump 23cm
   Density 2560kg/m^3

3 Mechanical characteristics

The specimens were removed from their formwork after 24 hours and cured in water (temperature 20°C). The mechanical tests were carried out at 28 days on cylinders Ø 160x320 mm capped with sulfur for HP concrete (Paris area materials) and ground for VHP concrete (Canadian materials).

The results are as in Table 1 (average values for 3 test pieces and standard deviation).

Table 1: Compressive strength and modulus

	HP concrete (Paris) capped cylinders (Tested at ENS Cachan)	VHP concrete (Canada) machined cylinders (Tested at LCPC Paris)	
		1st composition	2nd composition
σ_R [MPa]	79.8 (1.27)	120.7 (1.0)	120.0 (0.8)
E [MPa]	43 620 (320)	49 751 (-)	50 314 (-)

Concurrently, we are conducting mechanical tests to characterize concrete made before at the ENS, LCPC and at the University of

Sherbrooke. Table 2 presents the results of tests on cylindrical test pieces and the average of the results for the two series.

Table 2:

N° of Concrete	Size (cm)	Surfacing	Instrumentation	Test	Result
1) HPC Cachan 7 days	13x32	/	without	brasilian test	$\sigma_R=$ 4.8MPa
2) VHPC LCPC	16x32	/	without	brasilian test	$\sigma_R=$ 5.8MPa
3) VHPC LCPC	16x32	machined	extensometer J2P	5 cycles in compression	E = 47200 MPa
4) HPC Cachan 90 days	16x32	sulfur	without	compression	$\sigma_R=$ 77.4MPa
5) VHPC LCPC	16x32	machined	gauges εl εt	compression	$\sigma_R=$ 120MPa
6) VHPC Sherbrooke	11x22	machined	without	compression	$\sigma_R=$ 145MPa

Finally, in order to illustrate the behaviour, Figure 5 is the recording of test N°5 described in Table 2.

4 Conclusions

The objectives have been reached. The concretes have a very high workability and their strengths are near the predictions. Only the gravels of the Paris area (silico calcareous) have had some problems. We had to prewet these to achieve a good and relatively constant workability.

The strength of the Canadian concretes, measured on cylindrical test pieces (Ø 16x32 cm), which are bigger than those used in Europe (Ø 10x20 cm), conform with predictions and other results. The characterization of these concretes requires press capacities which are uncommon in France (typically 5000kN), so it is for this reason that, without doubt, we will reduce the dimensions of the test pieces in the future.

Last, we can observe that the failure surfaces (type slow) present only trans-granular failure and no detachment for all concrete studied (Figures 1-4).

Figure 1: From left to right: broken samples of HP concrete, VHP concrete (Sherbrooke, granitic gravel), and VHP concrete (Sherbrooke, dolomitic limestone gravel) after crushing.

Figure 2: Enlargement of Paris HP concrete (Seine gravel magnifying 2.5).

Figure 3: Enlargement of VHP Sherbrooke concrete (granitic aggregates magnifying 2.6).

Figure 4: Enlargement of VHP Sherbrooke concrete (dolomitic limestone aggregates magnifying 2.4).

Figure 5: The chemical characteristics of VHPC (see N°5, Table 2)

Instantaneous Young's modulus and Poisson's ratio:

$E_{i\ tangent}$ = 48000MPa
υ_i = 0.23 under 30MPa
υ_i = 0.26 under 60MPa
υ_i = 0.26 under 90MPa

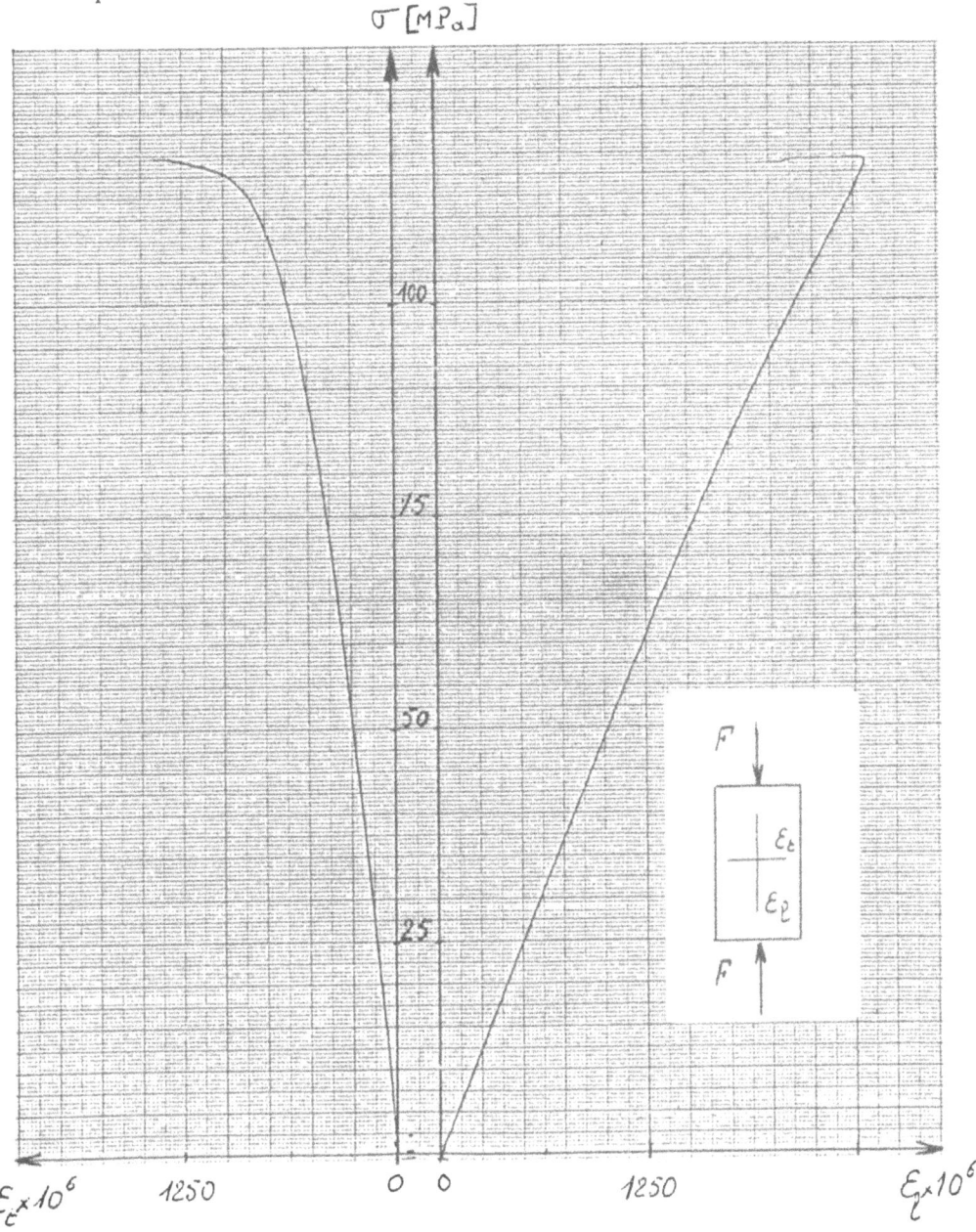

PART TWO
DURABILITY

14 DURABILITY OF HIGH PERFORMANCE CONCRETES: ALKALI - AGGREGATE REACTION AND CARBONATION

M. MORANVILLE-REGOURD
ENS de Cachan, France

1 Alkali - Aggregate reaction

The degradation of concrete due to alkali - aggregate reactions was identified for the first time in the USA in 1940. Then cases were detected in many countries, and in France recently, in dams, bridges, roads, buildings. Disorders occur at various ages of structures, two or ten years or more. They appear as :

- map cracking (Fig. 1). Cracks can develop rapidly i.e. 0.5 mm/year width,
- exudations of calcite and alkaline silicate gel,
- pop-outs due to superficial reactive aggregates,
- movements and displacements of structures like the swelling of a dam crest
- coloration or discoloration mainly along cracks (Fig.2).

1.1 Mechanism of alkali - Aggregate reactions
A humid environment, a high concentration of alkalis in the interstitial liquid phase of concrete, and alkali - reactive particles in the aggregates are the main factors responsible for alkali - aggregate reactions. Reactions are complex and heterogeneous. They occur between a liquid localized in capillary pores and solid particles irregularly distributed in the material. Chemical and physical processes interact in the formation of an alkaline silicate gel able to swell and crack concrete. The alkali - silica reaction is the most frequent case.

1.1.1 The alkali - silica reaction
The mechanism has several contributing factors but the main parts are, aggregate attack and expansion.

High Performance Concrete: From material to structure. Edited by Yves Malier. © 1992 Taylor & Francis.
Published by Taylor & Francis, 2 Park Square, Milton Park, Abingdon, Oxon, OX14 4RN. ISBN 0 419 17600 4.

Figure 1: Cracking due to alkali-silica reaction in a dam.

Figure 2: Concrete discoloration and dark zones along cracks in a bridge pile.

Aggregate attack (1):

- migration of Na$^+$, K$^+$, OH$^-$ ions from the interstitial liquid phase or pore solution towards the reactive silica particle: a physical process,
- reaction with aggregate and formation of a gel by flocculation of a colloidal solution: a chemical process,

Expansion (2):

- gel hydration and local expansion due to an osmotic pressure enhanced by Ca^{2+} ions,
- dissipation of gel in the cement paste as a function of the gel viscosity and the alkali/silica ratio.

1.1.2 Evolution of the interstitial pore solution
The pore solution squeezed from cement pastes, mortars and concretes can be considered as a soda or potash solution with traces of calcium and sulfate. In the presence of reactive aggregates the formation of an alkaline silicate reduces the concentration in Na$^+$, K$^+$, OH$^-$ but the equilibrium of electrical charges is maintained.

1.1.3 Products of the alkali - silica reaction
The alkali - silica reaction produces gels (Fig. 3) and crystals (Fig. 4) found in all deteriorated concrete structures either around aggregates or in cracks and pores of the cement paste, or in veins and cleavage planes within aggregates or at the concrete surface as exudations (3). Secondary products able to amplify the material degradation are ettringite and thaumasite.

1.2 Role of cement
Cements play a role by their effect on the composition of the pore solution and the matrix porosity.

1.2.1 Composition of the pore solution
In the alkali - aggregate reaction cements contribute a large proportion of the alkalis. A Portland cement liberates on average 70% of its alkalis as Na$_2$O eq (4). Blended cements with slag, fly ash, reactive pozzolans can minimise or eliminate the expansion in the presence of reactive aggregates. This is due to the reduction of alkalis in the pore solution and a refinement of the porosity in the cement paste and at the matrix-aggregate interface. If all pozzolans and more particularly fly ashes are not always efficient, silica fume on the contrary greatly reduces the OH$^-$ concentration in the pore solution (5) (Table 1).

Figure 3: Gel of alkali-silica reaction around an aggregate. From an EDAX spectrum, this gel is a potassium and calcium silicate.

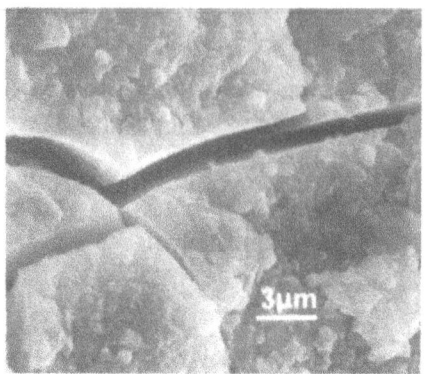 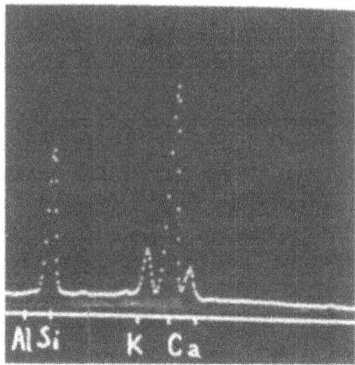

Figure 4: Crystals, products of alkali-silica reaction. From an EDAX spectrum crystals are higher in potassium but lower in calcium than the gel.

 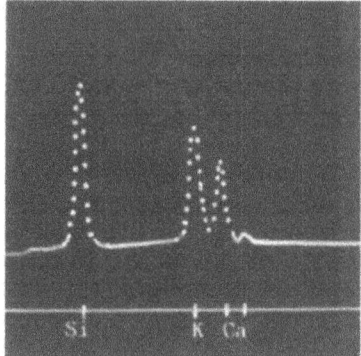

Table 1 : Concentration in the pore solution of cement pastes with increased amount of silica fume (5).

Silica fume % by weight	0	5	10	20	30
OH⁻ ions eq/l	0.50	0.30	0.15	0.10	0.02

1.2.2 Porosity
The ionic diffusion from the pore solution to reactive sites in aggregates (amorphous or cryptocrystalline silica) is a function of the matrix and interfaces cement - aggregates porosity. In a Portland cement paste with w/c = 0.5, C - S - H (37 %) and total porosity (31 %) are the main constituents (6). In the short term cements with silica fume have at the same total porosity more fine pores than Portland cements due to the filler effect and the initiation of the pozzolanic reaction of silica microspheres (7). If capillary pores remain large their lattice is however discontinuous (8) and in concrete the cement paste - aggregate interface is greatly improved i.e. there is no oriented portlandite.

1.3 Behaviour of silica fume concretes
The good behaviour of silica fume concretes has been reported by many authors (9, 10). In Iceland a cement with 7.5 % silica fume reduces the expansion to an acceptable limit (Fig. 5). In the presence of hornfels (Fig. 6) or bronzite or andesite, all reactive aggregates, the expansion has been eliminated by using 15 % silica fume. A pavement in Canada built in 1980 with siliceous limestones and a cement high in alkalis show no signs of deterioration in the concrete containing 20 % silica fume and 2.2 kg $Na_2 O$ eq/m³, 15 % silica fume did not stop the formation of gels but no sign of external deterioration was observed (11).

2 Carbonation

The carbonation of concrete may be either beneficial by increasing compressive strength due to the formation of a dense superficial layer of calcite microcrystals or detrimental by decreasing the pore solution pH and depassivating the reinforcement. The destruction of the passive film on steel occurs at pH between 9 and 10. The rate of reinforcement corrosion depends then on the electrical resistivity of the concrete.

Contradictory results have been published on the resistance of silica fume concretes to carbonation. From these studies it appears however that concrete curing plays an important role in the carbonation depth (Fig. 7) (2, 13). At equal 28 day compressive

Figure 5: Expansion versus time of mortar bars with Portland cement and rhyolite R and silica fume SF. The addition of 7.5 % SF is enough to reduce the expansion down to the limit $\Delta l/l = 0.05$ % at 6 months and 0.1 % at one year (9).

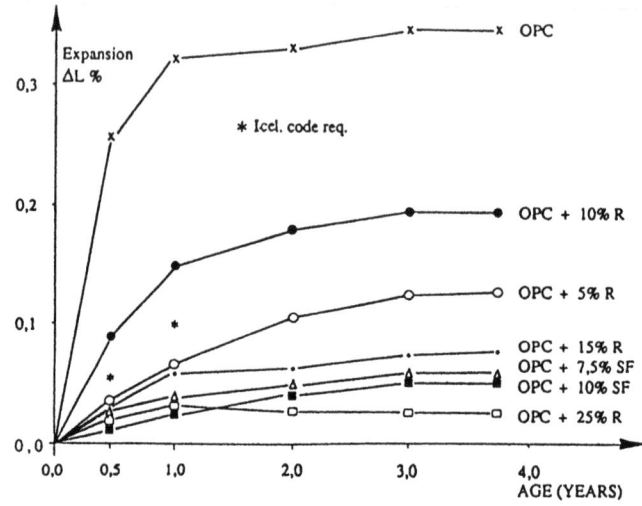

Figure 6: Effect of Portland cement at 0.97 % Na_2O eq replacement by a Portland cement at 0.16 % Na_2O eq or by blast furnace slag or pozzolans on the expansion at 15 months of mortar bars containing reactive hornfels aggregates. Silica fume is one of the most effective addition from the 15 % value (10).

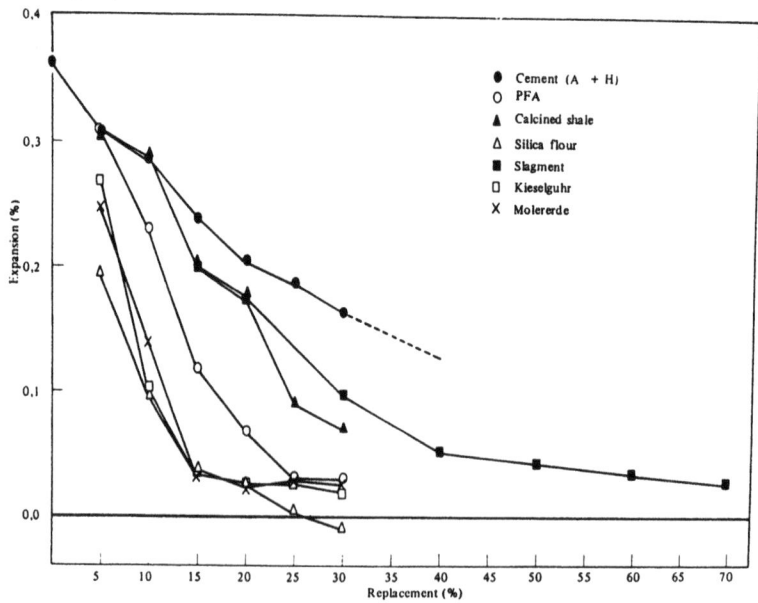

Figure 7: Carbonation depth of concretes after six years at 20°C and 50 % RH (12). Before exposure to carbonation the curing was :
a) one day in the mould,
b) one day in the mould then 26 days in water.

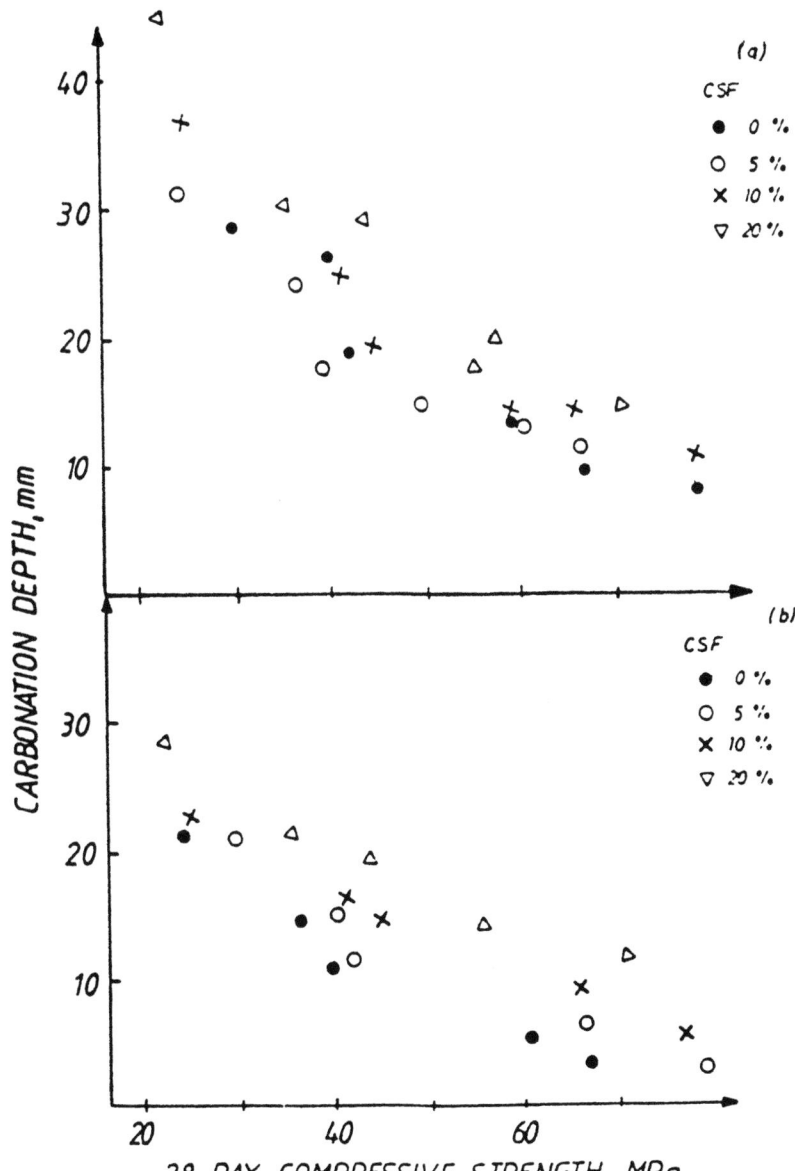

strength the higher the amount of silica fume the higher degree of carbonation of concrete due to a lower pH and amount of Ca(OH)$_2$ (14). For concretes with compressive strengths higher than 70 MPa silica fume does not seem to improve the resistance to carbonation (15).

If silica fume does not reduce significantly the carbonation depth it however increases the electrical resistivity (16) in relation with a discontinuous pore structure and a strong matrix-steel interface (17). The high electrical resistivity limits the galvanic current and thus the steel corrosion.

3 Conclusion

The resistance of high strength concretes to aggressive chemical agents is generally higher than that of normal concretes. This is the case for the alkali-aggregate reaction. This improvement is related to :

- a fine and discontinuous pore structure which reduces ionic diffusions through the matrix and cement paste-aggregate interface,
- a reduction of Ca(OH)$_2$ due to the pozzolanic reactivity of silica fume,
- a reduction of ions OH⁻ concentration in the pore solution.

Carbonation appears more complex. The carbonation depth is equal or superior to that of normal concretes and more particularly when curing time is too short in silica fume concretes. However the electrical resistivity of silica fume concrete is higher than that of normal concrete and prevents reinforcement corrosion even if steel is depassivated.

4 References

1. L.S. Dent-Glasser and N. Kataoka. The chemistry of alkali-aggregate reaction. Paper S 252/23, V[th] Int. Conf. AAR in concrete. Cape Town 1981.
2. S. Diamond, R.S. Barneyback and L. Struble. On the physics and chemistry of alkali silica reactions. V[th] Int. Conf. on AAR in concretes. Cape Town 1981.
3. M. Regourd and H. Hornain. Microstructure of reaction products. V[th] Int. Conf. Concrete Alkali Aggregate Reactions Ottawa 375-380, 1986.
4. S. Diamond ASR. Another look at mechanisms, 8th ICAAR, Kyoto, 83 - 94, 1989.
5. S. Diamond. Alkali reactions in concrete. Pore solution effects. 6th Int. Conf. Alkali in concrete. Copenhagen 155-166, 1983.

6. A.M. Harrisson, H.F.W. Taylor and N.B. Winter. Electron optical analyses of the phases in a portland cement clinker with some observations on the calculation of quantitative phase composition. Cem. Concr. Res. 15, 5, 775 - 780, 1985.

7. H. Uchikawa. Effect of blending components on hydration and structure formation. Principal Report, Theme 3.2, 249 - 280, 8th Int. Congress on the chemistry of cement. Rio de Janeiro, Vol I, 1986.

8. R.F. Feldman. Pore structure, permeability and diffusivity as related to durability. Principal Report. Theme 4.1., 8th Int. Congress on the chemistry of cement. Rio de Janeiro, Vol I, 336 - 358, 1986.

9. H. Olafsson. AAR in Iceland. Present state. 8th ICAAR, Kyoto, 65 - 70, 1989.

10. R.E. Oberholster and W.B. Westra. The effectiveness of mineral admixtures in reducing expansion due to alkali-aggregate reaction due to Malmesbury Group Aggregates. Vth Int. Conf. on AAR in concrete. Cape Town. Paper S 252/31, 1981.

11. P.C. Aitcin and M. Regourd. The use of condensed silica fume to control alkali-silica reaction. A field case study. Cem. Concr. Res. Vol 15, 711 - 719, 1985.

12. E.J. Sellevold. Condensed silica fume in concrete : a world review. Int. Workshop on condensed silica fume in concrete. Montreal, Paper n° 1, 77 pages, 1987.

13. M. Maage. Efficiency factors for condensed silica fume in concrete, 3rd CANMET, ACI Int. Conf. SP 114 - 38, Vol II, 783 - 798, 1989.

14. T. Yamato, M. Soeda and Y. Emoto. Chemical resistance of concrete containing condensed silica fume. 3rd CANMET, ACI Int. Conf. SP 114 - 44, Vol II, 899 - 913, 1989.

15. M. Maage and E.J. Sellevold. Feltmalinger av karbonatisering i betong med og uten silikastøv. Journal of the Nordic concrete Federation, 21 - 22, 1985.

16. N.S. Berke. Resistance of microsilica concrete to steel corrosion, 3rd CANMET, ACI Int. Conf. SP 114 - 42, Vol II, 861 - 886, 1989.

17. G. Grimaldi, J. Carpio and A. Raharinaivo. Effect of silica fume on carbonation and chloride penetration in mortars. 3rd CANMET, ACI Int. Conf. Supplementary. Paper, 320 - 334, 1989.

15 DURABILITY OF HIGH PERFORMANCE CONCRETE IN RELATION TO 'EXTERNAL' CHEMICAL ATTACK

J. GRANDET
INSA, Toulouse, France

In the most general of circumstances, we may define concrete durability as: the material's capacity to maintain its physical characteristics and mechanical performance in conditions of satisfactory safety throughout the service life of the construction (1).

The effect of purely external chemical attack (from acid solutions, water containing Cl⁻ ions, SO₄⁻⁻, sea-water, etc) on concrete is intimately connected with physico-chemical, physical, and even mechanical activity. It is not, in practice, possible to isolate aggressive agents, which act simultaneously or in succession, and to take account of only one of these whilst ignoring the others. Thus, the problem of durability is one of great complexity.

The purpose of this paper is to try to analyse the behaviour of HP. Is it more, or less, susceptible to chemical attack than ordinary concrete (30 MPa) ?

1 Outline of problem

Our knowledge concerning the durability of ordinary concrete has been build up over the years, and there is considerable literature on the subject. Although certain points remain obscure or doubtful, results from laboratory tests have been able to be confirmed by on-site observations of works constructed several decades ago.

HPC containing ultra fine materials and prepared with organic additives are of too recent date to allow reliable conclusions to be drawn from on-site behaviour analysis. The results of corrosion testing carried out in the laboratory can only be generalised with caution, as the test pieces used are small in size, and subjected in all cases to highly concentrated amounts of aggressive agents.

One of the possible ways of investigating HPC behaviour in relation to chemical attack is to extrapolate (with all the risks that may be entailed) from current data available on 30 MPa concrete.

High Performance Concrete: From material to structure. Edited by Yves Malier. © 1992 Taylor & Francis.
Published by Taylor & Francis, 2 Park Square, Milton Park, Abingdon, Oxon, OX14 4RN. ISBN 0 419 17600 4.

2 Chemical attack on concrete

Any chemical attack requires the presence of water, or at least a high level of humidity.There are generally acknowledged to be two main types of attack:

- the dissolution of concrete compounds (cement paste, aggregates) by the action of aggressive water solutions,
- expansion due to the crystallization of new compounds, which causes structural damage.

The essential factor in concrete durability is considered, by the authors of all studies, to be the penetration rate of water and gas, which depends on the structure's porosity and, above all, on the permeability of the cement paste, and perhaps also of the aggregates.

2.1 HPC microstructure

Comparing the microstructures of HPC. and ordinary concrete provides some interesting results (Fig. 1).

Concrete microstructure is modified by ultrafine particles, even though their presence is not absolutely indispensable for reaching 80 MPa.

Figure 1 : HPC microstructure compared with ordinary concrete

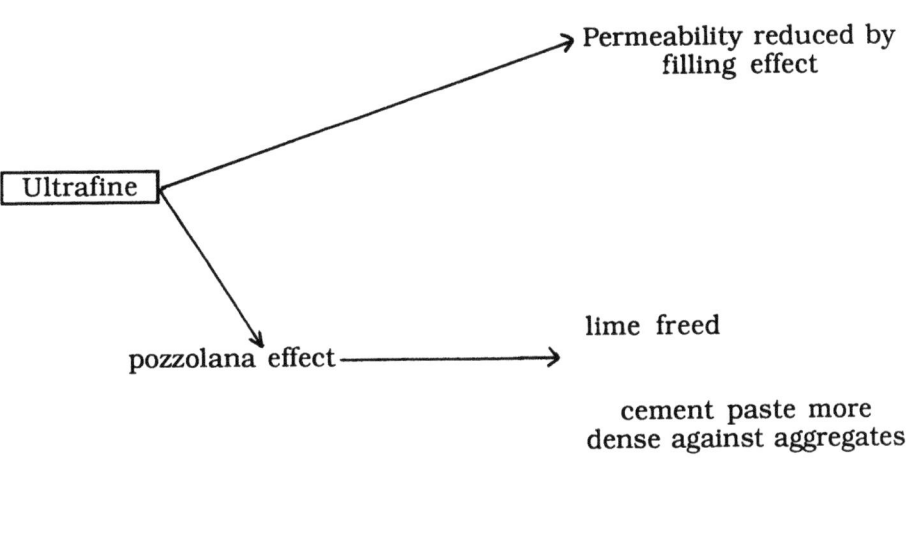

Filling effect

These ultrafine particles are minuscule spheres, usually of condensed silica fume (CSF); size of grain is between 0.02 and 1 µm, the average diameter being approximately 0.4 µm, compared with cement particles whose size varies between 1 and 80 µm, with an average diameter of 15 µm. The initial porosity of cement paste can thus be greatly reduced by incorporating CSF.

Pozzolanic effect

The amorphous silica in CSF combines with lime freed during cement hydration resulting in particularly dense and amorphous CSH which progressively fills up the remaining interstices in the structure. Furthermore, the "auréole de transition" which appears at the paste/aggregate interface, is thin and only slightly porous. This accounts for the often cross-granular appearance of the rupture surface. The absorption by CSF of lime, which is the most soluble hydrated constituent of cement, is certainly a guarantee of high durability.

Addition of water-reducing agent

CSF particles are flocculated by calcium ions freed by the cement. The addition of a wetting agent allows for the total dispersion of silica particles, whilst at the same time reducing the water/cement ratio necessary for the correct application of concrete.

In conclusion, the combined use of ultrafine particles and a water-reducing wetting agent results in concrete with a very low permeability. This is an essential factor governing durability.

2.2 Effect of different aggressive agents on concrete

The action of aggressive agents will, in each case, be investigated at the surface of a concrete structural element, and afterwards in relation to possible effects on the reinforcement.

2.2.1 Action of CO_2

Concrete structures are constantly affected by CO_2 in the atmosphere. In the presence of CO_2 which in aqueous solution is a weak acid, the different hydrates in portland cement paste (portlandite, CSH, hydrated calcium aluminates) become carbonated. They produce calcium carbonates (calcite, aragonite, vaterite) which partially close the concrete pores and so improve the mechanical behaviour of concrete.

On the other hand, this carbonation is accompanied by a reduction in the pH of the binding phase, from 12.8 to 9. At pH 12.8, the protective film covering the reinforcement is dissolved, leaving the steel surface unprotected. When the cement paste in contact with the reinforcement becomes carbonated, corrosion begins. The frontal advance of carbonation in terms of the water/cement mixing ratio is shown schematically in Fig. 2, and is governed by the following relationship :

$$\mathbf{d} = K\sqrt{t}$$

where t is the time expressed in years, and K is a coefficient dependent on CO_2 diffusion conditions in the concrete.

Taking into account their low permeability, curves relating to HPC can be found in the hatched bands (Fig. 2). At a given age, carbonation levels for HPC are roughly twice as low as for W/C 0.55 concrete.

Provided there is no cracking, and the thickness of covering is the same, the reinforcement of HPC elements should be much better protected from the effects of CO_2 than that of normal concrete.

2.2.2 Chloride action
Chloride ions come from the external environment (sea air, factory smoke, fluxing agents) and although their effects on concrete cannot be ignored, these are nothing compared to the devastation produced on steel. The rate of diffusion of chloride ions is greater than that of CO_2. When they reach the reinforcement, they start by corroding the anodic zones and pitting the metal, which can eventually lead to tensile steel failure.

Figure 2 : Carbonation rate - Permeability effect

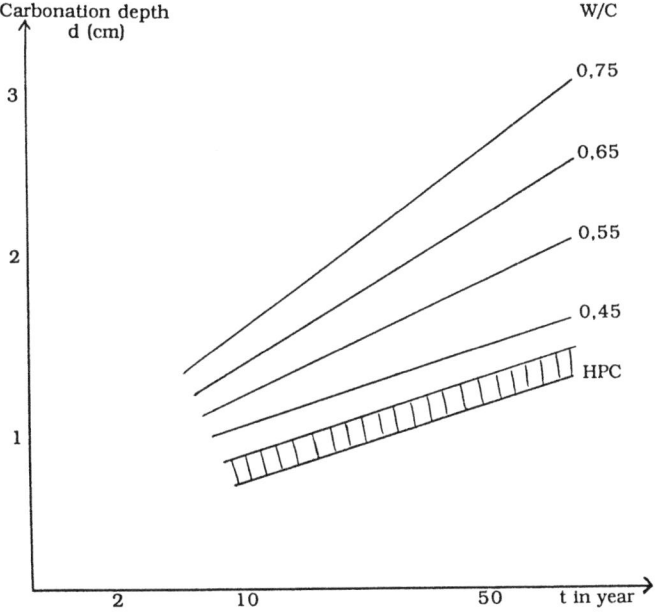

Reinforcement protection depends again, therefore on concrete permeability. Chloride ion penetration has furthermore been shown to be 2-5 times greater in portland cement than in containing additives cements (2). HPC containing condensed silica fume, should therefore be more resistant to chloride ions than ordinary concrete.

2.2.3 Sulphate action

Sulphate ions essentially react with the aluminate phase of cement producing expansive salts of the ettringite type which can lead to surface degradation of the concrete and thus to progressive sulphate attack throughout the construction.

HPC should also be more resistant to chloride ion attack, which is also linked to permeability, than ordinary concrete.

2.2.4 Action of sea-water

Concrete which has been alternately immersed and uncovered as a result of tidal action, is above all subject to physical attack, rather than chemical action alone, which only takes place when immersion is constant. Its overall effect is much less than the corresponding sum of effects of ions in sea-water solution acting individually (Cl^-, SO_4^{--}, Mg^{++}, etc) Some authors have suggested that the formation of calcium or magnesium carbonate deposits can block up surface pores and thus reduce the rate of ion penetration.

It can therefore be concluded that the overall behaviour or uncracked HPC in relation to aggressive ions should be very satisfactory. This is due to its low free lime content and permeability.

Microcracking and cracking, due to shrinkage and distortion in reinforced concrete, are inevitable. The presence of cracks greatly encourages the penetration of corrosive agents down to the reinforcing steel. The durability of HPC in relation to external chemical attack will thus be conditioned by the size of the existing crack network.

References

1. J. Calleja, Rapport principal (VII-2) 7ème Congrès International de la Chimie du Ciment, Paris 1980.
2. P. Schiessl, CEB-RILEM, International Workshop "Durability of Concrete Structures", Copenhagen, May 1983.
3. F. De Larrard, "Formulation et propriétés des bétons à très haute résistance", Thèse Paris, 1987.

16 FROST DURABILITY OF HIGH PERFORMANCE CONCRETES

R. GAGNE and P.C. AITCIN, Université de Sherbrooke
M. PIGEON and R. PLEAU, Université Laval, Québec, Canada

1 Introduction

It is now possible, with the use of superplasticizers, mineral additives and high performance Portland cements, to produce in plant concretes with very low water to binder ratios (<0.30) and compressive strengths of 100 MPa and more (Aïtcin, 1987). The users of these high performance concretes need to know if air-entrainment is necessary to protect them against freezing and thawing cycles and scaling due to deicer salts. The use of air entrainment reduces compressive strength significantly, and this decrease must be compensated by a lower water to binder ratio, which obviously increases production costs. It is presently in high-rise buildings, where durability problems are not very common, that most commercial uses of high performance concrete can be found. Durability problems cannot, however, be neglected in the construction of bridges and similar structures, and for concrete pavements, because in Northern climates these structures are exposed to freezing and thawing cycles and also generally to deicer salts. If high performance concrete is to used in such structures, the durability problems must be investigated.

High performance concrete is a new material, and therefore little information exists concerning its frost resistance in actual field conditions. Laboratory studies have been undertaken recently at Laval University to analyze the influence of various parameters (such as length of curing, air content, type of cement and aggregate characteristics) on the frost resistance of high performance concrete. In this paper, the results of the first part of this research program will be described. Laboratory concretes were submitted to freezing and thawing cycles in water (ASTM Standard C 666, procedure A) as well as to freezing and thawing cycles in the presence of deicer salts (ASTM Standard C 672).

High Performance Concrete: From material to structure. Edited by Yves Malier. © 1992 Taylor & Francis. Published by Taylor & Francis, 2 Park Square, Milton Park, Abingdon, Oxon, OX14 4RN. ISBN 0 419 17600 4.

2 On-going frost durability research at Laval University

The resistance of concrete to freezing and thawing cycles in air and in water and to scaling due to freezing in the presence of deicer salts has been the subject of laboratory investigations at Laval University for many years.With the use of the critical spacing factor technique, the influence of various parameters (water/cement ratio, length of curing, the use of silica fume, freezing in air versus freezing in water) has been determined quite accurately. In this technique, the same concrete mix is fabricated a number of times with different air void systems and specimens of each of these mixes are then submitted to the same freezing and thawing test. At the end of the tests, the value of the spacing factor required to protect the concrete against deterioration due to freezing and thawing cycles (L_{crit}) is determined. The spacing factor of the air voids is defined by ASTM Standard C 457 and represents approximately the average half-distance between two air voids walls. The higher the value of the critical spacing factor, the better the frost resistance, because the concrete then needs less protection against frost attack.

Table 1 shows certain values of the critical spacing factor obtained with various water to binder ratios, with two different cements (one normal Portland cement and one silica fume cement) and in two different freezing conditions (in air and in water). This table shows that there is little difference between the values obtained from freezing and thawing tests in air and in water. It also shows clearly that normal concretes are more resistant to freezing and thawing cycles than the silica fume concretes since the critical spacing factors are significantly higher for the normal concretes.

Table 1: Critical spacing factors obtained for different test conditions

$\dfrac{W}{C + SF}$	with SF		without SF
	cycles in air	cycles in water	cycles in water
0.50	400 µm	250 µm	500 µm
0.30	400 - 500 µm	300 µm	400 µm
0.25	-----	-----	750 µm

The influence of the water to binder ratio on frost durability (Table 1) is not very clear. For the silica fume concretes, tested in air or in water, the critical spacing factor increases slightly with the decrease of the water to binder ratio (from 0.50 to 0.30). For the normal concretes, the critical spacing factor decreases from 500 μm at 0.50 to 400 μm at 0.30. This represents a reduction in durability. It is only if the water to binder ratio is reduced to 0.25 that the critical spacing factor increases, to 750 μm.

Normally, a decrease of the water to binder ration would be expected to increase durability. Various hypothesis, based on the role played by this influence of the water to binder ratio (Pigeon et al., 1987, Foy et al., 1988). The research project on the durability of high performance concrete initiated recently should help to verify these hypothesis and also to better understand the action of frost on these concretes.

Little is known presently about the deicer salt scaling resistance of low water to binder ratio concretes. At Laval University, a first series of tests has shown that the scaling resistance of non-air-entrained concretes with a water to binder ratio of 0.25, normally cured for 14 days, could be as good as that of air-entrained normal concretes. It is probable that this excellent durability is simply due to the very low permeability of the paste (Foy et al., 1988). Further research will be carried out and will help to determine more precisely the influence of such parameters as the water to binder ratio, the spacing factor, the length of curing and the use of silica fume on the salt scaling durability of high performance concrete.

3 High performance concretes

Although the fabrication of high strength concrete is relatively well known nowadays, more research is needed before this new concrete can be used commercially at the lowest possible cost. The technology of high performance concrete is based on two fundamental principles : the cement paste must be as resistant as possible, but it is also extremely important to choose correctly the coarse aggregate that will be used to produce the concrete.

To obtain the most resistant possible cement paste, it is well known that the lowest possible water to binder ratio must be used. If the amount of mixing water is kept to a minimum, the cement particles are close to each other and the paste is dense and has a low porosity. There is, however, a limit to the reduction in the amount of the mixing water because the fluidity of the paste decreases with the amount of mixing water and the concrete becomes so stiff that it cannot be placed properly. With superplasticizers, concretes with water to binder ratios of the order of 0.20 to 0.25 (Aïtcin, 1987) and slumps of 100 mm and more can be produced in plant. The mechanical strength of the paste is also a function of the chemical and the physical characteristics of the cement. Unfortunately, the characteristics of the cement that are required to obtain the highest possible mechanical strengths are not really well known. It seems that fine grained cements with a high C3S content give the best results. The use of mineral additives, such as fly ash and silica fume, is not absolutely necessary, but helps to enhance the mechanical properties as well as the workability of high performance concretes (Aïtcin, 1987).

The choice of the coarse aggregate is particularly important. It must have a high mechanical strength and its modulus of elasticity must be as close as possible to that of the surrounding mortar in order to reduce as much as possible the differential deformations in the concrete. In North America, crushed hard dolomitic limestones with cubic particles of 10 to 12 mm maximum size tend to give the best results (Aïtcin, 1987).

4 Test program

It was decided, for the first part of the research program on the frost durability of high performance concretes, to study the durability of concretes made with three different water to binder ratios : 0.30, 0.26 and 0.23. The same high early strength Portland cement (ASTM type III, CSA type 30) was used for all mixes. This cement can be considered as a high performance cement, since it allows the production of concretes with significantly higher strengths than those made with normal Portland cements (Aïtcin, 1987). In all mixes, in order to obtain compressive strength values close to 100 MPa, 6% of the cement was replaced with silica fume. For each water to binder ratio, various air-entraining admixture dosages were utilized to obtain low air void spacing factor values (100-250µm, with high dosages), moderate values (250-500µm, with low dosages) and high values (500-800µm, no air-entraining agent). Four different curing periods were selected : 1,3,7 and 28 days. To simulate as much as possible actual field conditions, the specimens were cured in the following way : immediately after demolding, they were soaked in water for two to three minutes and then sealed in a number of layers of water-tight plastic film. In all, 27 concretes were submitted to the freezing and thawing tests and to the deicer salt scaling tests (see Table 2).

The freezing and thawing tests were carried out in accordance with ASTM Standard C 666 (procedure A, freezing and thawing in water), and the deicer salt scaling tests in accordance with ASTM Standard C 672. For the salt scaling tests, a 2.5% sodium chloride solution was used. Sodium chloride is now the most commonly used deicer salt in the province of Québec and the 25% concentration is generally considered to be the most damaging (Verbeck and Klieger, 1957). The air void spacing factor of all concretes tested were determined in accordance with the "modified point count" method of ASTM Standard C 457).

Table 2 : Summary of the 27 tests conditions.

$\dfrac{W}{C + SF}$	0.30			0.26			0.23		
$\overline{L}$ (µm)	100 to 250	250 to 500	500 to 800	100 to 250	250 to 500	500 to 800	100 to 250	250 to 500	500 to 800
curing (days)	1,3,7, 28	7,28	1,3,7, 28	1,3,7, 28	3,7, 28	1,3,7, 28	3,7	3,7	3,7

5 Materials, mix characteristics and experimental procedures

As previously mentioned, all mixes were made with the same high early strength Portland cement. The two main characteristics of this cement which allow it to develop very rapidly high mechanical strengths are its high C3S content (63%) and its very high Blaine fineness (5450 cm^2/g). The C2S content is 12%, the C3A content, 8%, and the C4AF content, 5%. The silica fume that was used as partial cement replacement (in the amount of 6% of the dry cement mass) contains more than 95% SiO2 and less than 1% carbon. The coarse aggregate was a very hard crushed dolomitic limestone from the Montréal area and the fine aggregate a natural granitic sand containing mostly particles of quartz and feldspar. The fineness modulus of this sand is 2.1.

Two admixtures were used to prepare the various mixes : a naphthalene based superplasticizer (Disal from Handy Chemicals) and a synthetic detergent based air-entraining agent (Microair from Master Builders).

The main characteristics of the three types of concrete that were made are presented in Table 3. As explained previously, for each water to binder ratio, many mixes were prepared in exactly the same way, except for the dosage of the air-entraining agent which was varied in order to obtain different values of the air void spacing factor. In Table 3, the range of values for the dosage of the air-entraining agent, the slump and the air content is indicated. Due to the low values of the water to binder ratio, the dosage of both admixtures (air-entraining agent and superplasticizer) is very high (many times the normal dosage).

Table 3 : Characteristics of the concrete mixes

$\frac{W}{C + SF}$	Water kg/m^3	cement kg/m^3	SF kg/m^3	fine ag kg/m^3	coar ag kg/m^3	slump mm	air %	SP ml/kg	AEA ml/kg
0.30	144	451	27	791	1041	45-90	2.8-4.9	28	0-2
0.26	148	546	32	703	1040	75-130	2.6-4.9	38	0-2
0.23	158	645	38	591	1025	30-35	2.5-3.9	52	0-3

Note : the admixture's dosages are expressed in ml per kg of cementitious material.

For each of the 27 concretes tested (see Table 2), three 75 x 100 x 400 mm prisms were cast : two for the freezing and thawing tests in water and one for the determination of the air-void characteristics of the hardened concrete. Two 75 x 225 x 300 mm plates were also cast for the deicer salt scaling tests. The surface of these plates was simply leveled-off with a wooden trowel. After curing, a rubber edge was fixed close to the edges of this surface, in order to keep a minimum of 3 mm of the 2.5% sodium chloride solution on the surface to be tested.

The freezing and thawing tests in water were carried out immediately after the end of the curing period, at a freezing rate of approximately 8°C/h (6 cycles per day with a maximum temperature of 4°C and a minimum of -18°C). Length change and ultrasomic pulse velocity measurements were carried out every 50 cycles to determine the degree of deterioration (of internal microcracking). The 75 x 225 x 300 mm plates were dried for 28 days in the laboratory before being submitted to the daily cycles of freezing and thawing (from +20°C to -20°C) in the presence of deicer salts. The scaled-off particles were collected and weighed every 5 cycles.

The compressive strength after 28 days of curing was measured for each mix on six 100 x 200 mm cylinders (Table 4). For each water to binder ratio, as expected, the strength was found to vary somewhat with the air content.

Table 4 : Compressive strength of the concrete mixes (at 28 days)

$\frac{W}{C + SF}$	compressive strength MPa	air content %
0.30	78 to 87	2.5 to 4.4
0.26	90 to 93	2.6 to 3.6
0.23	91 to 97	2.5 to 2.8

6 Freezing and thawing tests results

The results of the freezing and thawing tests in water for each of the 27 concretes used are summarized in Table 5, where the length change and the value of the ultrasonic pulse velocity at the end of the freezing and thawing cycles are shown.

Length change is usually considered to be the best indicator of the internal microcracking that can be due to freezing and thawing cycles (Pigeon et al., 1985). Normally, if, after 300 cycles, the length change is higher than approximately 200×10^{-6}, this indicates that a certain amount of microcracking has occured and that complete disruption will eventually happen if the number of cycles is sufficient (Pigeon et al., 1986). A value of the length change lower than 200×10^{-6} does not generally indicate the onset of microcracking because of the normal swelling of concrete in water (that can occur when the temperature is higher than 0°C) and because the difference between the values of the coefficient of thermal expansion for cement paste and aggregate can also cause a small innocuous length change. Of course, the value of 200×10^{-6} is a function of the mix characteristics and the materials used. In this series of tests on high performance concrete, because of the high cement content of the mixes, it can be considered that no micracking has occured if the length change is lower than approximately 500×10^{-6}.

Microcracking reduces the measured value of the ultrasonic pulse velocity, which is therefore also a good indicator of the internal deterioration that can be due to freezing and thawing cycles. The pulse velocity, expressed as a percentage of the value measured at 0 cycle, can thus help to confirm the results obtained with the length change measurements. Usually, if the velocity after 300 cycles is less than 90% of the initial value, it can be considered that a certain amount of microcracking has occured (Pigeon et al., 1987).

As can be seen in Table 5, the highest length change value is 542×10^{-6}, but, in most cases, the specimens were submitted to much more than 300 cycles, and, in certain cases, to more than 1000 cycles. It is thus clear that none of the specimens tested was affected by the freezing and thawing cycles. This conclusion is confirmed by the results of the pulse velocity measurements : all values at the end of the tests are higher than 100% of the initial value. This increase means that, not only has no microcracking occured, but that hydration has slowly continued during the cycles.

Table 5 : Freezing and thawing tests results

$\dfrac{W}{C+SF}$	curing period (d)	spacing factor (μm)	number of cycles	length change ($\Delta L/L \times 10^{-6}$)	ultrasonic pulse velocity ($\% V_o$)
0.30	1	255	760	275	103
	1	802	821	425	106
	3	255	750	259	104
	3	802	809	542	103
	7	216	608	226	102
	7	318	486	201	101
	7	817	1013	184	104
	28	216	300	319	100
	28	318	300	293	101
	28	817	304	378	102
0.26	1	182	760	259	103
	1	730	719	227	104
	3	285	679	228	101
	3	384	1151	258	105
	3	744	966	217	103
	7	285	657	283	103
	7	384	934	302	104
	7	744	944	265	103
	28	285	895	390	102
	28	384	1100	398	103
	28	744	295	326	100
0.23	3	170	1162	317	105
	3	457	898	338	103
	3	778	906	396	104
	7	170	1137	403	102
	7	457	872	347	103
	7	778	1171	413	103

7 Deicer salt scaling tests results

The results of the deicer salt scaling tests are summarised in Table 6, where the mass of scaled-off particles (in kg/m^2) after 50,100 and 150 cycles respectively is shown. These results clearly indicate that the deicer salt scaling resistance of all concretes tested is very good. In all cases, the mass of scaled-off particles is extremely low, since the highest value is 0.49 kg/m^2 after 150 cycles, i.e. significantly lower than the Norwegian standard limit of 1 kg/m^2 after 56 cycles (Petersson, 1986).

After 50 cycles, the surface of all specimens tested was almost completely intact. Very often, the small amount of deterioration that was observed was concentrated over a coarse aggregate particle situated just below the surface.

Table 6 : Scaling tests results

W / (C + SF)	curing period (d)	spacing factor (μm)	scaled-off	particles	(kg/m2)
			50 cylces	100 cycles	150 cycles
	1	255	0.08	0.13	0.19
	1	802	0.04	0.06	0.09
	3	255	0.04	0.05	0.10
	3	802	0.02	0.03	0.05
0.30	7	216	0.15	0.30	0.43
	7	318	0.15	0.18	0.25
	7	817	0.06	0.11	0.17
	28	216	0.12	0.32	0.49
	28	318	0.05	0.13	0.18
	28	817	0.09	0.20	0.32
	1	182	0.05	0.09	0.14
	1	730	0.02	0.06	0.08
	3	285	0.02	0.03	0.06
	3	384	0.05	0.09	0.16
	3	744	0.05	0.10	0.12
0.26	7	285	0.03	0.04	0.05
	7	384	0.04	0.08	0.15
	7	744	0.05	0.09	0.15
	28	285	0.04	0.08	0.17
	28	384	0.04	0.06	0.09
	28	744	0.03	0.05	0.07
	3	170	0.07	0.12	0.23
	3	457	0.01	0.03	0.04
0.23	3	778	0.05	0.15	0.25
	7	170	0.03	0.09	0.14
	7	457	0.02	0.04	0.05
	7	778	0.02	0.05	0.08

Figure 1 shows, for each water to binder ratio, the mass of scaled-off particles as a function of the number of cycles for the least and the most deteriorated concrete. For all mixes represented on this graph, regardless of the degree of deterioration, the damage due to scaling increases almost linearly with the number of cycles. This is typical of deicer salt scaling test results (Pigeon, 1989).

Figure 1: Scaled-off particles versus number of freeze-thaw cycles.

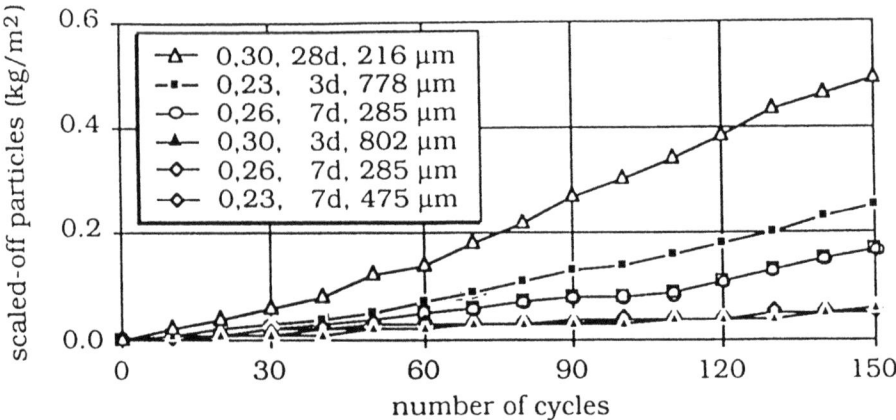

Legend:
- ─△─ 0,30, 28d, 216 µm
- ─■─ 0,23, 3d, 778 µm
- ─○─ 0,26, 7d, 285 µm
- ─▲─ 0,30, 3d, 802 µm
- ─◇─ 0,26, 7d, 285 µm
- ─◇─ 0,23, 7d, 475 µm

y-axis: scaled-off particles (kg/m²), scale 0.0 to 0.6
x-axis: number of cycles, 0 to 150

8 Discussion

None of the specimens that were fabricated for this first part of the research program on the frost durability of high performance concrete was damaged during the freezing and thawing tests in water, regardless of the water to binder ratio, the air-void spacing factor or the length of curing. Even the non-air-entrained concrete with a water to binder ratio of 0.30 was able to withstand without any significant deterioration 750 cycles of freezing and thawing in water after only one day of curing. It is thus clear that it is possible to fabricate non-air-entrained high performance concretes that are extremely resistant to frost.

All concretes with low water to binder ratios do not necessarily have such a good frost resistance. In a recent publication (Foy et al., 1988), it was shown that a reduction of the water-cement ratio from 0.30 to 0.25 could increase significantly the resistance to freezing and thawing cycles of ordinary Portland cement concretes cured for 14 days, but the results of this study also indicated that some concretes had been significantly damaged by the freezing and thawing cycles. It was therefore suggested by the authors that air-entrainment was still needed for these concretes to be frost resistant.

It is clear from what has just been said that the frost resistance of concrete is not only related to the water to binder ratio, but also to many other parameters (such as mix composition, curing, material characteristics, etc) and to the overall quality of the paste. The most important factor for frost durability is not the total porosity of the concrete or of the paste, but the size distribution of the capillary pores, since it is the size distribution that determines the amount of freezable water and the paste permeability (which are the two most important parameters with respect to frost resistance). At similar values of the water to binder ratio, the concretes fabricated by Foy et al., (1988) had a relatively lower compressive strength (60 to 70 MPa) than that of the concretes prepared for this study (80 to 90 MPa). This fact could explain their lower frost resistance. Even if, theoretically, compressive strength is not a very important factor for frost resistance, it is still a good indication of the overall quality of the paste. For the concretes analyzed in this study, it seems clear that the use of a high early strength Portland cement (ASTM type III) was much better than the use of a normal Portland cement (ASTM type I).

Natural exposure conditions are usually less severe than those used for laboratory tests, and some concretes that are not durable in the laboratory can be durable under field exposure conditions. Unfortunately, there is very little data available on the field performance of high strength concrete, and laboratory tests are still the only way to assess their durability and are the only basis on which guidelines for their use can be prepared. In the present state of knowledge, these guidelines can be summarized in the following way : concretes with water to binder ratios higher than 0.30 must be air-entrained and those with ratios less than 0.25 do not require air-entrainment. For intermediate values (0.25 to 0.30), no rules can be stated and laboratory tests should be carried out. These rules, of course, are only valid for concretes made with Portland cements (with or without silica fume) and good aggregates. Research is needed to determine the influence of mineral additives such as fly ash and granulated blast-furnace slag which are commonly used in France. The durability of the high performance concretes made with these additives cannot be predicted before this.

For all the specimens tested,the loss of mass due to deicer salt scaling was much lower than the limit of 1 kg/m^2 after 56 cycles recommended in the Norwegian standard. This excellent scaling resistance is related to the very low permeability of the paste which reduces the degree of saturation of the surface layer as well as the penetration of chloride ions. As in the freezing and thawing cycle tests, none of the variables examined (water to binder ratio, air-void spacing factor and curing) had any very significant influence on the test results. It is thus possible to fabricate high performance concretes that are frost resistant as well as deicer salt scaling resistant without using air-entrainment. Even the 0.30 water to binder ratio concretes cured for 24 hours only had an excellent performance. It even seems that, when air-entrainment is not required, it should not be used, because, although all values of the loss of mass are low (Table 6), the loss of mass is generally higher for the air-entrained concretes.

To fabricate a deicer-salt scaling-resistant high performance concrete, the 0.25 maximum water to binder ratio rule seems to be applicable, although the resistance to scaling appears to be less critical than the resistance to internal cracking. Recent results have indicated that non-air-entrained silica fume concretes with a water to binder ratio significantly higher than 0.30 could have an excellent deicer salt scaling resistance (Hammer and Sellevold, 1990). The tests by Foy et al. (1988) have shown that non-air-entrained 0.25 water-cement ratio concretes made with ordinary Portland cement and cured correctly can have an excellent deicer salt scaling resistance. Of course, since field exposure conditions can be quite different from laboratory testing conditions, this rule has to used with caution. In the laboratory, the minimum temperature of the cycle is generally very low (around -20°C) and the specimens are constantly covered with salt water, which is not the case in the field. Moreover, the temperature of the initial drying period before the tests is approximately 20°C in the laboratory, but can be much higher in the field. This, according to certain test results (Sorensen, 1983), can reduce significantly the scaling resistance of certain types of concrete. And, again, not much is known of the influence of the use of fly ash or slag and this remains to be investigated.

9 Conclusion

The main conclusion that can be drawn from the test results described in this paper is that it is possible to produce non-air-entrained high performance concretes that are frost resistant as well as deicer-salt scaling-resistant. It seems that non-air-entrained high performance concretes can be durable if they are made with Portland cements (with or without silica fume as partial cement replacement) and have a water to binder ratio of less than 0.25. The use of air entrainment, however, appears to be necessary if the water to binder ratio is higher than 0.30. For intermediate values (between 0.30 and 0.25), it is not possible to define a precise rule and laboratory tests are required in each individual case.

The limiting value of 0.25 for the water to binder ratio must be used with caution, since field exposure conditions are quite different from laboratory exposure conditions, and because many other factors, such as the minimum and maximum temperature of the cycles,the length of the freezing period, the temperature during the drying period after curing, the degree of saturation of the aggregates at the time of mixing, can have a significant influence on the frost and salt scaling resistance of high performance concrete. Very little is known presently of the effect of the use of mineral additives, such as granulated blast-furnace slag and fly ash, on the frost and salt scaling durability of high performance concrete. It is thus very difficult to evaluate the durability of these concretes, and more research in this area is clearly necessary.

References

1. Aïtcin, P.C., 1987, La technologie des bétons à très haute performance en Amérique du Nord, Materials and Stuctures/Matériaux et Constructions, vol. 20, pp. 180-189.
2. ASTM, 1988, Annual Book of ASTM Standards (1988) vol 04.02, section 4 : Concrete and Aggregates, American Society for Testing and Materials, Philadelphia.
3. Foy, C., Pigeon M and Banthia, N., 1988, Freeze-thaw Durability and Deicer Salt Scaling Resistance of a 0,25 Water-Cement Ratio Concrete, Cement and Concrete Research, vol. 18, no. 4, pp. 604-614.
4. Hammer, T.A. and Sellevold, E.J., 1990, Frost Resistance of High Strength Concrete, Second International Symposium on Utilization of High Strength Concrete, Berkekey, California, 23 p.
5. Petersson, P.E., 1986, the Influence of Silica Fume on the Salt Frost Resistance of Concrete, International Seminar on Some Aspects of Admixture and Industrial By-Products on the Durability of Concrete, Goteburg, Norway, 10 p.
6. Pigeon, M., 1989, la durabilité au gel du béton, Materials and Structures/Matériaux et Constructions, vol. 22, pp. 3-14.
7. Pigeon, M., Gagné, R. and Foy, C., 1987, Critical Air Void Spacing Factor for Low Water-Cement Ratio Concretes With and Without Silica Fume, Cement and Concrete Research, vol. 17, no. 6, pp. 896-906.
8. Pigeon, M., Pleau, R. and Aïtcin, P.C., 1986, Freeze-Thaw Durability of Concrete With and Without Silica Fume in the ASTM C 666 Procedure A Test Method : Internal Cracking versus Scaling, Cement, Concrete, and Aggregates, vol. 8, no.2, pp. 76-85.
9. Pigeon, M., Prévost, J. and Simard, J.M., 1985, Freeze-Thaw Durability Versus Freezing Rate, ACI Journal Proceedings, vol. 82 no. 5, pp. 684-692.
10. Sorensen, E.V., 1983, Freezing and Thawing Resistance of Condensed Silica Fume (microsilica) Concrete Exposed to Deicing Chemicals. Proceedings of the CANMET/ACI First International Conference on the Use of Fly-Ash, Silica Fume, Slag and Other Mineral By-Products in Concrete, ACI Special Publication SP-79, pp. 71-86.
11. Verbeck, G.J. and Klieger, P., 1957, Studies of "Salt" Scaling of Concrete, Highway Research Board Bulletin no. 150, pp. 1-13.

17 PERMEABILITY, AS SEEN BY THE RESEARCHER

D. PERRATON, P.C AITCIN
Université de Sherbrooke, Québec, Canada
A. CARLES-GIBERGUES, INSA, Toulouse, France

1 Introduction

It is now well known that the durability of concrete is closely related to its permeability. Why is it then, as stated by many researchers (recently, Hooton [1]), that we have not received any great changes to our knowledge of this subject since 1930? There are several causes. The following are some of them:

- Measurements of permeability generally have a high degree of uncertainty;
- There are, for a given concrete, as many permeability coefficients as there are for the fluids concerned (i.e. air, oxygen, water and saline solutions);
- For a given fluid, permeability of concrete is closely dependant on the hygrometric state of the material, itself difficult to quantify and control;
- The test can modify the structure of the concrete, and its permeability, thus running the risk of not accounting for the real properties of an actual concrete to be used on a construction;
- Finally, the permeability of some concrete, like high performance concrete, can not be measured by the classic water permeability measurement.

In this paper, we will show the diversity of results obtained on concrete according to the technique chosen (we have limited ourselves to three processes); determine if these techniques are applicable to high performance concretes, and, in the affirmative, expose the general tendency to noticeable variations for normal and high performance concrete.

High Performance Concrete: From material to structure. Edited by Yves Malier. © 1992 Taylor & Francis.
Published by Taylor & Francis, 2 Park Square, Milton Park, Abingdon, Oxon, OX14 4RN. ISBN 0 419 17600 4.

2 Permeability-basic equation

2.1 Definition

Permeability is the intrinsic property of a body which allows a fluid to pass through it. Classically, the coefficient of permeability is defined from the basic equation developed by Darcy, and enunciated in 1856 [2]. The permeability coefficient k, defines a permeable material of section A, which allows through it a flow Q of a fluid of viscosity η, and this under the effect of a pressure gradient dP/dZ. Darcy's equation is:

$$Q = -k \cdot \frac{A}{\eta} \cdot \frac{dP}{dZ} \qquad (1)$$

The coefficient of permeability was initially a term dependant upon the material and the fluid involved. Actual usage tends to take into account a coefficient, characteristic of the material only.

The complexity of a fluid transfer in a porous material can not be completely defined by an over simplification of the Darcy's law. One might especially look at the phenomenon of diffusion (i.e. for vapor) defined by Fick's law:

$$j = -D \cdot \frac{dC}{dL} \qquad (2)$$

where j = flow

$\frac{dC}{dL}$ = gradient of concentration

D = coefficient of diffusion

The transfer of a liquid at a velocity v in a unsaturated capillary network is given by Washburn's law:

$$v = \frac{r \cdot \gamma}{4 \cdot d \cdot \eta} \cdot \cos \theta \qquad (3)$$

where r = radius of the capillaries
γ = superficial tension
θ = contact angle
d = depth of penetration
η = viscosity of the fluid

Further on, we shall see how these different phenomena can occur in the permeability of concrete.

2.2 Units of measurement

Strictly, the use of the basic equation (1) leads us to express the permeability coefficient k in surface terms: m^2 for the S.I. system. The practice of using a sub-multiple has been established, which is the Darcy, valued at 10^{-12} m^2 or the Millidarcy which is valued at 10^{-15} m^2.

Confusion could arise by the use of Darcy's law in its initial form: $v = K. i$ (with v apparent velocity, under a hydraulic gradient $i = dH/dL$). In this case, the coefficient K, is homogeneous to a velocity and is expressed in m/s (or in cm/s, or even in ft/day!).

3 Relationships between the structure of two-phase concrete and its permeability

The permeability of a concrete structure evidently depends upon:

- the intrinsic permeability of its constituents,
- their geometric arrangement

The considerations of overall permeability of a construction are treated in another work by G. Ithurralde [3].

Concrete has for a long time been seen as a purely two-phase material: *paste and aggregate.* However, a more realistic approach now takes into account the existence of a particular zone of hydrated paste in contact with the aggregate called the "transition zone".

3.1 Hydrated cement paste

The permeability of a paste is a function of its intrinsic porosity. In an hydrated paste, we can distinguish gel pores, and capillary pores.

3.1.1 Gel pores

The C-S-H's major constituents of hydrated cement paste are microporous. The gel pores are very small, between 15 and 20 Å in diameter. Although they constitute a network of communicating pores, it seems by the calculation of Power's law that the permeability of this network is very low: $7 . 10^{-16}$ m/s.

3.1.2 Capillary pores

Capillary pores are the spaces, initially existing between the cement grains , and which have been incompletely filled by the hydrates. They can be partially or entirely filled with water, according to the state of saturation of the paste.

Hydrated cement paste being an "aging" material, it can be conceived that there would be, from an initial stage (itself strongly dependent upon the initial w/c ratio, or from the concentration of the aggregate), a modification of the pore network with time. The average volume and size of the pores decreases, and the network progressively breaks down (segmentation of the capillaries). This latter fact has been demonstrated by numerous studies, Powers being a pioneer in the field [5, 6]. This is to be explained by:

a) permeability at a given age, increases with the w/c ratio (Fig. 1).
b) permeability to a given w/c ratio decreases with time of curing (Table I).
c) the time of curing required in capillary segmentation is dependent upon the initial w/c ratio (Table 2).

Figure 1: Relation between permeability and water/cement ratio for mature cement paste (ref. [6]).

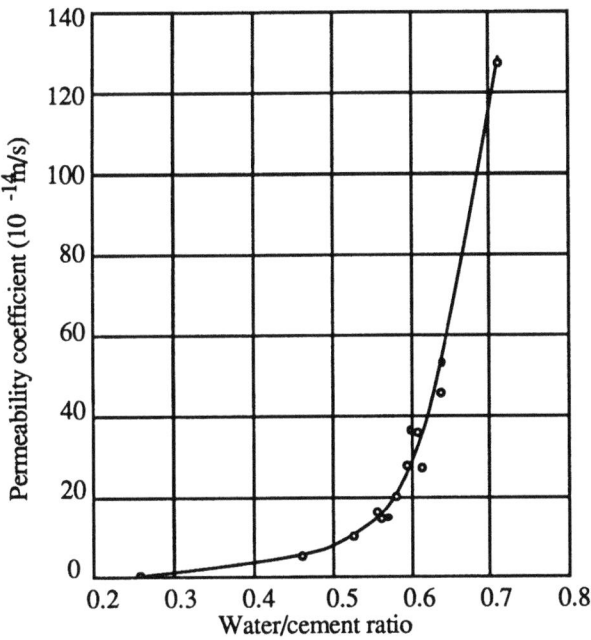

Table 1: Curing effect on the coefficient (cement paste ref. [5])			**Table 2:** Minimum curing time, before network of capillary pores breaks down (cement paste: ref. [5])	
Days	K(m/s)		w/c	Days
Fresh	2.10^{-6}		0.40	3
5	4.10^{-10}		0.45	7
8	4.10^{-10}		0.50	28
13	5.10^{-12}		0.60	180
24	1.10^{-12}		0.70	365

To sum-up, it may be said that the permeability of hydrated cement paste is controlled by capillary permeability.

3.2 Aggregate
Perhaps it would be interesting to remind ourselves of the now quite dated results presented by Powers [5].

Table 3 : Compared permeability of different rocks and cement pastes. Reading this table we can conclude that contrary to certain opinions, the aggregate cannot be taken to be impermeable compared to the binding matrix (hydrated cement paste) as soon as curing is in an advanced state.

Type of rock	Permeability coefficient (m/s) of same permeability	w/c ratio of cement paste entirely hydrated
Trap rock	$2.47 . 10^{-14}$	0.38
Quartz (diorite)	$8.24 . 10^{-14}$	0.42
Marble	$2.39 . 10^{-10}$	0.48
Marble	$5.77 . 10^{-10}$	0.66
Granite	$5.33 . 10^{-11}$	0.70
Gray	$1.23 . 10^{-10}$	0.71
Granite	$1.56 . 10^{-10}$	0.71

Moreover, it may be pertinent to complete the results in Table 3, by some results imputed to the work of Thenoz [7]. They bring to light the considerable influence of granite aggregate on the range of it's permeability coefficient:

- air permeability coefficient (8 samples), using sound granites, is between 2.10^{-3} to 8.10^{-2} 'Millidarcy'
- air permeability coefficient (7 samples), using deteriorative granites, is between 0.1 to 2.0 'Millidarcy'.

3.3 Transition Zone : permeability

In the literature, we notice that the permeability coefficient of a given concrete, is always largely superior to that of a hydrated cement paste having the same characteristics (w/c ratio, degree of hydration, etc.). For instance, a hydrated cement paste (w/c = 0.60) of permeability 40, corresponds to mortar of 'permeability 80' and concrete of 'permeability 600 to 1800' (for aggregates of 38 to 113 mm respectively). In a purely two-phase model, one might expect a reverse classification, because the addition of the aggregate increases the tortuosity of the fluid's path in the porous material and reduces the amount of paste, which ought to lower permeability. But, in fact, we well know now, that in a fresh concrete the structure of the paste is modified at the proximity of the aggregate, brought on by an effect of partitioning, which happens over a few hundredths of a millimeter. We observe at this interface, a zone where the porosity of the paste is increased. Consequently in this space, the formation of hydrates is very particular (thus, the portlandite appears in large crystals, preferentially oriented). These conclusions, based upon the study of simplified concrete tests [4, 8 and 9], have been confirmed from later studies on ordinary [10], and high performance concrete [11].

Figure 2: Variation of the porosity in the transitional interface (ref. [10])

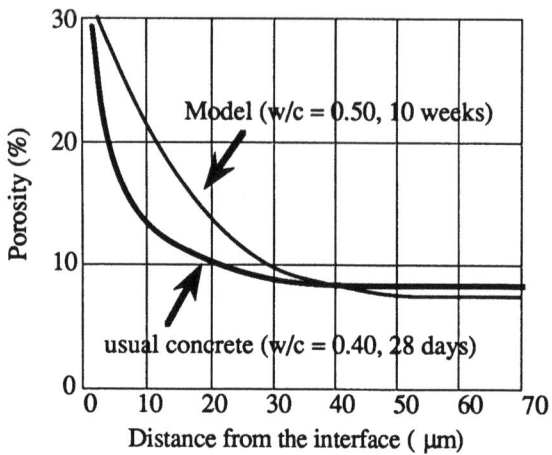

Model (w/c = 0.50, 10 weeks)

usual concrete (w/c = 0.40, 28 days)

Figure 3: Relation between permeability and capillary porosity of cement paste (ref. [5])

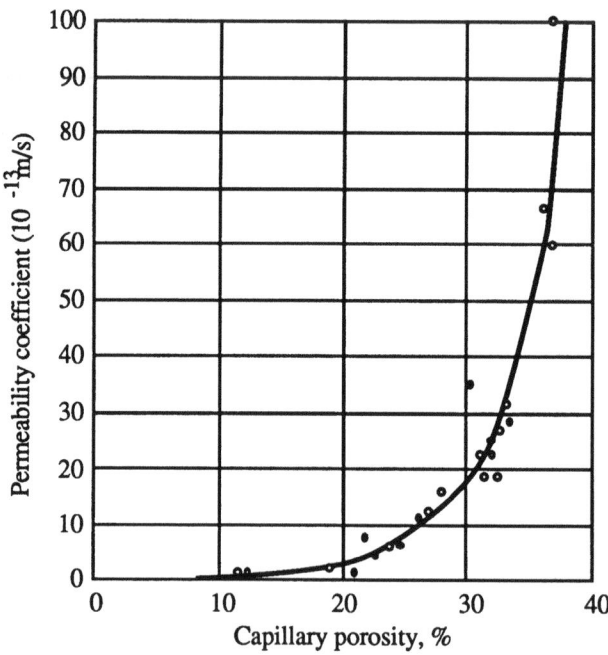

Figure 2, shows well that the porosity at the interface is significantly higher than that in the rest of the paste. In accordance to the relationship presented by Powers [5] between porosity and permeability (Fig. 3), we estimate that the permeability coefficient is increased at the transition zone by a factor 10 caused by the change in the porosity from 10 to 30%.

Accordingly, one may conclude that the interface surrounding the aggregate constitutes a preferential zone of drainage.

4 What kind of test is suitable to evaluate the concrete permeability?
This is a fundamental question which needs a satisfactory solution, if we want a measurement taken in the laboratory to take effective account of the real properties of concrete used for a construction. The following scheme (ref. [12]) illustrates well the multiplicity of the transfer processes of fluids in a real construction.

Table 4 : Transportation mechanisms

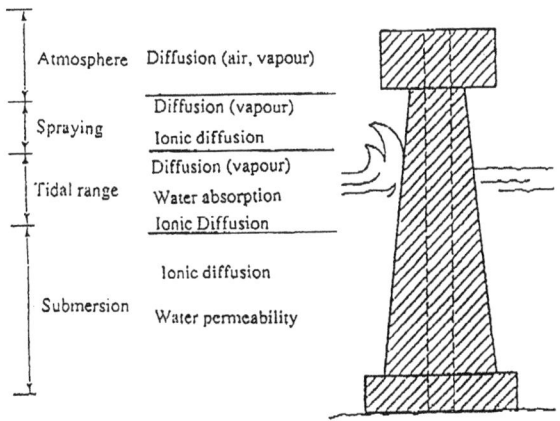

Primary transportation mechanisms for different exposure zones of an offshore concrete structure (ref. [12])

Experience has shown that results of tests on a given concrete with differing techniques will give as many permeability coefficients as mechanisms analyzed. In fact, it is important to evaluate the permeability of a given concrete according to the real site condition and, moreover, subject to the same type of fluid. Thus, in vain, one would try to define a concrete by an intrinsic coefficient of permeability.

We shall instead list our own test results, on chloride, water, and air permeability.

5 Chloride permeability test

5.1 Principle of the test method

The object of the AASHTO specification T277-831,'Rapid Determination of the Chloride Permeability of Concrete'. describes the procedure of the chloride permeability test. The test consists of determining the amount of electric current passed through a concrete sample, 95 mm dia. and 50 mm thick, of which one end is immersed in a solution of sodium chloride, the other in a sodium solution. A potential differential of 60 V is maintained for 6 hours. The total charge passed through the sample is expressed in Coulombs, and represents the chloride permeability.

5.2 Apparatus

The apparatus used is presented in Fig. 4. The concrete test sample undergoes the following preparations, before testing:

1 Held in a vacuum of at least 1 torr for 3 hours
2 Saturation by distilled water (that has been de-aired) in applied vacuum for 1 hour.
3 Held in water, at normal pressure, for18 hours.

The cylindrical surface of the sample is made water-proof by a coating of epoxy resin. The electrodes immersed in each cell are made up of wire-mesh.

Figure 4: Schematic of the ion-chloride permeability cell

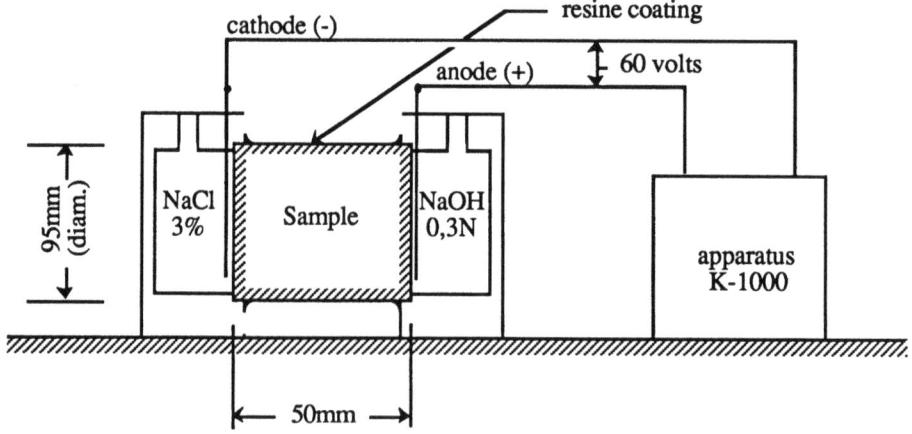

5.3 Results

For practicality, one follows the curve variations of the current intensity I in function of time. One can then calculate the total charge passed through the sample during the test (Coulombs). Figure 5 shows the diversity of the behavior for different concretes. Moreover, it is seen that the samples that are very permeable, undergoing an intensively high current, heat up a lot (their temperature may reach 75°C).

The tests undertaken at the Université de Sherbrooke have shown us the influence of the concrete's composition on chloride permeability.

5.3.1 Effects of the w/c ratio

Figure 5 shows that going from normal concrete (w/c 0.40 to 0.50) to high performance concrete (w/c = 0.24) brings about an important modification of the time/current diagram. For high performance concrete, the current remains constant throughout the test.

Figure 5: Typical curent-times relation for three different water/ cement ratios mesured on ion-chloride permeability test

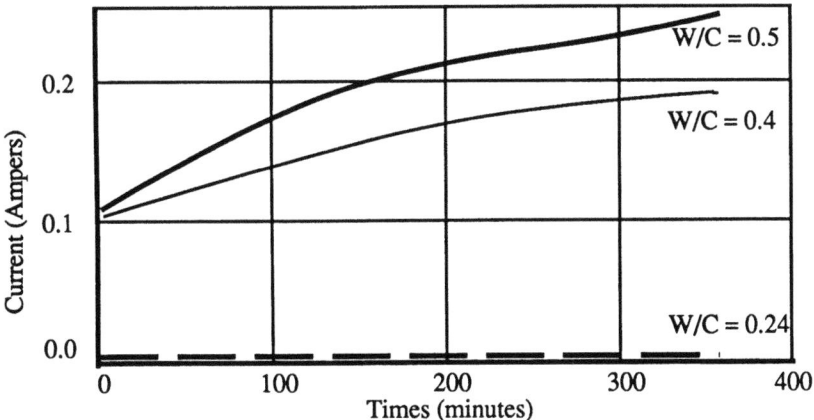

Another series of tests give us the results shown in Table 5. It appears that the reduction of the w/c ratio brings about an increasing impermeability in ordinary and high performance concrete. It is to be noted that this statement applies to cored concrete, checked after two years of aging on a construction (Lavalin building).

Upon analysis of Table 5 we can find out that the addition of silica fume has the same effect as the reduction of the w/c ratio.

Table 5 : The effect of the water/cement ratio on the chloride permeability test

	w/c	CURRENT (mA) INITIAL	FINAL	TOTAL ELECTRIC CHARGE (Coulombs)
	0.50	113	281	4 500
	0.40	101	197	3 500
	0.28	-----	-----	900
LABORA-	0.27*	6	7	125
TORY	0.16*	-----	-----	65
	0.24*	9	8	150
	LATEX	-----	-----	400/1 000
IN SITU	0.25	-----	-----	300

* Concretes containing silica fume

5.3.2 Effect of silica fume
To isolate the effect of the silica fume on permeability, we checked samples from two separate lots of concrete; one at w/c ratio of 0.40, the other at w/c ratio of 0.50; each lot tested with five different concentrations of silica fume. Fig. 6 clearly shows the following point:

a)The introduction of silica fume, at a given w/c, effectively makes a more impermeable concrete.;
b)Maximum effect is obtained with an 8-10% addition of silica fume.

Figure 6: Effect of silica fume dosage on chloride ion-permeability

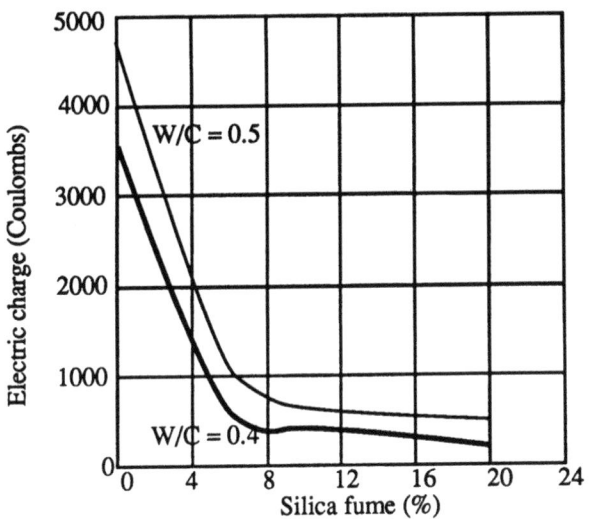

5.3.3 Effects of aggregates
According to results taken from the literature it seems that the aggregate must be taken into account at many levels.

Petrographic dispositions:
Comparative tests have resulted in total charges of 4500 Coulombs for limestone, 5,500 for granite and 6,000 for gravel.

Maximum size D:
The permeability varies in the same way as D.

Grain size distribution:
Discontinuity in the aggregate grain size distribution raises the permeability of concrete.

5.3.4 Effects of the air content
The air content and coefficient of permeability vary in the same way.

5.4 Conclusion
First, the method has drawbacks, not least of which is a partial incomprehension of the fundamental mechanisms the chloride permeability test. In particular, the following points have not yet received satisfactory answers:

- Would the initial amount of ions present in the solution be sufficient?
- How to quantify the motion of the ions brought about by the heating up of the sample during testing?
- What reactions are possible between the constituents of the concrete and the electrolyte?
- Is it possible that dissolution or local precipitation, which modifies pores arrangement, would bring about variations in permeability?

However, this method presents an important advantage, in that it can be applied on nearly all concretes, including high performance and latex modified concretes. The only exception is that it does not permit correct measuring on heavy concretes containing electrically conducting aggregate (i.e. minerals or metals). The test also can be used both in the laboratory, and 'in situ'.

To conclude, in spite of some adverse criticism, notably that the method really measures resistivity and not permeability, it does result in a sameness of classification of concretes to that given by a true permeability test. This latter is affirmed by Whiting [14] who proposes a scale of values reproduced in Table 6.

Table 6 : Scale of permeability to chloride ions proposed by Whiting (ref. [14])

Charge passed through the sample (Coulombs)	Permeability	Type of concrete (estimated w/c ratio)
More than 4000	High	Concrete w/c higher than 0.60
Between 2000 and 4000	Medium	Concrete w/c between 0.40-0.50
Between 1000 and 2000	Low	Concrete w/c less than 0.40
Between 100 and 1000	Very low	Latex concrete
Less than 100	Negligible	Concrete polymer

6 Water permeability test

6.1 Generalization
The permeability coefficient defined in Equation is an intrinsic characteristic of the material as long as a certain number of rules are respected:

- The material must be first saturated with de-aired water;
- The temperature of the water must be constant;
- The state of the flow must be steady and laminar;
- There must be no physical or chemical interactions between thewater and the constituents of the concrete.

This last point always causes problems when interpreting the results. And most researchers recognize that the water acts on the structure of the concrete, in which it percolates, either by dissolution of the hydrates (portlandite mainly) or opposingly by provoking a continuation of hydration of the anhydrides. These effects are furthermore amplified by the fact that the water is generally injected under pressure.

6.2 Test apparatus - preconditioning of samples
Reading review articles, 1 and 12, we see that different apparatus and procedures are in use. The one used at the Université de Sherbrooke is laid out in Fig. 7. The test samples are of cylindrical form, 150 mm dia. x 300 mm long, and axially perforated. The sample is placed in a water pressure chamber, which is then filled with pressurized water, and held in place with a pressure plate. Rubber discs are placed at each flat end of the sample to ensure sealing.The pressure on the water is raised gradually in the chamber, which forces a radial convergence flow. The volume of percolated water as a measured in function of time.

Figure 7: Schematic of water permeability cell used at Université de Sherbrooke

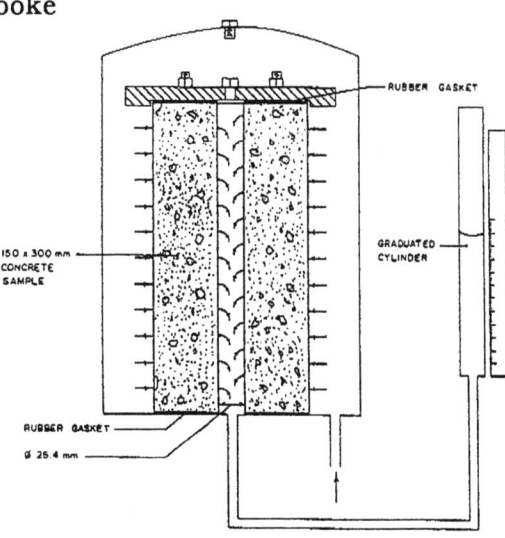

6.3 Results

The results obtained cross-check well with the literature, showing a reduction of the water permeability coefficient with their w/c ratio. They fit well with Fig. 8 which regroups values obtained by seven different researchers (Lawrence [12]). But the most important result, at least as concerns this book, is that we have not been able to measure significantly the percolation of water through concrete having a w/c ratio less than 0.40. So, water permeability can not be measured for high performance concrete.

Figure 8: Relation between permeability and water/cement ratio for mature cement paste (ref. [6])

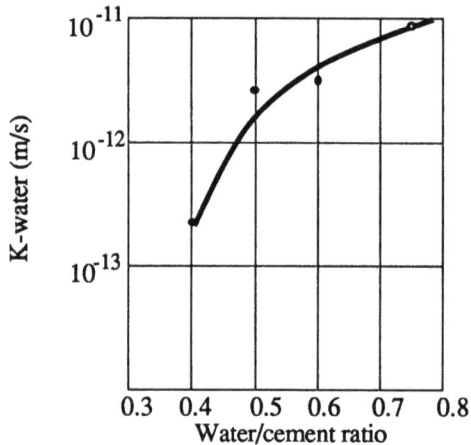

7 Air permeability measurement

7.1 Test method

Air permeability measurements of concretes are not as common as those of water permeability. However, the interest in such measurements is evident for certain types of works (nuclear reactors, gas tanks, etc.).

Like water permeameters, air permeameters are of variable design. Essentially they differ in their way of measuring under variable or constant pressure. The test apparatus used in both of our laboratories (Université de Sherbrooke and the LMDC de Toulouse) was conceived by Thenoz [15], from a prototype of the French Petroleum Institut [16].

7.1.1 Test apparatus

A section of this air permeater is shown in Figure 9. The test can be performed at various pressure gradients (33 to 93 kPa). The test specimens were 40 mm concretes and 50 mm in length. The vacuum inside the chamber (Fig. 9) creates a pressure gradient that forces air flow through the concrete. Mercury surrounding the cylinder walls forces the air to flow from end to end. Air penetration through the sample increases the pressure in the sealed chamber, which is indicated by the column of mercury. We measure the elapsed time to pass from the starting to the final level. For a given pressure gradient, permeability decreases as the required time increases. The test was performed in a room at controlled humidity and temperature.

Figure 9: Schematic of air permeability cell used at Université de Sherbrooke and at INSA de Toulouse.

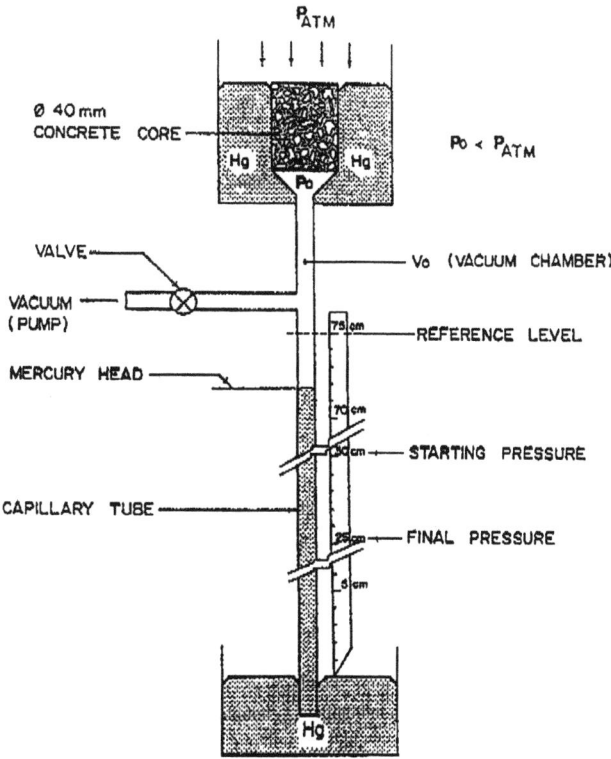

7.1.2 Precision

The experimental accuracy of the test is of ± 2% [15]. More important, however, is the disparity of the results obtained, due to the heterogeneous character of the material. Knowing the make-up of the latter, is in fact indispensable, to see if the observed differences between different concretes should be accounted for. To check these differences, we made three slabs of different concrete composition. Nine samples were cored from each slab and tested after storage in the controlled room for at least a month. The same samples were tested again after two additional days of drying in a ventilated oven at 60°C. The results are shown in Table 7.

One can draw from these values, the number of samples to test for a given concrete for a precision (10, 15 or 20%), and with a 95% confidence interval. These indications figure in Table 8.

Table 7: Precision of results, linked to the heterogeneity of the concrete

w/c	Drying	Coefficient of permeability (in 10^{-18} m^2)ka	
		Average	Precision (%)
0.33	laboratory	4.8	14
	oven	18.1	15
0.50	laboratory	7.1	10
	oven	30.3	9.5
0.67	laboratory	10.8	6
	oven	34.1	6

Table 8 : Number of samples to test for a given precision

w/c	Drying	Precision value of		
		10%	15%	20%
0.33	laboratory	11	6	3
	oven	13	6	4
0.50	laboratory	6	3	2
	oven	6	3	2
0.67	laboratory	3	2	1
	oven	3	2	1

Note that we may expect quite good accuracy from a reasonably low number of test samples.

7.2 Calculations for a more realistic coefficient of permeability

From the datum obtained in the test and the geometric characteristics of the test apparatus, it is possible to calculate an apparent permeability coefficient k_a, to use Houpert's equation [16]:

$$k_a = B \cdot G \cdot \frac{l}{t} \tag{5}$$

where
l = length of sample
t = dropping-time, from initial level to final level
B and G = coefficients, dependant on the geometry of the test apparatus and the flow conditions

The reader wishing to obtain the details of this equation may consult a previous publication [18]. Despite these improvments in the statement of k, it should be noted that Darcy'S law was developed for water, under laminar and steady flow conditions. However Maxwell [20] observed a molecular flow with a gas and consequently the air permeability coefficient calculated by the previous equation is not totally dependent on the material properties [15]. Thenoz [15] has a more realistic approach in calculating the modified air permeability coefficient k_c that applies solely to air:

$$k_c = \frac{k_a}{(1+D)} \tag{6}$$

As indicated by publication [18], the term D is a function of the porosity of the concrete (D is the ratio of the mean free path of the gas molecules to the radius of the pores). The term D may be evaluated by measuring the 'average' radius of the pores by Mercury porosimetry test. The corrective term is important (1 + D), having values 6 to 11 in our tests.

7.3 Results
Results obtained by the air permeability measurement are mainly consistent with those given by the 'water permeability test', though there may exist differences of 1 or even 2 orders of the coefficient values.

We shall examine the effect of the composition of a concrete upon its air permeability.

7.3.1 Effects of the w/c ratio
As shown by the data taken from the literature, and also according to our own tests (Fig. 10), the air permeability coefficient decreases with the w/c ratio.

7.3.2 The influence of an addition of silica fume
Nagataki and al. [19] have shown that the air permeability of a concrete is reduced by an addition of silica fume associated with a superplasticizer (Fig. 11). However, the opposite effect is observed for concretes mixed without superplasticizer. As when we tested for chloride permeability, it would be seen that the maximum effect would be obtained by a 10% addition of silica fume.

Figure 10: Effects of water/cement ratio on air permeability (ref. [19])

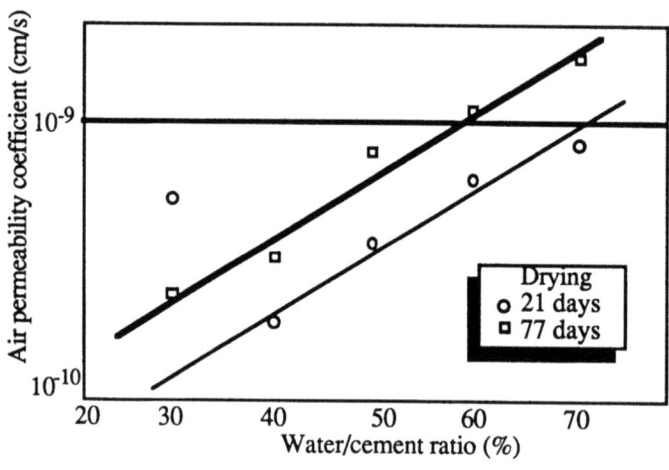

Figure 11: Coefficient of air permeability with condensed silica fume without high range water reducer admixture (ref. [19])

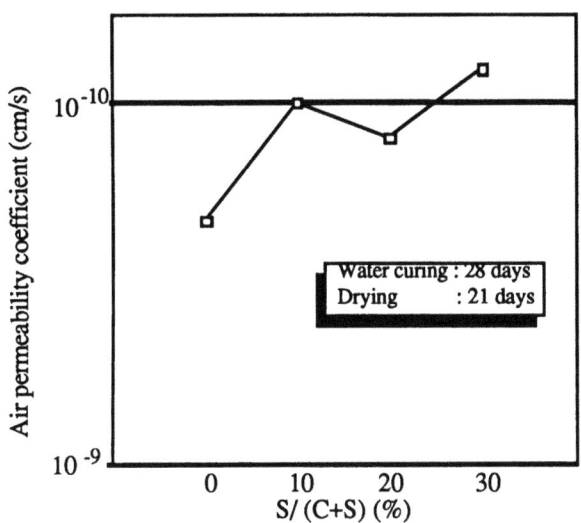

7.3.3 Effect of the air content
The results of [19] show that the increase of air content brings about a greater air permeability coefficient (Fig. 12).

7.3.4 Effect of the drying process
The moisture history influences the hardening time of concrete, and also its structure. At a given maturity, the degree of internal humidity of a concrete varies in the same way as that of the ambient environment. This explains our results at Fig. 13 which conform to those of the literature, where it appears that an oven treatment, at 60°C for 2 days, multiplies by a factor 3, the permeability coefficient of a concrete that has been preliminarily cured for 28 days at 20°C and dried in laboratory for at least a month.

Figure 12: Effects of air content on air permeability (ref. [19])

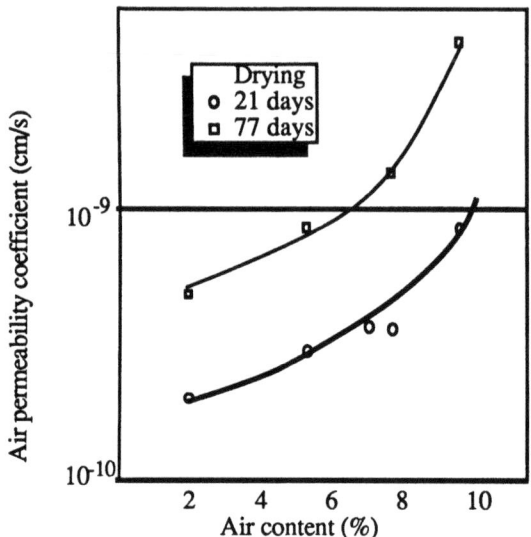

7.4 Conclusions

Measuring air permeability can be accomplished with rather unsophisticated apparatus. A measurement can be taken much quicker than as in the case for water permeability, as the steady flow condition is attained sooner.

Making use of the results is quite easy, if we content ourselves with the values of apparent permeability. However, things become more delicate if we wish to obtain the corrected permeability coefficient. In this latter case one can characterize the concretes of a low w/c ratio (< 0.40), the actual limitation of 0.30, on test samples of 40 mm dia. is imputable to the weakness of measured flows. This limitation figure will be lowered, probably allowing study of high performance concrete, by the use of a new apparatus on samples up to 100 mm in diameter.

Concerning the value of the coefficients of permeability calculated, we can be assured that contrary to what happens with water, the measuring fluid does not inter-react with the concrete.

Figure 13: Effects of drying on air permeability (ref. [18])

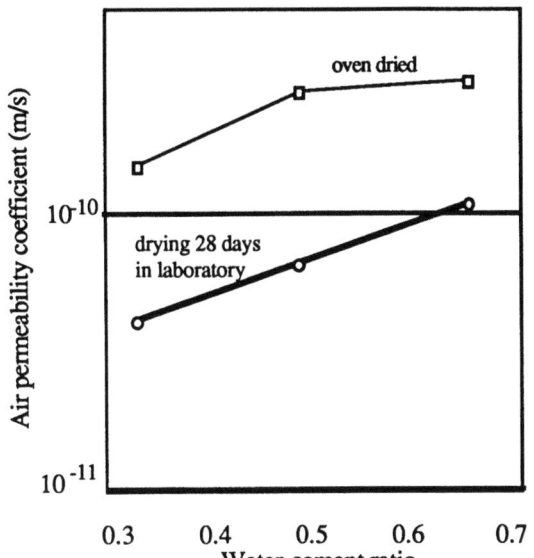

Fig. 13 - Effects of drying on air permeability (ref. [18])

8 Conclusion

We could conclude, like numerous other researchers, that there are many different procedures for measuring the permeability of concrete, and consequently, the correlation of the different results is not easy, if not impossible. However, in our point of view, it would be possibly more important to determine the phenomena relative to the fluid transfer, which establish themselves in the concrete in use, and to measure the corresponding coefficients of permeability (or diffusion).

The question is now: Do the tests on small samples in the laboratory give directly transposable information that can be used on actual constructions? Unfortunately, the answer is often no. We may say then, that much effort remains to be made in order to correlate the laboratory tests and the actual corresponding work on the site.

For high performance concrete the same conclusion applies as that drawn in the preceding paragraph, but in this case, because of their low permeability, special care must be taken in the permeability test and especially for the interpretation of the results.

References

1. Hooton, R.D. Concrete Permeability and the Search for the Holy Grail. Premier Colloque canadien sur le ciment et le béton - Université Laval, Québec, 1989.
2. Darcy. Les fontaines publiques de la ville de Dijon. Dalmont éditeur, Paris, 1856.
3. Ithurralde, A. La perméabilité vue par le maître d'ouvrage. Comptes rendus du Colloque bétons à hautes performances, Cachan, 1989.
4. Maso, J.C. Liaison entre les granulats et la pâte de ciment hydraté. 7e Congrès Int. Chim. Cim., Rapport principal, t.I, Ed. Septima, Paris, 1980.
5. Powers, T.C. Structures and Physical Properties of Hardened Portland Cement Pastes. J. Am. Ceram. Soc. (1), 1958, p. 1-6.
6. Powers, T.C., Copeland, L.E., Hayes, J.C., Mann, H.M. Permeability of portland cement paste. A.C.I. Journal, V.Cl, No. 3, Nov. 1954.
7. Thenoz, B. Communication personnelle.
8. Grandet, J., Ollivier, J.P. Orientations des hydrates au contact des granulats. Ve Congrès Int. Chim. Cim., tome III, Ed. Septima, Paris, 1980.
9. Grandet, J., Ollivier, J.P. Nouvelle méthode d'étude des interfaces ciment granulats. Ve Congrès Int. Chim. Cim., tome III, Ed. Septima, Paris, 1980.
10. Scrivener, K.L., Crumbie, A.K. A Study of the Interfacial Region Between Cement Paste and Aggregate in Concretes. MRS Symposia Proceedings, Ed. Mindess and Shah.
11. Scrivener, K.L., Bentur, A., Pratt, P.L. Quantitative Characterization of the Transition Zone in High Strength Concretes. Advances in Cement Research, Vol. 1, No. 4, October 1988.
12. Concrete Society Working Party. Permeability Testing of Site Concrete. A Review of Methods and Experiences. Concrete Society, London, 1988.
13. Whiting, D. Rapid Measurement of the Chloride Permeability of Concrete. Public Roads Magazine, Vol. 45, No. 3, 1981.
14. Whiting, D. Permeability of Selected Concretes. ACI SP-108, Permeability of Concretes, 1988.
15. Thenoz, B. Contribution à l'étude de la perméabilité des roches et de leur altérabilité : application à des roches granitiques. Thèse Doctorat ès Sciences, Faculté des Sciences, Toulouse, 1966.
16. Houpeurt, A. Revue de l'Institut Français du Pétrole. Vol. VI, No. 6, p. 180-90, 1951.
17. Schönlin, K. and Hilsdorf, H.K. Permeability as a Measure of Potential Durability of Concrete. ACI SP-108, Permeability of Concrete, 1988.

18. Perraton, D., Carles-Gibergues, A., Aïtcin, P.C. andThenoz, B. Air Permeability Measurement. Material Research Society, Boston, USA, november 1988, 11 p.
19. Nagataki, S. et Uzike, I. Air Permeability of Concretes Made with Fly Ash And Condensed Silica fume. ACI SP-91-52, Malhotra Ed., 1986.
20. Carman, P.-C. Diffusion and flow of gases and vapours through micropores - I slip flow and molecular streaming. Proceedings of the Royal Society of London, Serie A, Mathematical and Physical Sciences, vol. 203, 1950, p.56.

18 PERMEABILITY : THE OWNER'S VIEWPOINT

G. ITHURALDE
Electricité de France, SEPTEN, France

1 Introduction

For a number of years there has been growing interest in the properties of concrete other than the traditional compressive strength. The many studies and researches undertaken in this area are explained both by the increasing concern of owners with the service lives of existing or planned structures, and by the related need facing contractors to innovate in structural design to meet economic challenges.

In this respect, P. RICHARD (1) emphasizes that, in most cases, the material is not decisive in determining the overall shape of the structure : it is in fact possible to proceed quite far in the design of a structure without mentioning the material as such. The choice of material will be made when the cross-section of the structure is decided, or on the basis of secondary roles unrelated to the primary function of the structure.

The evolution of the design of containments,which must combine strength and impermeability under the most severe normal operating and accidental conditions, illustrates this point (Fig.1).

For such structures, as for offshore structures, or silos, impermeability and life may be the decisive factors in the economic choices once the general shape of the structure has been specified: for works of this type, the ageing of the structure and its long-term ability to perform its functions in all circumstances must also be taken into account.

The durability of the material is therefore of major concern to the owner, and may justify the extra cost of a material;

But how is the durability of a material to be judged ? Is its durability a sufficient criterion for judging the life of a structure ?

It is now acknowledged that the permeability of a concrete is a key indicator :

- of the water tightness of a concrete structure

High Performance Concrete: From material to structure. Edited by Yves Malier. © 1992 Taylor & Francis.
Published by Taylor & Francis, 2 Park Square, Milton Park, Abingdon, Oxon, OX14 4RN. ISBN 0 419 17600 4.

Figure 1:

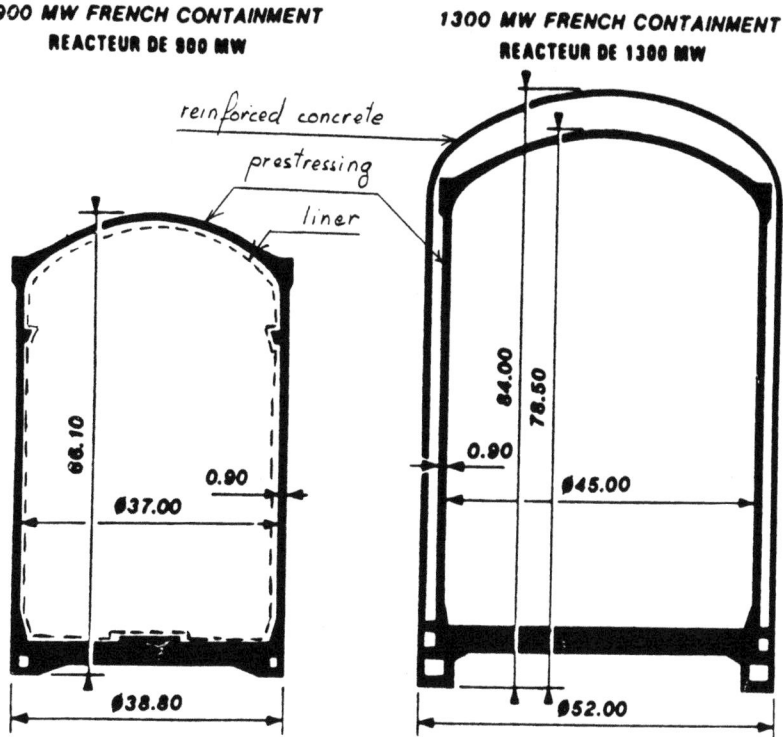

900 MW FRENCH CONTAINMENT
REACTEUR DE 900 MW

1300 MW FRENCH CONTAINMENT
REACTEUR DE 1300 MW

- and also of the durability of the material in the presence of physical and chemical aggression.

However, the relation between the permeability of a concrete and the desired durability performances is not always easy to put into quantitative form. The same applies to the relation between the permeability of a concrete and the water tightness of a stucture.

2 Relation between permeability and durability

Permeability is representative of the mass transfer of a liquid or gas through a porous medium (the term "diffusion" is also used in the case of gases).

2.1 Durability of a reinforced or prestressed concrete
With respect to durability, one of the main factors governing the resistance of a reinforced or prestressed concrete to corrosion of the reinforcements or tendons is the capacity of the concrete to maintain a passivating environment in their vicinity.

Carbonation of the concrete and the diffusion of chloride ions or of water vapour are so common physico-chemical phenomena that destroy this resistance.

The absorption of water is also a major source of damage to a stucture, not only through its effects on the durability of the reinforcements embedded in the concrete but also because of the risk of alkali-aggregate reactions and freezing-thaving cycles. It is especially important to evaluate this water transport mechanism in the case of concretes in offshore structures exposed to infiltration by seawater.

All of these processes of degradation are linked to a mechanism of diffusion, primarily in the surface (skin) concrete constituting the first few centimetres of a stucture : diffusion of chloride ions, water vapour, and atmospheric CO_2, which react with the hydrated compouds of the cement, or else of OH⁻ions or alkalis in the case of alkali-aggregate reactions.

To forestall these aggressions, the quality of the surface concrete covering the reinforcements is a better guarantee than its thickness, yet regulations specify only the thickness.

2.2 Laboratory tests

The many laboratory tests used to evaluate these diffusion and absorption mechanisms (2) do not strictly speaking measure the permeability of the concrete. But the potential degradation of a concrete exposed to these aggressions is directly related to the degree of freedom of certain gas or liquid transfer mechanisms to act in the concrete. In this sense, they may be regarded as a good indicator of the permeability of the concrete.

Measurements of permeability :

Measurements of the mass transfer of aliquid or gaseous fluid under a pressure gradient is, on the other hand, a true test of permeability (Fig.2)

Such a measurement makes possible in principle a more global assessment of the mechanisms of water and gas transport and, in consequence, of the ability of a concrete to withsland various forms of aggression.

However, the handful of theoretical considerations concerning the notion of permeability expounded in the appendix reveal the difficulties of interpretation that may arise with the many tests developed to evaluate the mechanisms of mass transfer of afluid in a concrete.

As with the diffusion and absorption tests mentioned earlier, the assesment of the permeability of a concrete is highly dependent on the microstructure of the material, on its porosity and pore distribution, and on the free water content of the concrete.

The tests conducted by J.C. Maréchal, at the request of EDF-SEPTEN, in connection with a study of the mass transfer of an air + water vapour mixture through permeable concrete subjected to

Figure 2: EDF/CEMETE permeability meter

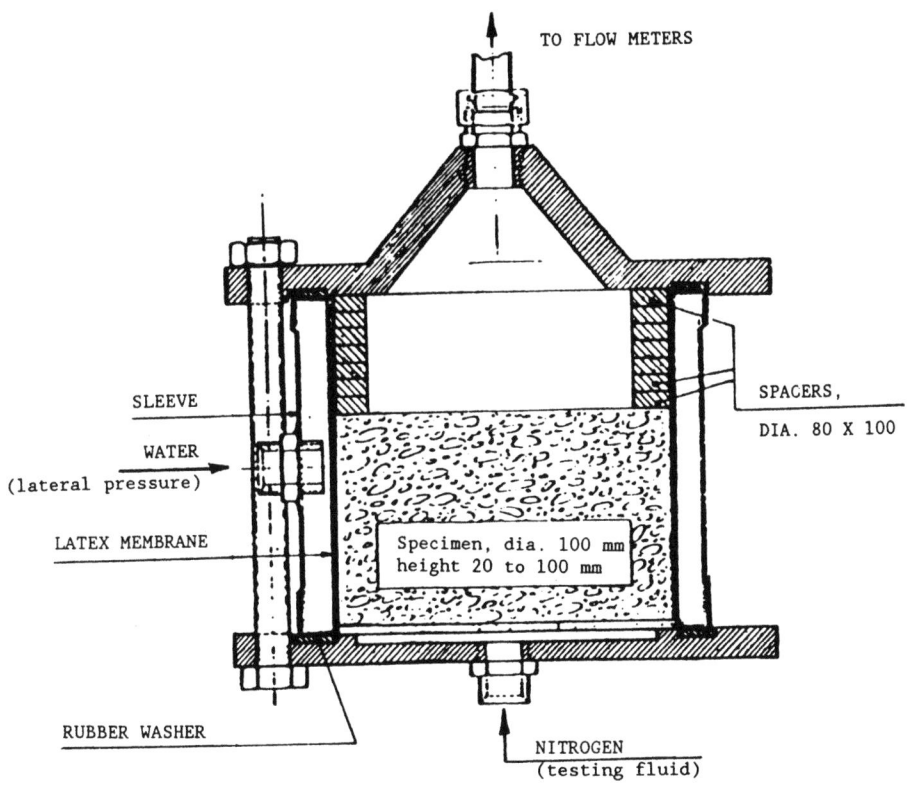

extreme thermodynamic stresses P = 0.6 MPa (87 psi), T = 150°C (302°F) explain (Figs. 3 and 4) the various phenomena that occur in the concrete (adsorption, absorption, diffusion of air, water, and water vapour) during a transfer phase.

This thinking highlights the difficulties encountered by researchers (3) in developing a representative test of permeability : the measured permeability depends on the thermodynamic and hygrometric state of the concrete during the test (and thus on how the specimens are dried and stored), which may substantially alter the microstucture of the concreta by affecting the chemical phenomena of hydration between the water and the cement.

It should also be pointed out that permeability tests may sometimes turn out to be impracticable or unrepresentative (steady state not established) at very low flow rates, such as are pratically always found in the case of concretes subjected to rather small water pressure gradients. Here, a more appropriate test is measurement of the depth of penetration of the liquid.

Figure 3: EDF/CEMETE and CEBTP experimental setup

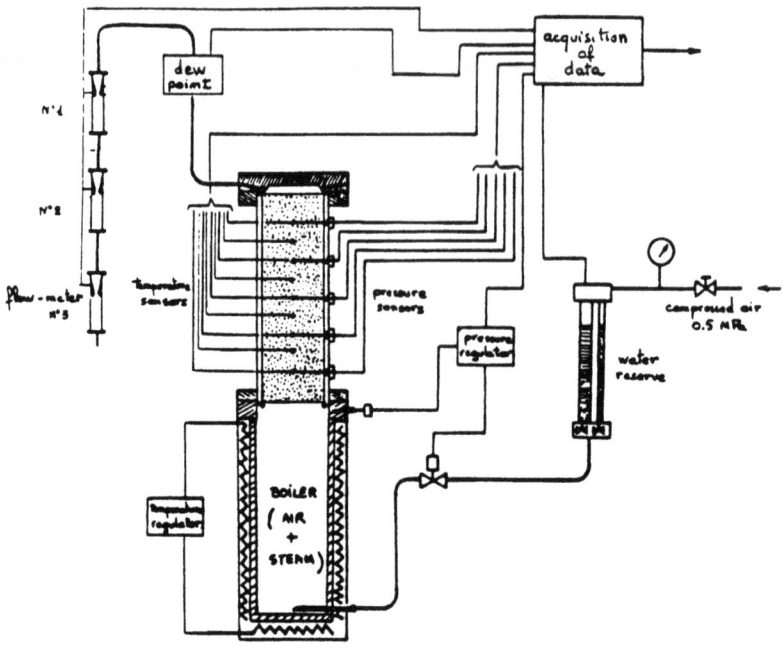

Figure 4: Tightness of concrete test specimens subjected to a loss-of-coolant accident

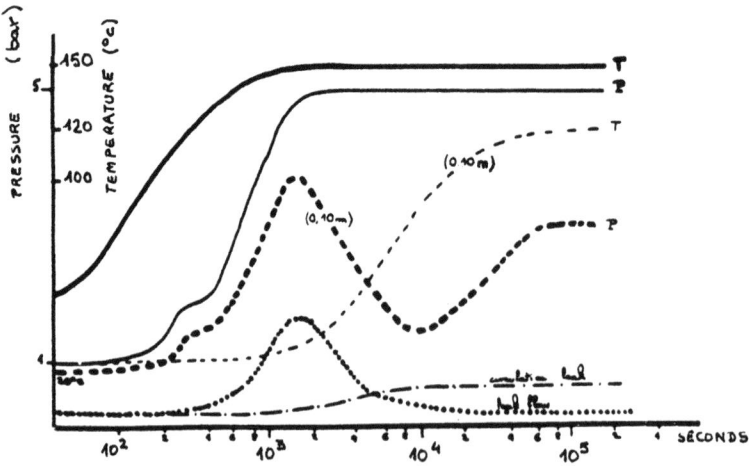

3 Relation between permeability and water tightness of structure

To illustrate this second thought on the relation between the permeability of the concrete and the tightness of the structure, Table 1 gives the results of tests of permeability to air carried out on concrete cores taken from existing structures and compares them with the overall air-tightness of the structures built with the materials.

Table 1: Permeability and other properties of existing nuclear containment structures, compared to material properties

	Structure		Concrete						
	Tight ness % day (1)	Crac king state	Permeability $(10^{-18}m^2)$ (2)			Other properties			
			satured	dried 24h at 40°C (104°F)	dry (3)	Porosity %	% satura tion	density kg/m3 (4)	W/C (5)
N°1	1,85	+++ fine	0,03	0,81	190	13,3	90,5	2,190	0,55
N°2	1,27	++ medium	0,02	0,07	210	16,1	95,4	2,150	0,58
N°3	0,69	+ coarse	0,03	0,035	35,5	10,6	96,7	2,240	0,48
N°4	0,57	none	0,06	0,53	60	13	97	2,230	0,51
N°5	0,31	none	0,03	0,04	98	14	96,2	2,130	0,54

(1) Rate of leakage from a containment in a test with air at approx. 0.5 MPa (73 psi), in % of mass of air in containment
(2) Permeability tests performed on 100-mm-dia (3,94 in) concrete cores taken from the structure
(3) Oven drying at 80°C (176°F) to constant weight
(4) Note : 1kg/m3 = 0,0624 pcf
(5) With a cement content c ranging from 350 (770 lb) to 375 kg (825 lb) according to the site.

The table reveals the importance of the open porosity of the concrete in the degree of tightness of the structure, see Fig. 5.

Figure 5: Illustration of permeability and porosity (after Bakker [(2)]

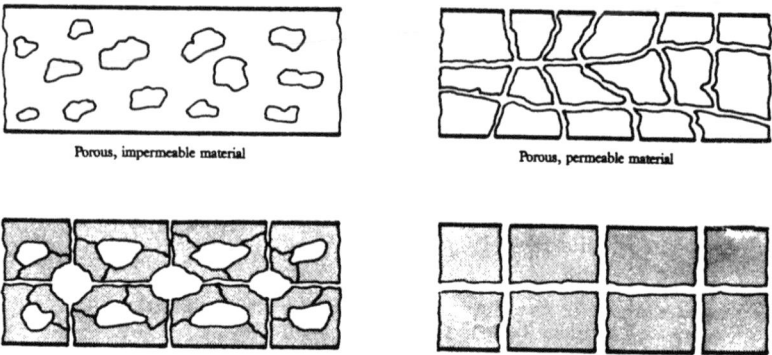

Porous, impermeable material

Porous, permeable material

High porosity, low permeability

Low porosity, high permeability

These results also highlight the fact that knowledge of the permeability of the material concrete alone is insufficient for an overall assessment of the tightness of a structure, which depends, notably, on the cracking state of the structure :

a dense concrete with fine, non-open prores ensures a low permeability : this is generally the case with concretes having a low water/binder ratio (w/c). But such concretes are also characterized by high endogenous shrinkage, which may cause cracking in the structure if deformations of the concrete are blocked.
conversely, the designer may attempt to make the structure thicker to increase its tightness accordingly. But massive parts are subjected to large thermal shrinkage when the concrete sets.

Therefore, even with a concrete of low permeability,the cracking induced by blocking its deformations may significantly impair the tightness of the structure.
These examples clearly illustrate the need to match the material and structure; the necessary match must be borne in mind in the design and in the construction to ensure that the structure will be tight and durable.

4 Advantages of high performance concretes

The risks of using a conventional concrete in massive structures that must remain tight are now well known ; they are related to early cracking, during the setting and hardening of the concrete. These risks (additional contribution of thermal and endogenous shrinkage) cannot be perfectly controlled in conventional concretes.

Because of their microstructure (transgranular fracture, low porosity of matrix), conventionally formulated high performance (HP) concretes having a low w/c have a much lower intrinsic permeability than conventional concretes. But they do not allow any control, or prevention, of early random microcracking.

On the other hand, a concrete formulation based on the principle of replacement of part of the cement by silica fume and fillers may lead to :

- a substantial reduction of the heat of hydration
- higher density
- higher tensile strength
- an intrinsic permeability at least 10 times better

4.1 Case studies

For 1988, EDF-SEPTEN's thinking about the design of nuclear containments has been along these lines.

A case study has been undertaken with the help of the LCPC.

4.1.1 Laboratory study

The performance of the concrete mix (Tables. 2 to 4) developed at the LCPC by F. de Larrard using local materials from the Civaux (Vienne) site is better than that of the conventional 36 MPa (5,220 psi) concrete :

- compressive strength of the order of 65 to 70 MPa (9,425 to 10,150 psi) at 28 days
- 20% increase in tensile strength (with respect to conventional concrete made with CPJ)
- permeability to air improved by a factor of 10 with an appropriate testing method (specimens not dried)

Numerical simulations representing a complete containment structure have made it possible to compare the blocked strains to the admissible strains of the concrete. This comparison has been applied to conventional concrete and to HP concrete (Figs. 6 and 7).

The results of the computations confirm the advantages of an HP concrete so formulated in terms of a lower risk of the appearance and opening of vertical cracks like those so far observed in most containments built with conventional concretes.

4.1.2 Experiments in the field

In a second stage, the laboratory work on HP concrete was continued at the Civaux site, with the Fougerolle-France construction company to :

- confirm the results of the LCPC studies (concerning cracking)

Table 2: Mix designs

Study	Concrete	Aggregate 12,5/25 mm kg	Limestone 4/12,5 mm kg	Sand 0/4mm kg	Airvault CPJ 55 Cement kg	Silica Fume kg	Filler kg	Plasticizer % kg	Added Water kg
Stage 1 (1)	Control	890	208	725	375	-	-	(5,6)	190
	HP 1	944	221	769	266	40,3	66,4	(34,6)	127
Stage 2 (2)	Concrete B 1*	759	333	749	350	-	-	0,3%	195
	HP 2**	891	209	873	266	40,3	-	1,4% (9,08)	161

(1) Mix design developed using materials supplied by contractor (05/88), except for the sand (Boulonnais sand)
(2) Revised mix design of B1 concrete (Contractor's study concrete as of 8/7/88) and HP concrete (LCPC) with non-defillerized sand and use of a new plasticizer (Rheobuilt 1000) for the HP concrete.

Table 3: Summary of convertions and HP concrete

	Conventional Concrete B1	HP Concrete B10
In conformity with general specifications	No(sand grading being corrected)	No (justification file in accordance with document 65 in progress)
Compressive strength (28 d)	45 to 47 MPa (6,500 to 6,800 psi)	65 to 70 MPa 9,400 to 10,100 psi)
Bending strength (28 d)	4,6 MPa (667 psi)	5,5 MPa (798 psi)
Admissible strain (28 d)	$155,10^{-6}$	$160,10^{-6}$
Permeability, air (.meth 1 (. meth 2	3 to $14,10^{-18}$ m2 $5,10^{-20}$ m2	0,1 to $0,8.10^{-18}$ m2 $5,10^{-20}$ m2
Porosity	13,6%	12,1%
Basic creep Desiccation creep (μm/MPa at 2 months)	19,2 26,7	13,5 0,5
Modulus	32,000 Mpa (4.64×10^{6} psi)	36,000 MPa 5.25×10^{6} psi)
Occluded air	1,6 %	0,8 %
pH on cores from model	12,4	12,3
θ max	40°C (104°F) (37° with CPJ/LCPC)	30°C (86°F°
Max temperature rise	60° (140°F), 28 h after start of concreting	48° (118°F), 38 h after start of concreting
Bleeding	Yes	No
Treatment of construction joints	delicate washing 2/3 h after concreting	. curing with water before desiccation . then cleaning with pressurized water the next day
Slump	10 cm (3.94 in.)	20 cm (7.88 in.)

Table 4: Comparison of concrete properties

	Conventional Concrete B1	HS Concrete B10
Pumping to great high	Ok (dia. 125mm/4.925 in.)	Ok (with dia. 150mm/5.91 in. and staged reductions of pipe diameter)
Digital simulations : blocked deformations of structure	> admissible deformations	< admissible deformations
Model		
Number of cracks per side	8	1
Crack opening*	(1x40 μm) +(4x100 μm) +(2x200 μm) +(1x500 μm)	100 μm
Air flowrates at 0,5 bars (7.25 psi) : N1/h.m (cubic ft/ h.yd*) -construction joints - body	47/200　(1.5/6.4) 225/350　(7.2/11.2) flowrates unacceptable	24/0　　　(0.77/0) 4/26　　(0.13/0.83) Ok
microbubbles between pourings	Yes, at 2 bars (29 psi)	Yes, at 2 bars (29 psi), on outside surface only (effect more marked and continuous)

Figure 6: Results of thermal calculations, HSC2 concrete

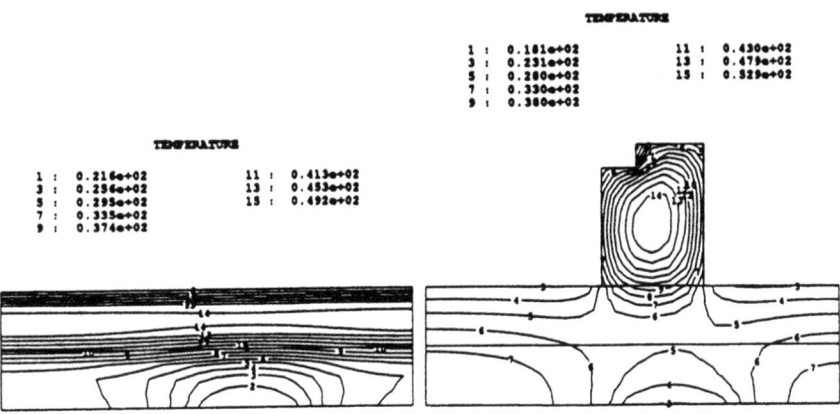

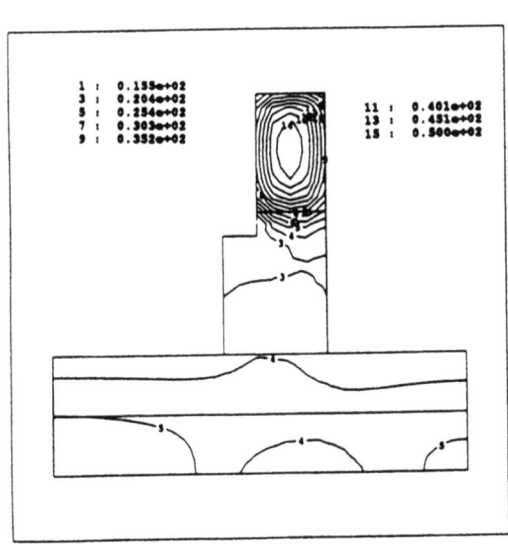

Figure 7: Result of mechanical calculations, OC2 and HSC2. Crack pattern as predicted by the numerical computations

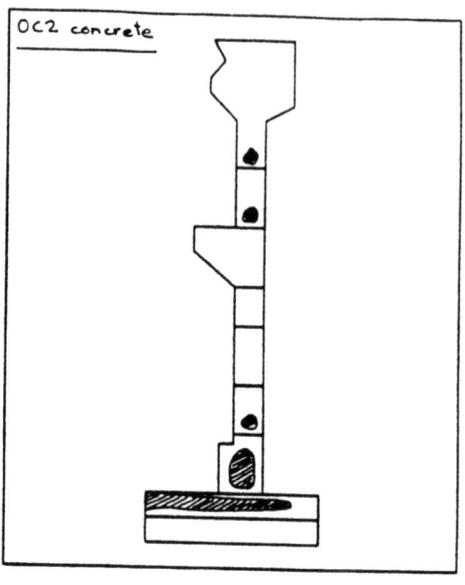

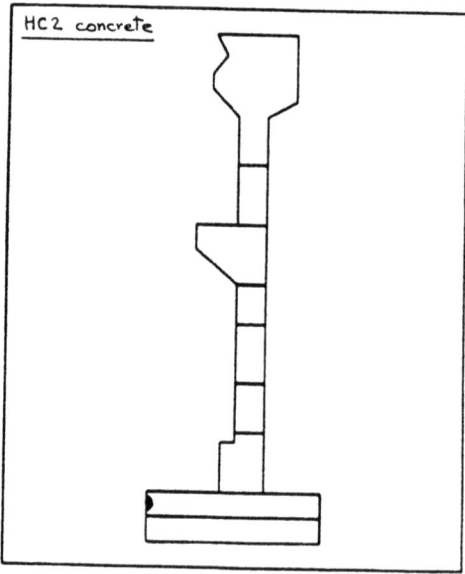

- compare conventional B1 concrete to the HP concrete under realistic conditions of placement (mixing, pumping, vibration, usable life, interval between pourings, handling of construction joints, curing)

This second stage of the study consisted of :

- mix design and field feasibility tests (conducted in the same spirit as those required by the specifications for conventional concretes)
- the construction of a model representing a containment

(a) Preparation, pumping
The field development of the HP concrete formulation studied inthe laboratory did not in the end pose any real problems : after a few hitches, in particular in pumping to a great height (70 m = 77 yd), the concrete prepared exhibited a consistent slump (about 20 cm = 8 in) and no segregation at the mixer or pump outfeed.

(b) Model (fig. 8)
The conreting of a section of tunnel not classified as a safety element provided an opportunity to produce a model representinga containment 1.20 m (1.31 yd) thick, with its reinforcements and prestressing ducts; two 20-m (22 yd) lengths were built, making it possible to compare the two types of concrete (conventional and HP) in an actual structure.
The model clearly showed :

- a maximum temperature rise of 40°C (104° F) in the mass of the conventional concrete, against 30°C (86°F) in the HP concrete, a reduction of 25% (17%) with the latter (fig. 9).
- high air flowrates (during tests using testing networks embedded in the concrete) in the conventional concrete, both in the mass
and at construction joints

By comparison, the flowrates in the HP concrete shell are reduced by a factor of 10 and are quite satisfactory : the flowrates measured in the HP concrete shell during the construction of the containments, less than 25 Nl/h.ml at 0.5 bar (0.80 cubic ft/h. yd at 7.25 psi), are regarded as excellent.

- a strong tendency to vertical cracking of the B1 conventional concrete, with an interval of every other prestressing duct ; and the opening of three of the cracks exceeded 200 μm (7.880 10-6 in.)

Figure 8: Section of model of tunnel wall showing reinforcement

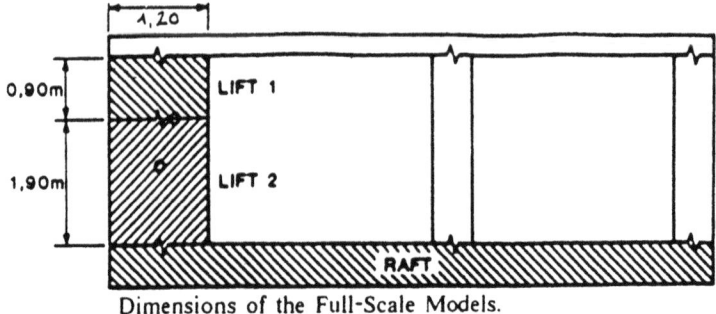

Dimensions of the Full-Scale Models.

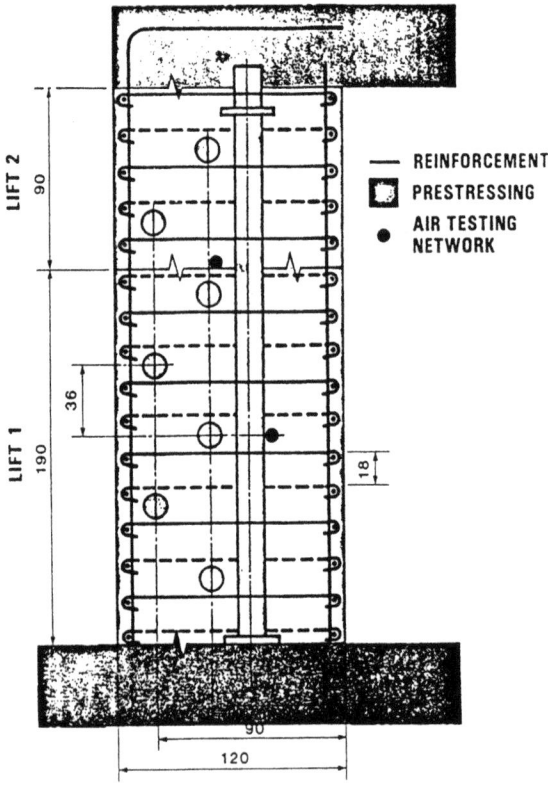

REINFORCEMENT
PRESTRESSING
AIR TESTING NETWORK

Figure 9: Site of Civaux. Recording of temperature of concrete

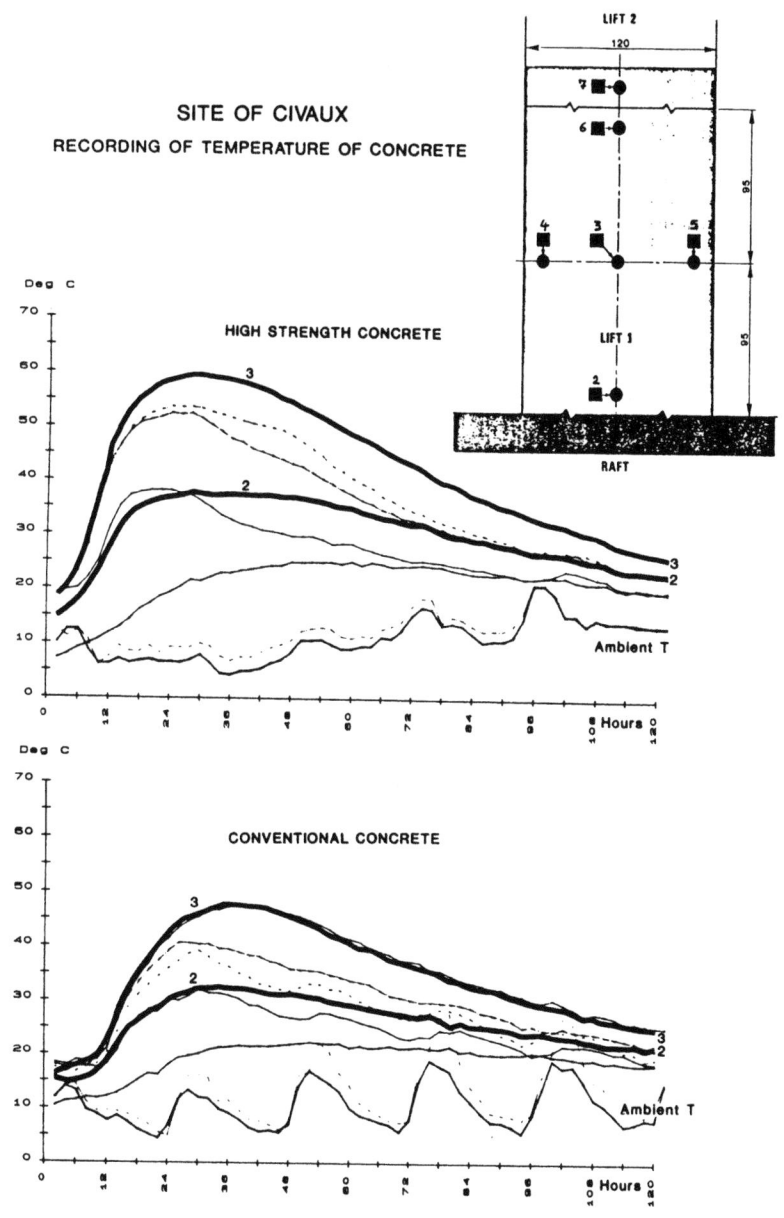

By comparison, there was only one microcrack, with an opening of 100 μm (3.940 10^{-6} in.) in the HP concrete shell.

- the susceptibility of the HP concrete to surface desiccation because of the absence of bleeding ; this desiccation made it necessary to cure the construction joint surface with water less than a half hour after the end of concreting.

4.1.3 Comments
This case study of a containment reveals the advantages of a special HP concrete formulation for the design of tight and lasting structures, together with a quality far superior to that obtained with conventional approaches.

5 Conclusion
The durability of the material is an essential factor in assessing the life of a structure. It is determined by many laboratory tests aimed at evaluating the risks of degradation.

Each of the tests provides an indication about the permeability of the material, just like actual permeability tests with a pressure gradient. The very diversity of these tests is a guarantee that the ability of the material to remain tight will be determined as accurately as possible.

But an impermeable material is not enough to guarantee the life of a structure, nor to guarantee that the structure will be tight when it is required to be.

The experience acquired and the analyses conducted in the course of the construction of 20 containments with double concrete walls have shown that a perfect match between the material and the structure must be achieved in the design and construction of a structure for it to be tight and durable.

Here, the properties of a judiciously formulated high performance concrete may help to improve the performance of a structure, both in terms of durability, by providing larger margins, and by making it possible to design tighter structures.

References

1. L'interaction matériau structure dans la conception des ouvrages, P. RICHARD, ENPC- Les bétons à hautes performances,15/17 mars 1989.
2. Permeability Testing Site Concrete- The Concrete Society, Technical Report n° 312 August 1988.
3. La perméabilité vue par les chercheurs, D. Perraton, ENSET - Colloque bétons à hautes performances, 11/14 sept. 1989.

Appendix:

Theoretical considerations related to the concept of permeability

The many experimental studies conducted so far have shown that Darcy's empirical law gives a good approximation of fluid flow in a porous medium such as concrete. But the assumptions on which this law is based must never be forgotten :

- laminar flow
- uniform, continuous, isotropic medium
- single fluid

$$\vec{v} = - k \overrightarrow{grad} \frac{P}{\gamma_w} \tag{1}$$

where: k, the Darcy permeability coefficient, depends on:

- the properties of the fluid (kinematic viscosity γ)
- the mean porosity and pore size (and thus on the components of the concrete and their arrangement in the structure)

and where: $\vec{v}$, the velocity of the fluid in the capillary interstices, is small (less than 1 cm/s, so that $\mathcal{R} < 10$), but great enough so that the mechanism of absorption of water molecules does not become preponderant, in the case of permeability to water.

It is also possible to define a permeability value K that is independent of the fluid : this is intrinsic permeability of a powdery medium in laminar flow (Hagen-Poiseuille's theoretical law).

$$\vec{v} = - K/\mu \overrightarrow{grad} P \tag{2}$$

where is μ the dynamic viscosity of the fluid.

Comparing (1) and (2) gives:

$$K = k \, \mu/\gamma_w = k \, \gamma/g \tag{3}$$

where γ is the kinematic viscosity of the fluid.

Dimensions

$$\left[L^2\right] = \left[L \; T^{-1}\right] \cdot \left[L^2 \; T^{-1}\right] / \left[L \; T^{-2}\right]$$

Units (Sl system)

$$(m^2) = (m/s) \cdot (m^2/s)/(m/s^2)$$

At 20°C, equation (3) gives the following :

Air	$K (m^2) = 15.392 \times 10^{-7} \cdot k (m/s)$
Water	$K (m^2) = 1.027 \times 10^{-7} \cdot k (m/s)$

In other words, for a concrete having a given intrinsic permeability K, Darcy's permeability k is 15 times as great for water as for air at 20°C.

To assess the tightness of a concrete, it is also possible to determine, from the continuity equation and Poiseuille's law, the relation between the flowrates of air (regarded as an ideal gas) and water (an incompressible fluid) through a concrete having a given intrinsic permeability K, under steady-state conditions with a pressure gradient :

$$Qwater/Qair = \mu air/\mu water \cdot (2Po/P+Po)$$

At 20°C, this gives $Qwater+Qair=6/1000$

19 Hpcs AND ALKALI SILICA REACTIONS. THE DOUBLE ROLE OF POZZOLANIC MATERIALS

A. CRIAUD, G. CADORET
Technodes SA Groupe Ciments Français, Paris, France

1 Introduction

The quality of an HPC formulation must guarantee long-term physico-chemical resistance of the material to its environment. One of the risks that is to be anticipated, the better to prevent it, is that of disorders due to alkali aggregate reactions. This is a slow process in human terms and is often associated with other causes of disorders, of mechanical or physico-chemical origin. There is much literature on the subject, and many lively debates have been devoted to it (see for example Hobbs (1988) and the acts of congresses devoted to these problems [Grattan-Bellew (1987), Okada et al. (1989)]).

Reactive mechanisms are reasonably well-known and relate to all the constituents of concrete as a whole. But certain points are still obscure: the pessimum effect, the role of calcium, the relationship between swelling and the appearance of 'gels' produced from alkali-aggregate reactions, and secondary reactions, etc. These are also questions which have been debated a great deal [Diamond (1989), Chatterji et al. (1987), Jones (1989)].

Alkali silica reactions in concrete are relatively easy to produce in the laboratory, and easily speeded up and amplified. The difficulty resides in the inevitable extrapolation of the conclusions from the test results, which are carried out in strictly controlled conditions, to the reality of building sites.

Effective preventive solutions against ASR [Alkali-Silica/Silicate Reactions exist (see [Hobbs (1989)] for a critical review) and are particularly compatible with, if not inseparable from, the notion of HPCs. Use of pozzolanic (in the broadest sense) additions, the quest for optimum compactness and reduced quantities of water contribute overall to obtaining a durable HPC as far as alkali aggregate reactions are concerned. Such considerations do not however constitute a guarantee of perfect properties in the long

High Performance Concrete: From material to structure. Edited by Yves Malier. © 1992 Taylor & Francis.
Published by Taylor & Francis, 2 Park Square, Milton Park, Abingdon, Oxon, OX14 4RN. ISBN 0 419 17600 4.

term. HPCs and their components are thus subject to the traditional assessment methods.

2 Alkali-Silica reactions in concrete

These require the presence of two conditions intrinsic to concrete:

- the presence of 'potentially reactive' aggregates, i.e. with siliceous or silicated minerals liable to be soluble reactive within the concrete,
- a sufficiently high concentration of alkalis in the capillary pore-fluids on contact with these aggregates.

The third necessary condition is linked to the environment of concrete. According to various authors, relative humidity must permanently exceed 80% for reactions to be triggered, persist and produce damage (Bérubé et al., 1989).

Finally, certain factors linked to the placing of the concrete in the construction contribute indirectly to the triggering or to the propogation of the reactions (cracking for example).

2.1 The interstitial solution

Solid species formed during the hydration of binders (CSH, portlandite, ettringite, AFm and AFt phases) incorporate a large proportion of initial mixing water into their structure. The composition of the residual fluid is rapidly stabilised (approx. one week) when Portland cement is used. Hydration of certain additions (natural or artificial pozzolanas) is slower, and corresponding soluble species are liberated more slowly in solution.

The chemistry of pore fluids is directly related to the nature of the binder, to its content in soluble alkali oxides and to the water/binder ratio [Diamond (1989), Taylor (1990)]. Experimental data has mainly been obtained with pure pastes or mortars and measured contents of OH^- ions were between 0.2 and 0.8 M under these conditions, corresponding to pHs of around 12.8 to 13.7.

The main cations are Na^+ and K^+, in equivalent amounts:

$$[Na^+] + [K^+] = [OH^-]$$

This is why the first member (alkali cations) is often confused - sometimes incorrectly - with the second (alkalis or OH^- ions).

The composition of the capillary solution of concrete is decisive for the triggering and maintenance of silica-alkaline reactions, because it is in contact with aggregates and governs their solubility. For a given cement, the lower the quantity of water, the more alkaline the pore-fluid is. We are of course talking here of **concentration** in alkalis, whereas the idea of total alkali content of concrete, expressed in kg of Na_2O eq per m^3, refers to the total **quantity**, independently of the w/c ratio.

However, it must be acknowledged that we have very little information relating to the concentrations of alkalis in capillary pores of concretes.

2.2 Aggregates: solubility and reactivity

Siliceous minerals have a certain **solubility** in an alkaline environment. For mineral species which are well characterised from a stoichiometric and crystallinity point of view, such as quartz or chalcedony, the theoretical solubility as a function of the pH is known at every temperature. It increases with the pH, the temperature and the 'disorder' of the silicate lattice.

$$SiO_2 \text{ (solid)} + 2 \text{ } OH^- \text{ -----> } H_2 \text{ } SiO_4^- \hspace{3cm} K_1 \text{ (T)}$$

Within the concrete, interactions, however, take place between the solution, the hydrates of the cement (possibly with the anhydrous coumpounds), the fillers and the aggregates. Potential reactivity is the consequence of a complex set of reactions of dissolution and precipitation of new phases. Portlandite, $Ca(OH)_2$, is the probable source of calcium and of OH^- required for the reaction, as the work of Chatterji et al. (1987) and Struble (1987) shows.

2.3 The environment of concrete

The reaction can only progress and become destructive if the concrete has intrinsic permeability and if the environment supplies the water needed for the reaction.

Transportation of ionic species by diffusion or dispersal is only possible if there is a certain continuity of the interstitial solution in the pores. In other words, the concrete must be water-saturated. Filling of the capillary pores also decreases the quantity of 'expansion cells' which could contain the gels. The reactions are in other cases local and of limited extent (therefore they cause no harm), since there is 'exhaustion' or sufficient lowering of the concentration of one or other of the reactants (SiO_2 , reactive, OH^- $Ca(OH)_2$, or water).

2.4 The reaction products

The reaction products have very variable morphologies and compositions [Regourd-Moranville (1989), Davies and Oberholster (1988)]. They are basically Ca-SiO_2^- Na-K bearing compounds with sometimes high porportions of water (10 to 25% by weight according to some authors, when in gel form). They are mobile to a greater or lesser degree. The symbolical notation CSHK (Dron 1990) or CSHN reflects the fact that normal CSH gels resemble these products though they are significantly richer in Na and K. Their stoichiometry is as indeterminate as the CSH.

According to some authors, such compounds should be present in all concretes, in conformitywith the theoretical considerations in regard to overall balance within concrete (thermodynamics and phase rule). These products would only be destructive in certain forms and in excessive quantities.

3 The case of HPCs: the role of pozzolanic materials

3.1 Influence on the chemistry of concrete

2.1.1 Acceleration of hydration/pozzolanic activity
Kinetics data of cement hydration show that the presence of pozzolana or microsilica accelerates the hydration of C_3S and the phenomenous of setting (Figure 1). Formed CSH silica gels have an average Ca/Si ration of 1.1, which trap or adsorb alkali cations more easily (K better than Na according to observations [Glasser and Marr (1984)] than traditional CSH.

The quantity of portlandite after 14 days is lower than in the reference paste, whatever the w/c ratio. The portlandite morphology is visibly affected (Figure 2, from Vernet and Noworyta, unpublished) in the presence of microsilica.

As Taylor (1990) has observed, the chemistry of alkali-silica reactions is the same as that of pozzolanic additions:

$$SiO_2 + Ca(OH)_2 -----> 'CSH'$$

'Pozzolanic' reactions are more rapid than ASR by vitue of the very small size of the reactive particles. Newly formed products are on the one hand more easily dispersed or better distributed within the material, and on the other also more apt to allow incorporation of alkali cations into their structure. Pozzolanic properties are also amplified through increasing the temperature.

3.1.2 Composition of pore fluids
In the absence of pozzolanic materials, low water/binder ratios (<0.4) of high performance concretes should lead to increases in alkalinity of the interstitial solution and hence favor alkali-silica reactions (Figure 3). This is indicated by expansion tests on concrete (see Bérubé et al., 1989).

In the presence of pozzolanic materials, and in particular of condensed silica fume, one initially observe (15 hours) an increase in the concentration of OH-, then a decline in pH relative to the reference, made even more dramatic if the proportion and reactivity of the addition are increased (Figure 4, from Larbi et al., 1990). The quantity of water which can evaporate (not chemically bound) increases in the presence of fly ash or silica fume (there is in effect less clinker to hydrate), and this explains the drop in concentration

298

Figure 1: Changing characteristics of conductivity of a suspension of C$_3$S in water (c/s = 4), showing that hydration of tricalcic silicate is accelerated in the presence of condensed silica fume.

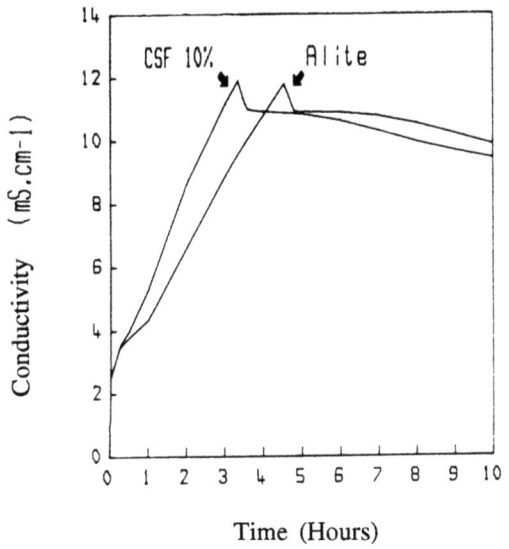

Figure 2: Paste consisting of alite and condensed silica fume. The alteration of portlandite can be seen.

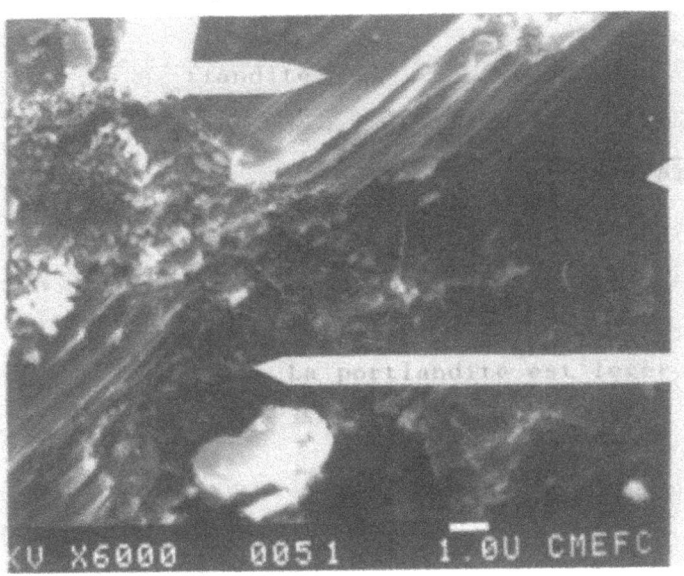

Figure 3: Influence of the w/c ratio on swelling (Québec reactive aggregate), from Bérubé et al., 1989.

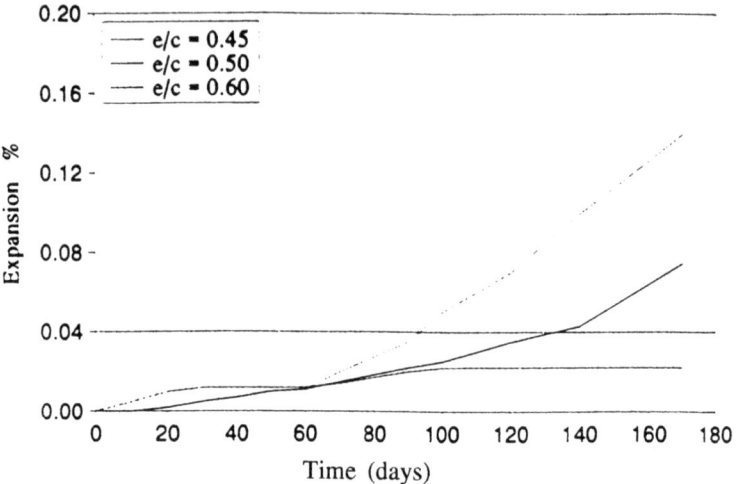

Acnor concrete test CAN3-A23.2-14A

Figure 4: Changing characteristics of OH- concrentration in interstitial fluid of pastes undergoing hydration, from Larbi et al., 1990

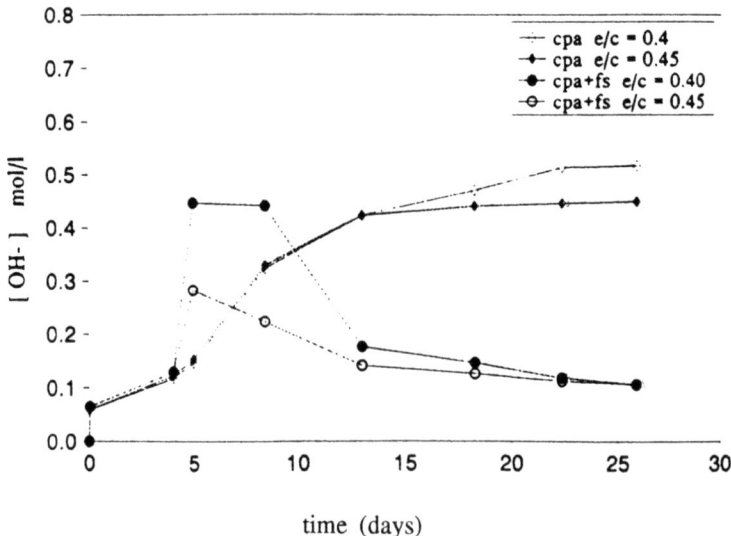

of OH⁻. The solubility of siliceous minerals of aggregates declines with pH, thus reducing the risks of ASR.

However, the partial solubility of alkali pozzolanic materials, sometimes helps to increase significantly the pH of pore fluids, which runs counter to one of the effects sought, the diminution of the pH of the environment. The use of certain additions (p.f.a. but also silica fume) requires a compromise between their contents in alkali cations presumed soluble, and the expected decrease of permeability (see paragraph 3).

3.2 Microstructure
Low water/binder ratios lead to a decline in the permeability of concrete. Furthermore, fine and ultra-fine additions help to increase the compactness through their effect on granular filling. Porosity measurements (Hg) show that the relationship between porosity and permeability of HPCs reuslts from the fact of the pores being discontinuous rather than from their varied size: the pores are indeed much coarser than in traditional concretes [Feldman and Huang (1985), Taylor (1990)]. In the presence of pozzolanic materials, permeability of pastes decreases as the portlandite is consumed.

The possibilities of transporting aqueous species (silicate ions, alkali cations, hydroxyl ions, calcium) required to achieve alkal-silica reactions, are much smaller since the discontinuity of the pores restricts their diffusion in HPCs.

3.3 Conclusion
Pozzolanic materials thus act both on the chemical nature of concrete (composition of interstitial fluid, nature of minerals formed) and on its microstructure (permeability and porosity). This double role is fulfilled to an even greater degree if there is a homogeneous distribution of constituents in the mixture. Depending on the nature and quality of these materials, or of their combination, the reactions will either be noticeable after a few hours only (in the case of silica fume or micro silica), or within the weeks or even months following.

4 Durability of HPCs with regard to ASR: experience and prevention

4.1 References from real-life applications
High performance concretes can be obtained with or whitout pozzolanic additions. The first HPCs were achieved by Freyssinet in 1949. He obtained 50 MPa in 1 hour and 100 MPa in a few days by using mechanical clamping and stoving without any additives.

Generally speaking however, there are few references from real-life applications as far as durability in relation to alkali silica reactions is concerned. This is due on the one hand to a relatively

301

recent introduction of industrial production of HPCs and on the other to the difficulty of reconstituting compositions actually used.

4.2 Test results

There are many more references relating to the properties of HPCs or 'HPMs' (for mortars), with or without pozzolanic materials, during laboratory ageing tests. We are obliged to use accelerated methods (using temperature, or/in enrichment in alkali ions)to be able to judge the effectiveness within reasonable times. Practical test include ASTM C441 (mortar + Pyrex glass), the NBRI test, and ACNOR Can 3A23.2-14A.

As for condensed silica fume, and given the industrial nature of this by-product, unexpected properties have been noted in the literature that could be explained in terms of variations in alkali contents (Bérubé et al., 1989), as well as in terms of variations in the amount (Perry adn Gillott, 1985). On the other hand, using it in circumstances where the quality and regularity of the quantitative analysis are controlled is a good means of preventing such occurrences (Aitcin and Regourd, 1985).

Test relating to concretes or mortars produced using water reducing admixtures - with or without silica fume - are inconclusive as far as the influence of the water/binder ratio is concerned [Figg (1975), Perry and Gillott (1985)].

As for p.f.a., the content in the mix is also critical in this case (30 to 40% is recommended) and depends on the reactive aggregate under consideration [see Bérubé et al. (1989) and Hobbs (1989)]. Partial solubility of alkali cations of pfa can help to increase the pH of the interstitial solution (Nixon and Page, 1978).

4.3 The prevention of reactions in HPCs

This applies only to HPCs containing potentially reactive aggregates.

The efficiency of the performances of additions can vary. Their ability to reduce expansion depends on their nature and composition, fineness, reactivity towards portlandite, and on their alkali oxides content, but also on the reactiviyt of the aggregate whose effect is to be controlled. Each concrete composition should thus be assessed experimentally in a systematic fashion.

Accelerated tests of the ACNOR or NBRI type are preferable to ASTM C441 test when it comes to judging the reactivity of silica fume or pfa.

There is currently no reliable and recognised method enabling the long-term reactivity of a concrete formulation to be judged, and research is currently under way to address this requirement.

To assess the reactivity of pozzolanic additions, we prefer a method enabling binder-aggregate couples to be tested (DMT test unpublished work) to the ASTM C441 method.

5 Conclusion

HPCs combine a number of qualities which must prevent disorders due to alkali aggregate reactions appearing. The low water permeability restricts the mobility of the ions in the pores of the concrete. The application conditions are generally excellent and are highly determinant in terms of the final quality of the material. Finally, when the binder includes an efficient pozzolanic addition (the efficiency being related to its reactivity and to the homogeneity of the mixture), the risks of alkali aggregate reactions are reduced through consumption of the portlandite and lowering of the pH.

References

1. Aitcin PC and Regourd M. (1985): The use of condensed silica fume to control alkali silica reaction. A field case study - Cement and Concrete Research, Vol 15, 711-719.
2. Bérubé M.A., Fournier B., Vezina D., Frenette J., Duchêne J., Choquette M. (1989): Les réactions alcali-granulats dans le béton. Cahier 1, Université Laval, 362 p.
3. Chatterji S., Thaulow N., Fersen A.D. and Christiansen P. (1987): Mechanisms of accelerating effects of NaCl and Ca(OH)2 on alkali silica reaction concrete alkali-aggregate reactions, Proc. 7th Int. Conf. Otawa, ed by P.E. Grattan Bellew, Noyes Publication, 115-119.
4. Davies G. and Oberholster R.E. (1988): Alkali-silica reaction products and their development - Cement and Concrete Research, Vol 18, 621-635.
5. Diamond S. (1989): ASR - Another look at mechanisms in alkali-aggregate reaction, ed by Okada and al., 83-94.
6. Dron R. (1990): Thermodynamique de la réaction alcali-silice. Bull. Liaison Labo. P. et Ch., n° 166, 55-59.
7. Feldman R.F. and Huang C.Y. (1985): Cement and concrete research, Vol. 15, p 765.
8. Figg J.W. (1975): Preliminary appraisal of problem areas and reactive aggregates with appropriate preventive measure - Proceedings of the 2nd Int. Conf. on Concrete Alkali-Aggregate reaction, Reykjavik, 245-258.
9. Glasser F.P. and Marr J. (1984): The effect of mineral additives on the composition of cement pore fluids. British Ceramic Proc. on "The chemistry and chemically related properties of cement", Vol. 36, 419-429.
10. Grattan-Bellew P.E. (1987): Concrete alkali-aggregate reactions. Proceedings of the 7th Int. Conf. 1986, Ottawa, Noyes Publications, 509 p.
11. Hobbs D.W; (1988): Alkali-silica reaction in concrete. Thomas Telford, Londres, 183 p.

12. Hobbs D.W. (1989): Effect of mineral and chemical admixtures on alkali-aggregate reaction. 8th Int. Conf. on Alkali-Aggregate Reaction, ed by Okada et al., 173-186.
13. Jones T.N. (1989): Mechanism of reactions involving british chert and flint aggregates. 8th Int. Conf. on Alkali-Aggregate Reaction, ed by Okada et al., 6 p.
14. Larbi J.A., Fraay A.L. and Bijen J.M. (1990): The chemistry of the pore fluid of silica fume-blended cement systems. Cement and Concrete Research, Vol 20, 506-516.
15. Nixon P. and Page C. (1987): Pore solution chemistry and alkali-aggregate reaction - Proc. of K and B. Bryant Mather Int. Conf. Concrete Durability, ACI (2), SP 100-1833-62.
16. Okada K., Nishibayashi S. and Kawamura M. (1989): Alkali-Aggregate Reaction Proceedings of the 8th Int. Conf. 1989, Kyoto, 857 p.
17. Perry C. and Gillott J.E. (1985): The feasibility of using silica fume to control concrete expansion due to alkali-aggregate reaction - in Durability of Building Materials, Elsevier Science, 133-146.
18. Regourd-Moranville M. (1989): Products of reaction and petrographic examination. 8th Int. Conf. on Alkali-Aggregate Reaction, ed by Okada et al., 6 p.
19. Sellevold E.J. and Nilsen T. (1986): Condensed silica fume in concrete: a world review. Publication Elkem Chemicals, 98 p.
20. Struble L.J. (1987): The influence of cement pore solution on Alkali Silica Reaction, Ph D thesis, Pendue University (NBSIR 87 - 3632).
21. Taylor HFW (1990): Cement Chemistry Academic Press, 475 p.
22. Thomas MDA (1990): A comparison of the properties of OPC and PFA concretes in 30-year-old-mass concrete structures, in "Durability of Building Materials and Components", ed by J.M. Baker, P.J. Nixon, A.J. Majumdar and H. Davies, 383-394.

20 ACCELERATED CARBONATION: COMPARISON BETWEEN THE JOIGNY BRIDGE, HIGH PERFORMANCE CONCRETE AND AN ORDINARY CONCRETE

C. LEVY
Lafarge Coppée Recherche, France

1 Purpose of the study

As part of the National Project on 'New Developements in Concrete', and more precisely, during the construction of the Joigny Bridge using high performance concrete, the Lafarge Coppee Recherche laboratory carried out a study on the carbonation of high performance concrete (HPC).

HPC, in addition to high mechanical strength, exhibits excellent durability. Numerous tests have confirmed this in respect of its resistance to frost, sulfates and as regards the diffusion of chloride ions.

Little is known about the durability of these concretes faced with the problems of carbonation (particularly for HPCs without silica fume). We therefore took advantage of the construction of the Joigny Bridge (built with B 60 concrete delivered by Béton Chantiers Bourgogne - Ciments Lafarge), to comparatively study, by an accelerated method, the advancement of carbonation of two concretes having the mix designs and characteristics (per 1 m³ of dry materials) shown in Table 1:

High Performance Concrete: From material to structure. Edited by Yves Malier. © 1992 Taylor & Francis.
Published by Taylor & Francis, 2 Park Square, Milton Park, Abingdon, Oxon, OX14 4RN. ISBN 0 419 17600 4.

Table 1 : Mix designs and properties

	Mixes in kg/m^3 of dry materials								
Concrete type	Ultra fine sand	0/4 Seine	6/20 Seine	HP Cor-maille	Eau	Mel-ment	Mel-retard	slump	Rc 28 j 16x32
Joigny HPC	105	648	1027	450	160	11.25	4.5	16 cm	64.6 MPa
Ordinary concrete	105	698	1077	350	180	-	-	16.5 cm	41.4 MPa

The concretes were on the one hand the HPC used for pouring the Joigny Bridge deck, and secondly an ordinary concrete (OC) consisting of the same aggregates and the same cement (but at a lower dosage) and without admixtures.

The purpose of this study was to compare the depth and speed of carbonation (which depends on the aptitude to air penetration) in these concretes. It should be noted that the control concrete ("Ordinary concrete") is representative of concretes currently in use in structures and that it is thus interesting to put it in competition with a B 60 (these results obtained on the Joigny construction site are in fact better than those obtained in the laboratory on the occasion of this study).

2 Operating method

Initially, our task was to develop equipment and an operating method capable of causing accelerated carbonation of concretes. This is because, although the chemical reaction of carbonation is relatively fast in itself, the in-depth progress of this phenomenon is rather slow and particularly when the density of the concrete increases.

With this in mind, we created an accelerated carbonation enclosure with forced carbon dioxide circulation at constant hygrometry (carbonation is optimum between 50 and 70 % relative humidity) in which we placed the concrete test pieces. Dr. Piguet has shown that the acceleration in time of a test of this sort is equal to the ratio of the CO_2 concentration between the test conditons and the normal atmospheric conditions. In this text:

$$\frac{\text{concentration } CO_2 \text{ pure}}{\text{concentration } CO_2 \text{ in the air}} = 3000$$

Theoretically the holding time of 36 days in the enclosure should thus correspond to 300 years of natural carbonation.

2.1 Equipment
We used a 200 l PVC tank fitted with a cover closing hermetically. The carbon dioxide, supplied in bottles under pressure is introduced into the tank after passing through a wash bottle (containing ethyleneglycol). The same device is installed at the output from the tank. At the bottom of this enclosure, there was a basin containing a saturated solution of ammonium sulfate first to keep the relative humidity roughly constant. A small fan (with timer and potentiometer) provided uniformity of the environment. A hole was made in the tank in order to check the ambient humidity in it. Three levels of removeable grills enabled a maximum of space to be used in the tank and 72 cylinders to be positioned. See the photos of the installation in the Appendix.

2.2 Concrete test pieces
The concretes were made in the laboratory in accordance with standards in an Eirich mixer (in batches of 40 l). In the case of HPC, the superplasticizer was introduced in two parts (as on the Joigny Bridge site):

- 0.5 % of the weight of cement in the mixing water
- 2 % in the mixer after homogenization of the materials.

We made concrete cylindrical samples in steel molds having a diameter of 11 cm and a height of 22 cm. After 24 hours, they were removed from the molds, and placed in the moist chamber (20°C, 100% RH). After the determined storage time (2 or 28 days) the test pieces were sawn in the middle and thus gave two small cylinders (diam=h=11 cm). They were then placed in the accelerated carbonation enclosure for a certain duration (e.g. 36 days). Then, a splitting test was used to separate each cylinder into two samples (half cylinder diam=h=11 cm). For each test (concrete type, type and duration of storage, carbonation duration), two 11 x 22 cm samples were made, which gave 8 test pieces per test.
A 1 % solution of phenolphthalein was sprayed onto the surfaces of the eight samples. The carbonation depth was then measured 1/2 hour afterwards, on the four sides of the 11 cm x 11 cm square. A fixer was then sprayed in order to take photos later, if required.

3 Tests and results

3.1 Comparative test OC/HPC
We compared the carbonation depths of the two concretes, after :

* normalized storage at 20°C and 100 % RH for 28 days
* then various hold times in the accelerated carbonation tank of 9,18,27,36 and 72 days.

All the carbonation depth readings are given in the tables in the Appendix, which also include :

- the splitting strengths of the 11 x 11 cm test piece.
- the carbonation depth, specifying on each of the four sides :

* the form of attack by carbonation
* the mean dx
* the maximum dx max

The curves in Fig.1 well illustrate the very clear difference between the risk and the speed of carbonation of ordinary concrete OC and that of high performance concrete.

Figure 1 : Carbonation of HPC and OC

| Depth of carbonation dx | | dx OC
dx max OC
dx HPC |

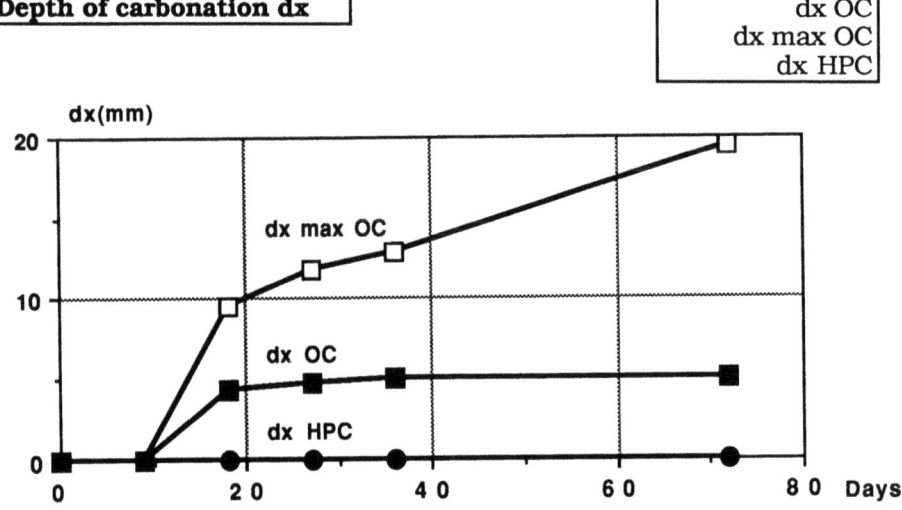

In fact, even after 72 days of accelerated carbonation, there was no progress made by carbonation in the HPC whereas the ordinary concrete was carbonated over several millimeters as from eighteen days. The photos in the Appendix clearly show these phenomena. It should be noted that carbonation is virtually nil for the two concretes after 9 days holding in the accelerated carbonation tank. This may be explained by the rather high relative humidity at the beginning (80-90% RH), which made the penetration of the CO_2

molecules difficult. Several days development was necessary to stabilize the relative humidity in the tank around 65% RH.

We observed that even after 7 days of accelerated carbonation, the HPC was not carbonated except on the leveled surface: the upper trowelled part of the 11 x 22 test piece. This zone consists of cement laitance, rich in water, and thus less compact and more sensitive to the advance of carbonation.

3.2 Supplementary studies

From the basic study (28 days curing in a moist room at 20°C and 100 % RH, split test pieces, use of phenolphthalein) we varied certain parameters. The table of the results from these tests and the photos of samples are in the Appendix .

a) Storage in a damp room for 2 days instead of 28 days

The value of storing concrete test pieces in a damp room is, in addition to getting closer to the compression test standards, to carry out carbonation tests on well matured concrete that have already reached the major part of their intrinsic strengths (at an infinite age). This is more coherent as part as an accelerated test. On the other hand, after 28 days in a damp room, the first millimeters of the external layer of the test piece are saturated with water and diminish the carbon dioxide's aptitude for penetration. In addition, this is further away from reality (shuttering removed on site 1 to 2 days after pouring the concrete).

This is why we carried out tests only leaving the test pieces (OC and HPC) for two days in a damp room and then 36 days in the accelerated carbonation tank. The difference between the two concretes shows up still further (dx = carbonation depth), see Table 2. It is noted that the HPC still does not carbonate (except on the trowelled surface).

Table 2 : Carbonation after storage for 2 and 28 days

Conservation in damp room 20°C - 100 % HR	Concrete type	dx (mm)	dx max (mm)
28 days	O.C.	5	12.9
	HPC.	0	0
2 days	OC	8.3	19.8
	HPC	0	0

b) Storage for 28 days in a dry room at 20°C and 50% instead of the damp room 20°C and 100% RH

This test was capable of working with sufficiently old concretes, for which the penetration of carbon dioxide is easier than when the concrete is stored in a humid room. This is why the carbonation depth measurements (see table in the Appendix) are greater for the two concretes, after 36 days of accelerated carbonation, see Table 3.

Table 3 : Carbonation after storage in dry conditions for 28 days

Storage in damp room 20°C - 100 % HR	Concrete type	dx (mm)	dx max (mm)
humid (100 % RH)	OC	5	12.9
	HPC.	0	0
dry (50 % RH)	OC	8.7	22.9
	HPC	1.6	1.6

It is important to note that, in this case, the carbonation depth in the HPC concrete is not zero. It will be observed that the carbonation depth is the same as in the case of storage in a damp room for the leveled surfaces (1 mm) and the base (0 mm), but has increased (3 mm) on the cylindrical face (sides 2 and 4 in the tables in the Appendix). The results obtained are even so quite in favor of HPC (see photo in the Appendix). The two following photos in the same appendix clearly show the influence of relative humidity in the place where the concrete test pieces underwent curing.

3.3 Test for optimizing the operating method

A set of 8 samples of HPC (28 days in damp room +36 days of carbonation) was obtained by sawing, and then drying with compressed air. The results are the same as those obtained by splitting. Reading the results is a little easier, but preparation of the test pieces is less so.

To show up the progress of the carbonation (thus nearer the core of the concrete), we sprayed a solution of thymolphthalein onto a set of 8 HPC test pieces (28 days in damp room + 36 days of carbonation). Unfortunately, the grey blue color of the thymol is virtually impossible to see on concrete. Similarly, a test on two samples of OC (28 days in damp room +72 days carbonation) was carried out with Alizarin yellow without success.

Different methods of preparing the test pieces before reading were tested:

* reading 30 minutes after spraying with phenolphtalein
* reading 1 hour after spraying with phenolphthalein
* reading 24 hours after spraying with phenolphthalein
* spraying of distilled water after splitting, then spraying of phenolphthalein and reading 30 minutes afterwards.

It results from this that 30 minutes wait is enough to read the carbonation depth under good conditions. Spraying distilled water improves this reading.

4 Conclusions

A reliable operating method was developed to characterize the durability of concretes with respect to carbonation. We were able to update the influence of certain parameters on the results of these accelerated carbonation tests.

It appears clearly that durability with respect to carbonation of high performance concretes (here Rc 28 = approx. 65 MPa) is very substantially greater than that of an ordinary structural concrete (here Rc 28 = approx. 40 MPa). The penetration of the molecules of carbon dioxide is virtually nil in the case of the HPC, a concrete whose compactness is much better than that of a more current concrete.

Such a difference is particularly surprising (and satisfactory !) in that it was an HPC whose performance is not exceptional and which was manufactured industrially for the construction of the Joigny Bridge.

References

1. Piguet. Bulletin du Ciment, Switzerland Ang. 1988.

Appendix

Figure 2: Test set up

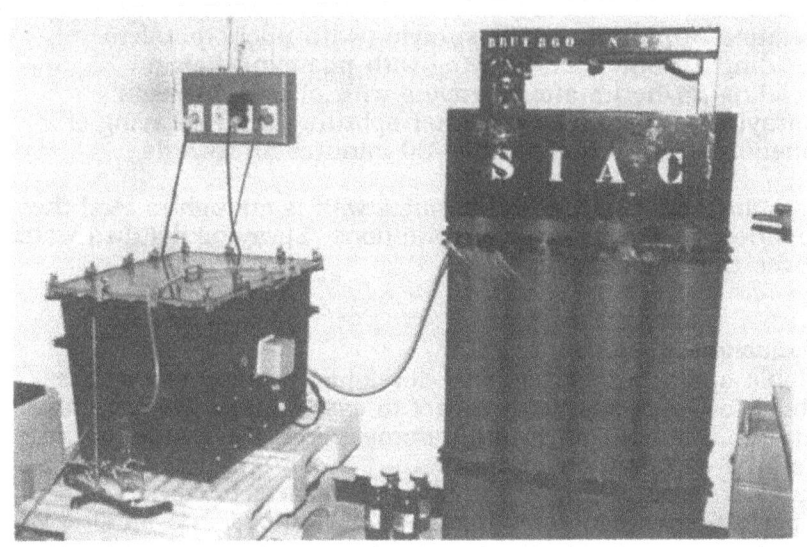

Figure 3: Forms of progress of carbonation from RILEM recommendation

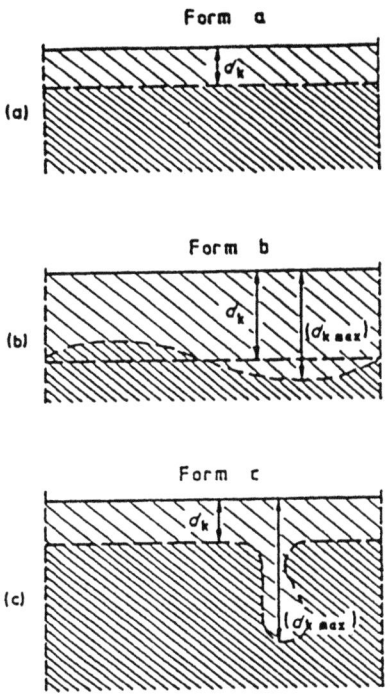

Figure 4: Numbering of sides

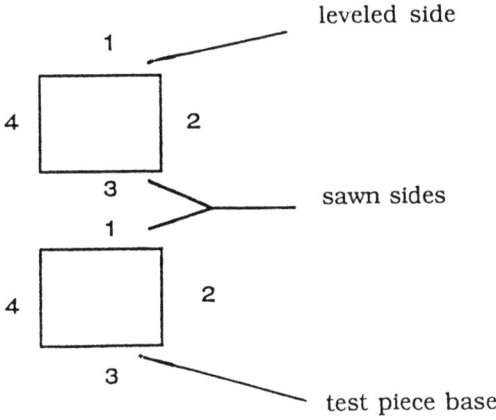

Table 4: Comparison between carbonation of HPC and OC stored for 28 days damp room + 9 days carbonation

splitting (kN)	height (mm)	Rf (Mpa)	Form	dx 1	2	3	4	m	dx max. 1	2	3	4	m	concrete type	Observations
76	105	4.2	a	1	0	0	0	0	1	0	0	0	0	OC	Carbonation on leveled face
			a	1	0	0	0	0	1	0	0	0	0		
72	106	3.95		0	0	0	0	0	0	0	0	0	0	OC	
				0	0	0	0	0	0	0	0	0	0		
84	105	4.6	a	1	0	0	0	0	1	0	0	0	0	OC	Carbonation on leveled face
			a	1	0	0	0	0	1	0	0	0	0		
78	107	4.2		0	0	0	0	0	0	0	0	0	0	OC	
				0	0	0	0	0	0	0	0	0	0		
mean OC		4.2				0						0			
100	107	5.4	a	1	0	0	0	0	1	0	0	0	0	HPC	Carbonation on leveled face
			a	1	0	0	0	0	1	0	0	0	0		
90	106	4.9		0	0	0	0	0	0	0	0	0	0	HPC	
				0	0	0	0	0	0	0	0	0	0		
96	106	5.25	a	1	0	0	0	0	1	0	0	0	0	HPC	Carbonation on leveled face
			a	1	0	0	0	0	1	0	0	0	0		
88	110	4.65		0	0	0	0	0	0	0	0	0	0	HPC	
				0	0	0	0	0	0	0	0	0	0		
mean HPC		5.1				0						0			

Table 5: Comparison between carbonation of HPC and OC stored for 28 days 100% RH + 18 days carbonation

splitting (kN)	height (mm)	Rf (Mpa)	Form	CARBONATION (mm)										concrete type	Observations
				dx					dx max.						
				1	2	3	4	m	1	2	3	4	m		
82	107	4.43	b	4	6	5	5	5	8	12	9	11	10	OC	dx : 4.5 mm
			b	4	7	4	5	5	9	12	9	13	9		
76	108	4.07	b	2	4	4	5	3.7	8	10	9	11	9.5	OC	dx max : 9.5 mm
			b	3	5	4	6	4.5	8	9	9	9	8.7		
74	106	4.04	b	4	5	4	5	4.5	9	11	8	12	10	OC	dx : 4.3 mm
			b	3	5	4	5	4.2	8	9	9	10	10		
74	106	4.04	b	3	5	4	5	4.2	8	10	8	10	9.5	OC	dx max : 9.4 mm
			b	4	5	3	5	4.2	8	11	8	9	9		
mean OC		4.2						4.4					9.5		
86	106	4.7	a	1	0	0	0	0	1	0	0	0	0	HPC	Carbonation on leveled face
			a	1	0	0	0	0	1	0	0	0	0		
102	106	5.57		0	0	0	0	0	0	0	0	0	0	HPC	
				0	0	0	0	0	0	0	0	0	0		
94	108	5.04	a	1	0	0	0	0	1	0	0	0	0	HPC	Carbonation on leveled face
			a	1	0	0	0	0	1	0	0	0	0		
96	105	5.29		0	0	0	0	0	0	0	0	0	0	HPC	
				0	0	0	0	0	0	0	0	0	0		
mean HPC		5.2						0					0		

Table 6: Comparison between carbonation of HPC and OC stored for 28 days 100% RH +27 days carbonation

splitting (kN)	height (mm)	Rf (Mpa)	Form	CARBONATION (mm)										concrete type	Observations
				dx					dx max.						
				1	2	3	4	m	1	2	3	4	m		
68	106	3.7	b	3	6	5	6	5	8	14	12	16	12.5	OC	dx : 4.9 mm
			b	3	5	4	6	4.5	9	13	13	15	12.5		
78	107	4.2	b	4	6	4	6	5	7	15	12	17	12.7	OC	dx max : 12.6 mm
			b	4	5	5	6	5	8	15	13	16	13		
80	106	4.4	b	4	5	3	7	4.7	7	13	9	12	10.2	OC	dx : 4.5 mm
			b	3	5	4	5	4.2	9	13	12	14	12		
78	107	4.2	b	4	6	4	6	5	9	15	12	15	12	OC	dx max : 10.9 mm
			b	2	6	3	6	4.2	7	11	10	12	9.7		
mean OC		4.1						4.7					11.8		
84	107	4.5	a	1	0	0	0	0	1	0	0	0	0	HPC	Carbonation on leveled face
			a	1	0	0	0	0	1	0	0	0	0		
100	104	5.6		0	0	0	0	0	0	0	0	0	0	HPC	
				0	0	0	0	0	0	0	0	0	0		
84	107	5.2	a	1	0	0	0	0	1	0	0	0	0	HPC	Carbonation on leveled face
			a	1	0	0	0	0	1	0	0	0	0		
96	106	4.5		0	0	0	0	0	0	0	0	0	0	HPC	
				0	0	0	0	0	0	0	0	0	0		
mean OC		4.95						0					0		

Table 7: Comparison between carbonation of HPC and OC stored for 28 days 100% RH +36 days carbonation

splitting (kN)	height (mm)	Rf (Mpa)	Form	CARBONATION (mm) dx					dx max.					concrete type	Observations
				1	2	3	4	m	1	2	3	4	m		
78	106	4.26	c	7	11	6	11	8.7	13	25	12	18	17	OC	dx : 6.1 mm
			c	3	6	3	5	4.2	11	16	5	13	11.2		
78	106	4.26	c	5	9	3	11	7	12	19	5	23	14.7	OC	dx max : 13.7 mm
			c	4	6	3	6	4.7	10	12	11	16	12.2		
90	107	4.87	c	4	6	3	7	5.0	12	12	8	22	13.5	OC	dx : 3.9 mm
			c	3	5	4	5	4.2	9	24	14	16	15.7		
86	107	4.65	c	5	2	1	5	3.2	12	14	1	13	10	OC	dx max : 12 mm
			c	3	6	1	3	3.2	12	12	3	9	9		
mean OC		4.5						5					12.9		
102	108	5.46	a	1	0	0	0	0	1	0	0	0	0	HPC	Carbonation on leveled face
			a	1	0	0	0	0	1	0	0	0	0		
90	107	4.87	-	0	0	0	0	0	0	0	0	0	0	HPC	
			-	0	0	0	0	0	0	0	0	0	0		
90	107	4.87	a	1	0	0	0	0	1	0	0	0	0	HPC	Carbonation on leveled face
			a	1	0	0	0	0	1	0	0	0	0		
90	107	4.87	-	0	0	0	0	0	0	0	0	0	0	HPC	
			-	0	0	0	0	0	0	0	0	0	0		
mean HPC		5						0					0		

Table 8: Comparison between carbonation of HPC and OC stored for 28 days 100% RH + 72 days carbonation

splitting (kN)	height (mm)	Rf (Mpa)	Form	dx					dx max.					concrete type	Observations
				1	2	3	4	m	1	2	3	4	m		
77	107	4.1												OC	Test with Alysarine yellow not valid
76	105	4.2	c	7	3	3	7	5.0	17	19	18	17	17.8	OC	dx : 5.2 mm dx max : 19.3 mm
			c	7	5	2	7	5.3	21	28	15	19	20.8		
85	106	4.7	c	6	4	3	7	5.0	16	19	18	18	17.8	OC	dx : 5.1 mm
			c	7	7	2	5	5.3	21	28	15	20	21.0		
86	107	4.7	c	3	10	3	5	5.3	16	33	9	18	19.0	OC	dx max : 19.6 mm
			c	2	6	2	9	4.8	11	18	20	33	20.5		
mean OC		4.4						5.1					19.5		
98	107	5.3	a	1	0	0	0	0	1	0	0	0	0	HPC	Carbonation on leveled face
				0	0	0	0	0	0	0	0	0	0		
94	106	5.1		0	0	0	0	0	0	0	0	0	0	HPC	
				0	0	0	0	0	0	0	0	0	0		
108	108	5.8	a	1	0	0	0	0	1	0	0	0	0	HPC	Carbonation on leveled face
				0	0	0	0	0	0	0	0	0	0		
101	106	5.5		0	0	0	0	0	0	0	0	0	0	HPC	
				0	0	0	0	m	0	0	0	0	m		
mean HPC		5.4						0					0		

Table 9: Comparison between carbonation of HPC and OC stored for 2 days 100% RH + 36 days carbonation

splitting (kN)	height (mm)	Rf (Mpa)	Form	dx 1	2	3	4	m	dx max. 1	2	3	4	m	concrete type	Observations
66	105	3.6	c	9	13	14	13	12.2	12	16	12	15	13.7	OC	dx : 8.5 mm
			c	7	9	6	8	7.5	11	16	8	13	12		
80	108	4.3	c	6	9	7	10	8	10	25	21	33	22.2	OC	dx max : 17.7 mm
			c	6	7	5	8	6.5	12	32	21	27	23		
80	107	4.3	c	5	7	6	9	6.7	12	29	10	31	20.5	OC	dx : 8 mm
			c	6	8	5	9	7	13	34	13	32	23		
75	106	4.0	c	5	4	3	6	4.5	12	32	16	29	22.2	OC	dx max : 21.8 mm
			c	4	5	3	8	5	15	28	12	31	21.5		
mean OC		4						8.3					19.8		
88	107	4.8	a	1	0	0	0	0	1	0	0	0	0	HPC	Carbonation on leveled face
			a	1	0	0	0	0	1	0	0	0	0		
84	105	4.6		0	0	0	0	0	0	0	0	0	0	HPC	
				0	0	0	0	0	0	0	0	0	0		
84	107	4.5	a	1	0	0	0	0	1	0	0	0	0	HPC	Carbonation on leveled face
			a	1	0	0	0	0	1	0	0	0	0		
82	107	4.4		0	0	0	0	0	0	0	0	0	0	HPC	
				0	0	0	0	0	0	0	0	0	0		
mean HPC		4.6						0					0		

Table 10: Comparison between carbonation of HPC and OC stored for 28 days 50% RH + 36 days carbonation

splitting (kN)	height (mm)	Rf (Mpa)	Form	CARBONATION (mm) dx					dx max.					concrete type	Observations
				1	2	3	4	m	1	2	3	4	m		
78	107	4.22	c	10	15	9	23	14	26	22	12	29	22.2	OC	dx : 9.1 mm
			c	7	15	5	22	12	23	26	15	27	22.7		
78	106	4.26	c	3	7	4	5	4.7	14	15	12	12	13.2	OC	dx max : 17.7 mm
			c	3	7	4	8	5.5	14	12	12	14	13.0		
94	108	5.04	c	6	19	6	8	9.7	37	37	13	33	30	OC	dx : 8.2 mm
			c	6	18	7	9	10	35	38	15	35	30.7		
92	107	4.98	c	3	7	3	10	5.7	25	26	18	35	26	OC	dx max : 28 mm
			c	4	6	5	15	7.5	28	24	17	33	25.5		
mean OC		4.7						8.7					22.9		
94	108	5.04	a	1	3	0	3	1.7	1	3	0	3	1.7	HPC	dx : 1.6 mm
			a	1	3	0	3	1.7	1	3	0	3	1.7		
96	106	5.25	a	0	3	0	3	1.5	0	3	0	3	1.5	HPC	dx max : 1.6 mm
			a	0	3	0	3	1.5	0	3	0	3	1.5		
104	106	5.68	a	1	3	0	3	1.7	1	3	0	3	1.7	HPC	dx : 1.6 mm
			a	1	3	0	3	1.7	1	3	0	3	1.7		
102	106	5.57	a	0	3	0	3	1.5	0	3	0	3	1.5	HPC	dx max : 1.6 mm
			a	0	3	0	3	1.5	0	3	0	3	1.5		
mean HPC		5.4						1.6					1.6		

320

Figure 5: Comparison HPC - OC 28 days in damp room
9 and 18 days of accelerated carbonation

A - Storage 28 days 20°, 100% RH + 9 days carbonation HPC OC

B - Storage 28 days 20°, 100% RH + 18 days carbonation HPC OC

Figure 6: Comparison HPC - OC 28 days in damp room
27 and 36 days of accelerated carbonation

C - Storage 28 days 20°, 100% RH + 27 days carbonation HPC OC

D - Storage 28 days 20°, 100% RH + 36 days carbonation HPC OC

Figure 7: Comparison HPC - OC 28 days in damp room
72 days of accelerated carbonation

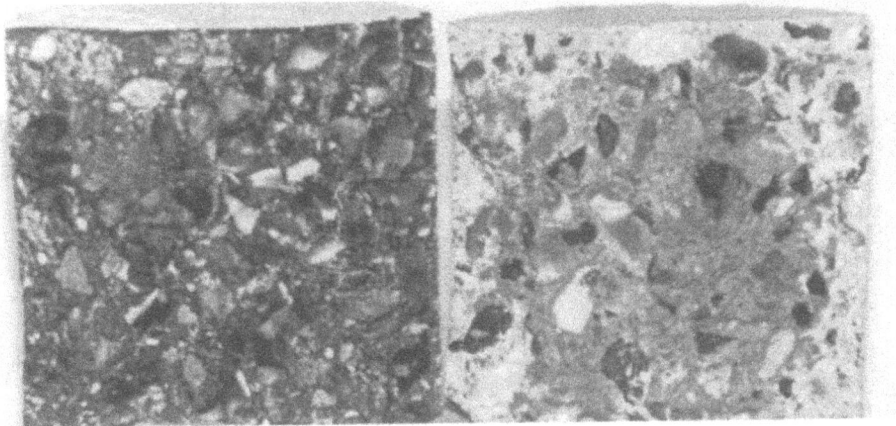

E - Storage 28 days 20°, 100% RH + 72 days carbonation HPC OC

Figure 8: Comparison HPC - OC 36 days in damp room
2 days of accelerated carbonation, 28 days in dry room

F - Storage 2 days 20°, 100% RH + 36 days carbonation HPC OC

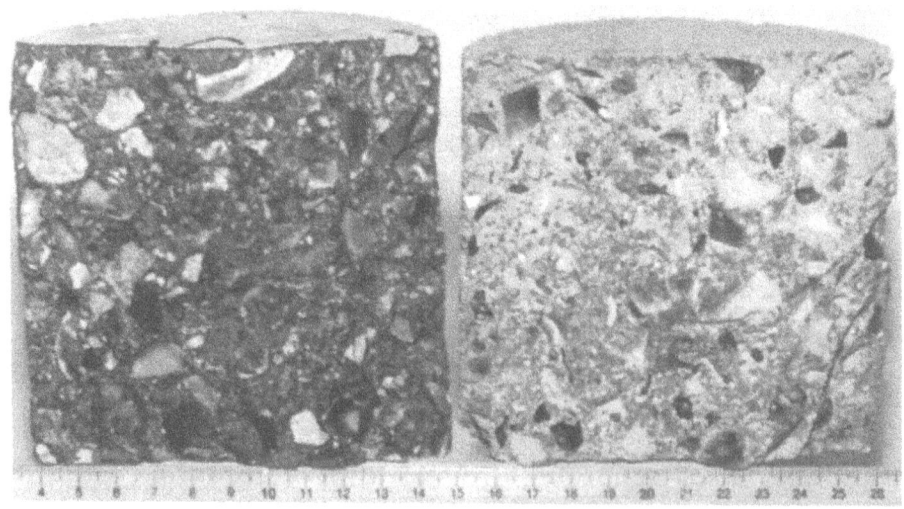

G - Storage 28 days 20°, 50% RH + 36 days carbonation HPC OC

Figure 9: Comparison in damp room - dry room - 36 days
carbonation - OC - HPC

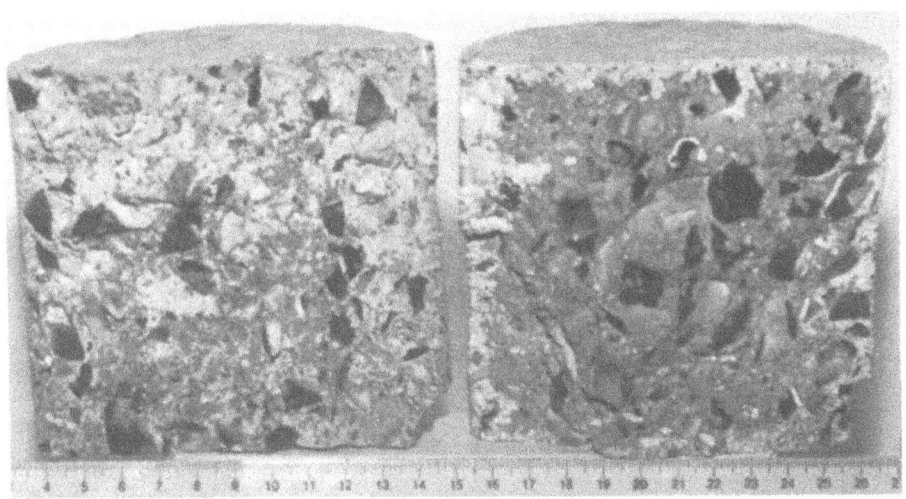

H - OC - Carbonation 36 days
28 days, 50% RH/storage 28 days, 100% RH

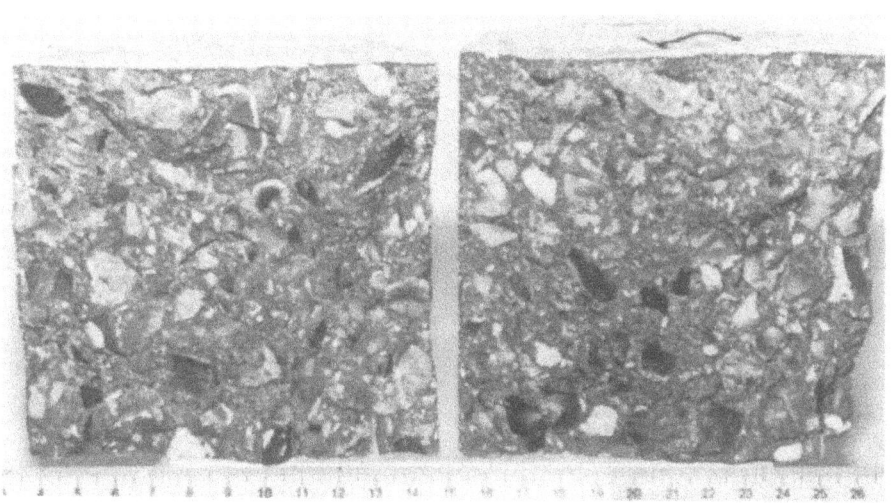

I - HPC - Carbonation 36 days
28 days, 100% RH/storage 28 days, 50% RH

PART THREE
CONSTRUCTION

21 THE FRENCH NATIONAL PROJECT 'NEW WAYS FOR CONCRETE' — OBJECTIVES AND METHODOLOGY

Y. MALIER
ENS Cachan, France

1 Origins

The origins of the French Project are to be found in the studies conducted by the national steering committee for research in civil engineering, (Conseil d'Orientation de la Recherche en Génie Civil, chaired by Jean Chapon), which resulted in the publication of the 1985 white paper on scientific and technical perspectives in civil engineering.

The project is implemented within the framework of the civil engineering research programme conducted under the joint supervision of the ministerial departments in charge of public works and of research.

2 One objective and four main ideas

The objective of the project is adequately defined by its title : it is to explore and develop new ways for concrete. In view of this objective, the following four main ideas have been defined:

2.1 To ensure the best possible continuity between fundamental research, applied research, experimental structures, codes and development. In the field of new developments for concrete, we think that the construction of experimental structures which really test the results of research is an absolute prerequisite to all development actions.

2.2 To promote the idea of closely associating the concepts of 'new materials', 'new designs' and 'new construction methods'.

The present situation has led to a paradox: the quality of the design studies made by engineers has led to dimensioning of structures which uses traditional concrete with a fairly remarkable optimisation level.

The consequence, one might almost say, the paradoxical result, is that, beside this propitious optimization, there has been too little

High Performance Concrete: From material to structure. Edited by Yves Malier. © 1992 Taylor & Francis.
Published by Taylor & Francis, 2 Park Square, Milton Park, Abingdon, Oxon, OX14 4RN. ISBN 0 419 17600 4.

apparent interest in questioning the traditional properties of concrete. Hence the reluctance so often expressed towards high performance concretes.

Our purpose is to contribute to reversing such a trend. New devices (external prestressing), new reinforcements (composites, fibres), new shapes (thin shells, triangulated and mixed structures), new production methods (pushing, components) may enable us to rethink partially the design of structures with respect to the new properties of materials and the evolution of the resulting methods of fabrication.

2.3 To highlight the notion of high performance', showing clearly that for various 'new developments', high or very high compressive strength, other important characteristics may be improved, like ductility, water and gas tightness, density, abrasion and impact resistance, resistance to frost, or in specific cases, for example the capacity to be pumped for long distances, to be projected, etc.

2.4 To stress the fact that the best methods of using high performance concrete corresponds most often to the notion of improved durability, a notion which is more and more essential and measurable for the owner.

3 The conditions necessary for the project

The following conditions in particular may be mentioned:

3.1 The collaboration of owners with a nationwide dimension, able to open up possibilities for the study and realization of innovative experimental structures and to accept the constraints specific to such operations (specific studies, follow-up,etc).

3.2 The effective participation of contracting firms, materials suppliers and civil engineering laboratories in the Project.

3.3 The acceptance by each partner of the sharing of research results (while naturally respecting confidentiality) and of the necessity to coordinate with the other members of the project an important part of its activity in the field of 'New ways for concrete'.

3.4 The implementation of complementary applied research on objectives adapted to the specificities of experimental structures, funded directly by the Project.

3.5 The implementation of instrumentation and follow-up programmes for these structures in the course of time.

3.6 The planning of accompanying actions: international relations, engineer and technician education and training, publications, etc.

4 The main themes

4.1 At the beginning, in 1986
We organized the themes with the help of a matrix of representation in which the columns represented the experimental structures (bridges offshore structures, roads, slender products, components, drainage structures, structures subject to shocks, etc) and the lines represented the finalised research necessary to these experiments, conducted in the continuation of the research programmes proposed to each partner (HS, HB and VHS concrete, HS/light coupling, metal fibres, glass fibres, HS components assembly, durability, etc).

4.2 Current situation
We have of course never claimed to be exhaustive in the field of new ways for concrete and, in the course of negotiations with the owners, the initial matrix evolved, as we wanted, and as was necessary, into a tighter one.

Thus the "Direction des Routes" (Road Direction) of the Ministry in charge of public works and the EDF (French Electricity Board) were the first owners to resolutely commit themselves to the National Project "New Ways for Concrete".

The construction of the Joigny experimental bridge and the preliminary studies on the Civeaux nuclear plant were the first experimental operations.

Other actions are currently in the course of study or already realised or advanced:

- prestressed structures by pre-stressing (VHP),
- university building (VHP),
- bridge (VHP),
- airport runways (fibres and HP),
- reinforced drainage networks (fibres and HP).

5 The partners

A list on 1st December 1990 would include, among others :

- owners (Direction des Routes, Ville de Paris, EDF, Aéroport de Paris),
- firms in the materials industry (CF, ELF, Saint-Gobain, Lafarge),
- contracting firms (SOGEA, SPIE, PPB, GTM, Bouygues, Freyssinet, Campenon Bernard),
- laboratories (LCPC, INSAL, INSAT, CEBTP, CERIB, CSTB, ENSC),

- professional bodies and institutions (FNTP, SNBATI),
- scientific associations (AFPC, AFREM).

6 The administrative organization

We want it to be as light as possible. All partners have signed the National Project Charter. An Executive Committee, chaired by Roger Lacroix, was set up, comprising all the partners and meeting twice a year, as well as a Technical Committee, now chaired by Lucien Pliskin, which together with Yves Malier as Director of the Project, determine the scientific policy of the National Project.The project is scheduled to continue until 1st January 1992.

7 Conclusion

We wish to say how, in the course of building experimental structures, an atmosphere of fruitful exchange has developed within the National Project between engineers and researchers from both the private and public sectors, an atmosphere which, we are sure, will be most beneficial to construction in the future.

22 FROM MATERIAL TO STRUCTURE

P. RICHARD
Groupe Bouygues, Direction Scientifique, Paris, France

1 Introduction

The debate is dominated to-day as it will be for years to come by the economic choice between steel and concrete or more precisely, between steel and prestressed or reinforced concrete, the performance of concrete alone not being comparable with those of steel.

New materials are neither technologically nor economically ready to replace conventional materials in the construction industry.

The choice between the two materials is actually an economic one, except when either is required for its intrinsic properties, concrete for foundations or piers, steel for bridge spans longer than 400 or 500 meters.

Efforts are made to combine steel and prestressed or reinforced concrete; an example will be given later. Except for some simple cases of short-span bridges, these efforts come up against two difficulties : expensive traditionalism with respect to design of connections between steel and concrete elements on the one hand and cooperation problems between the concrete and steel industries on the other hand.

There are however some experimental projects which in my opinion, are far from the mark from the economic viewpoint. When the two materials oppose each other, the general structural design is always determined by factors other than the material. The function of the structure will be a determining factor. The bridge is a typical example, its general structure depending on the function to be fulfilled: roadbridge, railway bridge or road-rail bridge. The type of crossing is then taken into account : maritime valley, or land and river valley for instance.

Moreover, environment has an impact on the erection method and the latter can affect or even determine the structural design. Though the economic result does not lead necessarily to the same span lengths for steel or concrete, the general design of the structure is often identical.

High Performance Concrete: From material to structure. Edited by Yves Malier. © 1992 Taylor & Francis.
Published by Taylor & Francis, 2 Park Square, Milton Park, Abingdon, Oxon, OX14 4RN. ISBN 0 419 17600 4.

The structure's length, directly linked to environment, will be a determining factor. Investment in erection equipment, in the case of a very long bridge for instance, will have little impact on the cost of the structure. These means may thus be sophisticated, allow high erection rates, resulting in the reduction of the site fixed costs. This investment will affect the concrete-steel economic choice only slightly; concrete, which provides heavier structures, will result in more expensive equipment but the extra cost will be of little consequence if the structure is long.

The cross section of a bridge can be either a solid web or three-dimensional truss box girder or beams, both for steel and concrete.

Only at the production and sometimes erection technology levels do the differences become apparent.

These statements are obvious, but I deem it necessary to recall them because if performances of materials are determining from the economic viewpoint and if improving them can be justified, it is worth while to add the price of a structure is also linked to many other factors; to-day, thanks to technological efforts for rational use of materials, structural designs encountered are the same, whether made of steel or concrete. Post-tensioning, irrespective of its theoretical aspects, provides elegant and excellent solutions to assembly which have endowed concrete with as much versatility as steel.

2 Case study

To illustrate my remarks, I will now present various solutions contemplated for the competitive bid for construction of a major structure in Denmark. It is an exceptional structure due to its length (6 km) and its rail-road function. It is maritime crossing with a great water depth over its whole length, equal to or greater than 7 m. The ground, well consolidated by glaciations, poses no problem.

We note on the figures that three solutions were contemplated by the Consultant: steel box girders with 105-m spans (Figure 1), prestressed concrete box girders with 105-m spans (Figure 2) and three-dimensional steel trusses allowing 144-m spans (Figure 3). It should be noted that the geometry of the prestressed box girder structure is not very clever since it combines relatively short spans (105 meters) with very wide piers, and is thus expensive with respect to dredging, foundations and structural reinforced concrete (Figure 4).

One of the competitors offered an alternative solution which, by joining the two road viaducts, allowed a decrease in pier size on the one hand and by using variable inertia box girders, enabled 120-m spans without increasing the deck cost on the other hand.

Figure 1

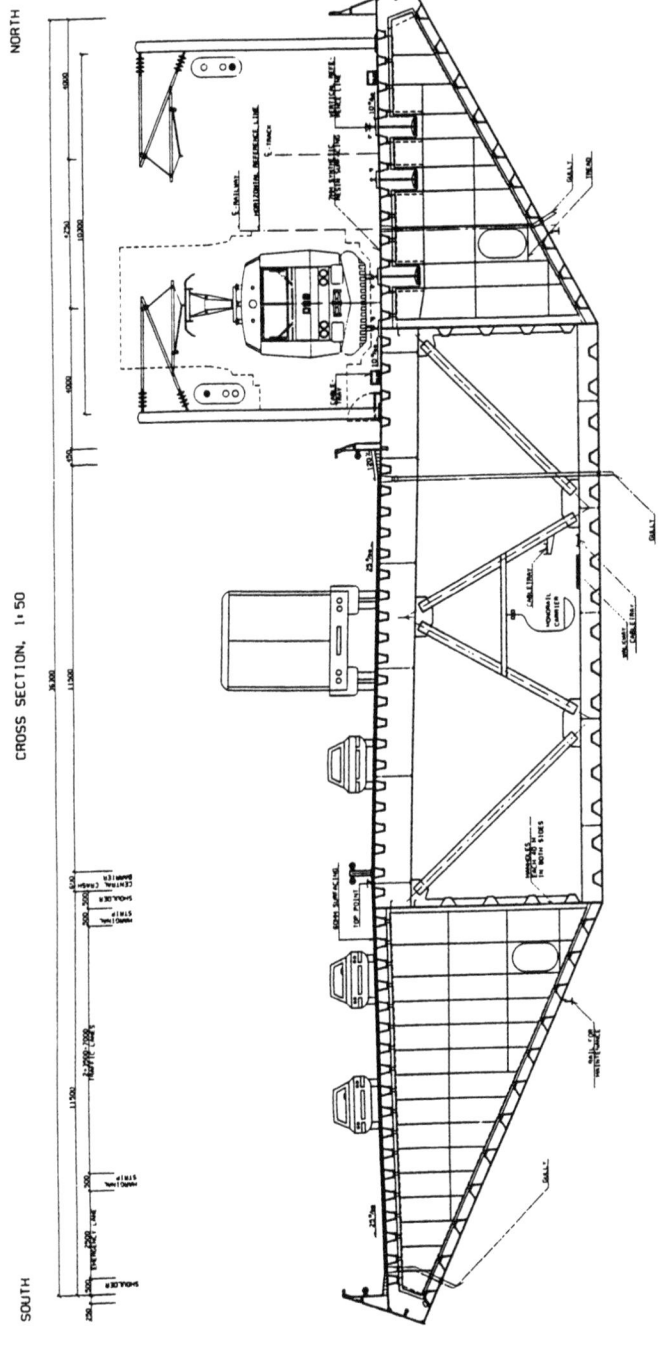

CROSS SECTION, 1:50

NORTH

SOUTH

Figure 2

CROSS SECTION

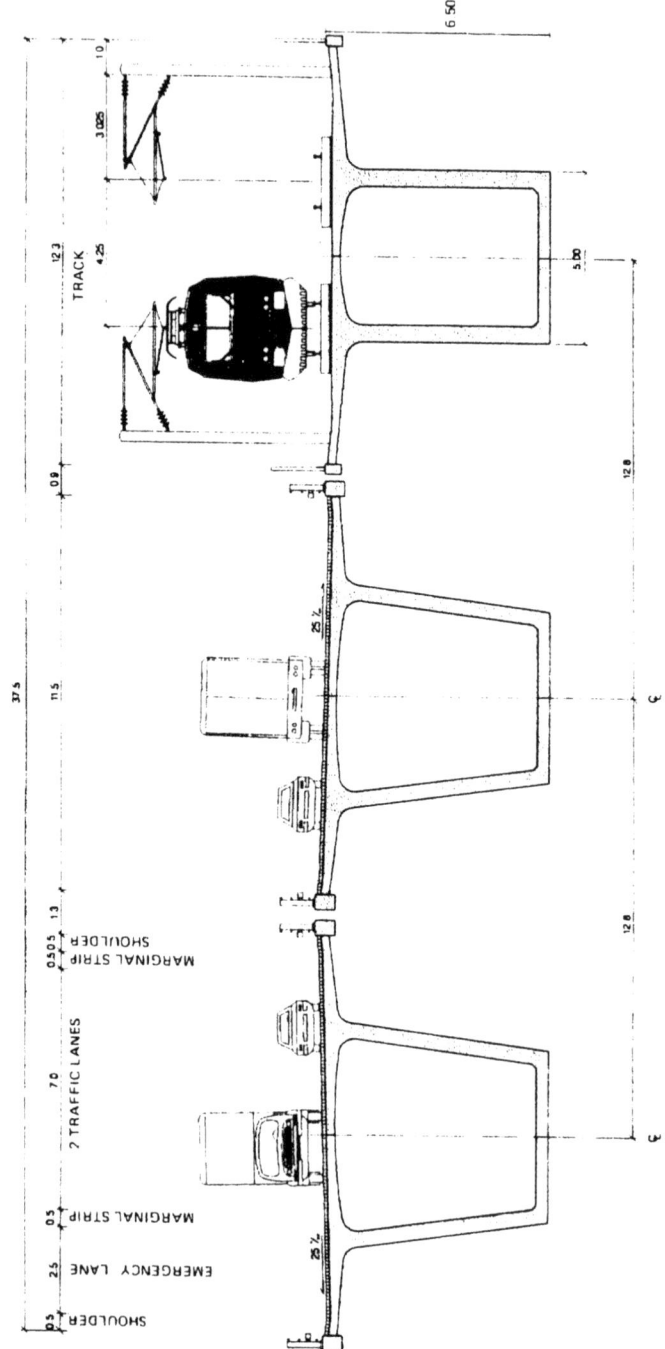

Figure 3

CROSS SECTION

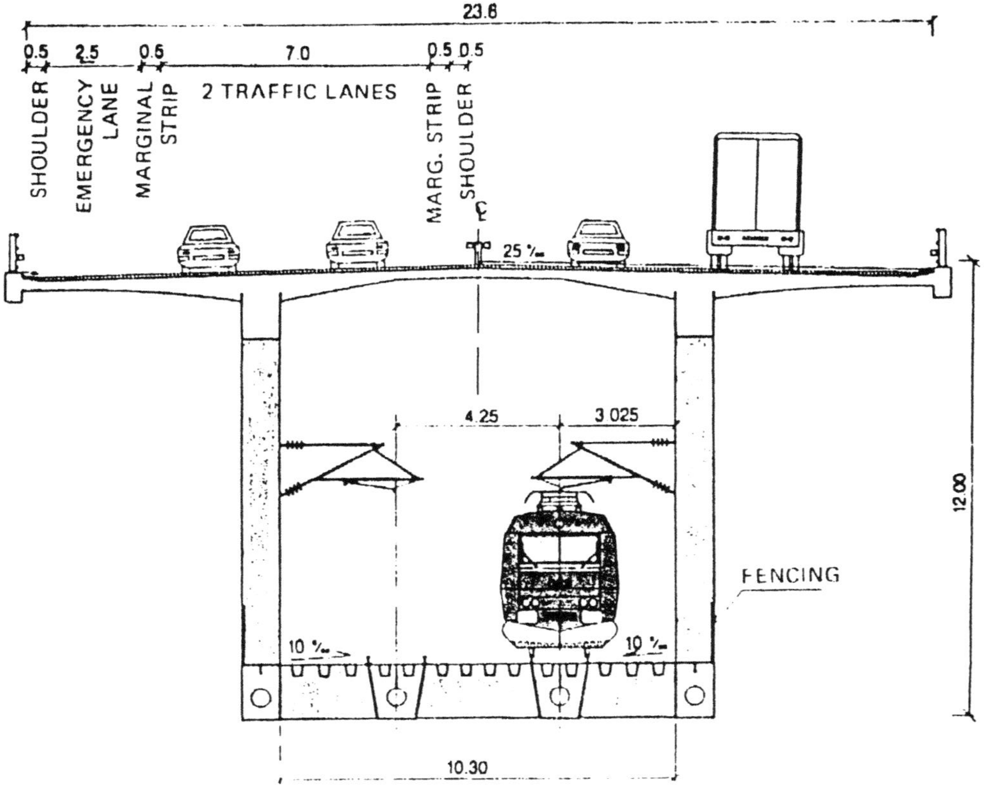

Figure 4

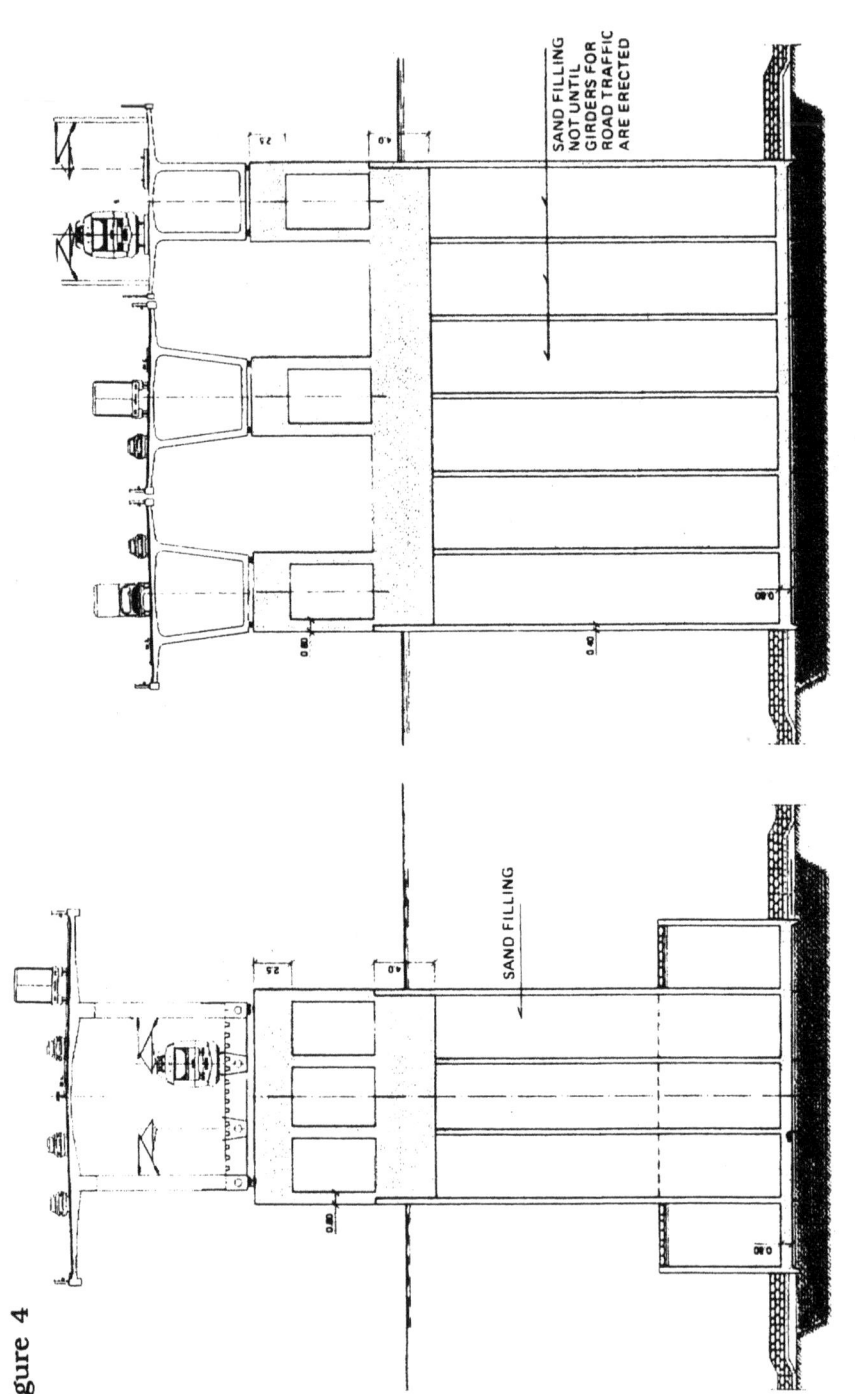

CROSS SECTION IN PRINCIPLE FOR
COMPOSITE AND STEEL SOLUTION.

CROSS SECTION IN PRINCIPLE
FOR CONCRETE SOLUTION.

Finally, we proposed the three-dimensional concrete truss solution whose general geometry is identical to that of the three-dimensional composite steel truss, having 144-m spans as well (Figures 5,6,7).

We thus note that different material lead to solutions of the same type, determined by environment. Moreover, it should be added that all of the bidders based their design on erection of entire spans, a method made necessary by a fully maritime crossing with sufficient draft and a bridge length such that mobilization of heavy erection equipment is justified (Figures 8 and 9).

The difference related to the nature of the material can be seen in the general designs since the design of a steel box girder is of course, very different from that of a concrete box girder. This explains the two single webs for the three-dimensional composite truss and two V-shaped double webs for the three-dimensional concrete truss.

Now, we will try to analyse these differences between steel and concrete box girders. As the strength of steel is about twelve times greater than that of concrete, the steel section strictly necessary with respect to longitudinal bending of the bridge, will be of lesser thickness.

Unlike concrete, steel thicknesses are unable to take transversal bending forces. It is thus necessary to create a cross beam system, diaphragms and a third beam system, from diaphragm to diaphragm for localized longitudinal bending. Moreover, instabilities under compression or shear forces require stiffeners. All these secondary functions entail considerable expense in terms of materials. Disregarding the road function and taking into account equivalent live loads, it can be shown that the quantity of steel necessary is reduced by more than half. So the quantity of steel is as though its strength were only six times greater than that of concrete.

As steel density is three times greater than that of concrete, the steel box girder deck should thus be half the weight of the concrete deck. But experience shows that the concrete deck is three to four times heavier because the concrete box girder also has its drawbacks, in particular it does not make optimum use of web concrete.

Lastly, to conclude, let us examine prices. A ton of steel for a box girder costs about FF 20 000, a ton of prestressed concrete about FF 3 000, i.e. 6 to 7 times cheaper than steel. Prestressed concrete box girders, though made of a material having lower strength than steel, will thus be more economical, their geometry being more adapted to the functional requirements.

These simple considerations show how functional constraints can make a material uncompetitive, despite its greater intrinsic properties. In the case involved, the steel box girder is unsuited to compete with the concrete girder. A geometry whose main advantage will be to combine functions is required. Roads and

Figure 5

CROSS SECTION SCALE 1/80

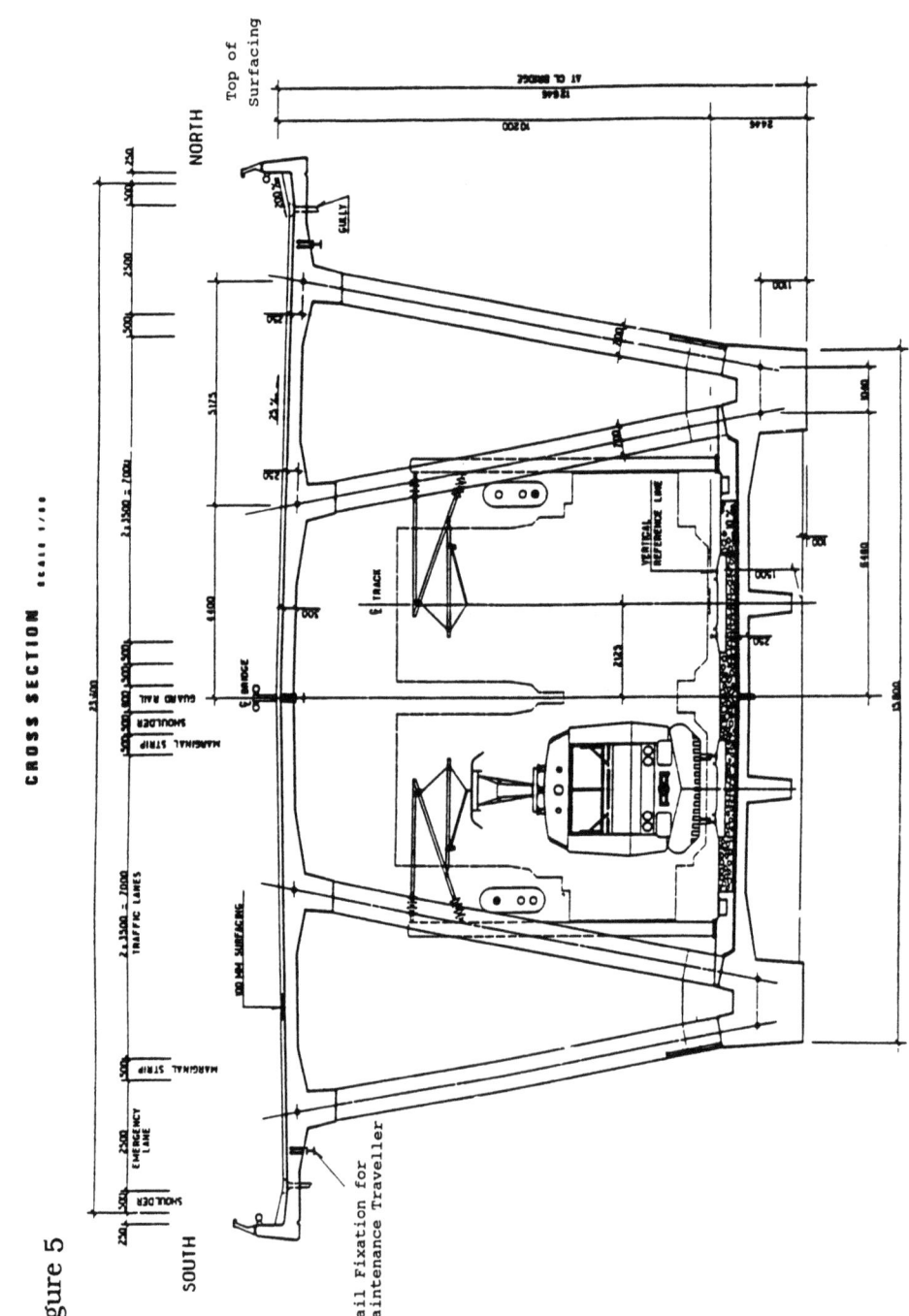

Rail Fixation for
Maintenance Traveller

Figure 6

SPAN : 144 m   ALTERNATIVE
number : 42

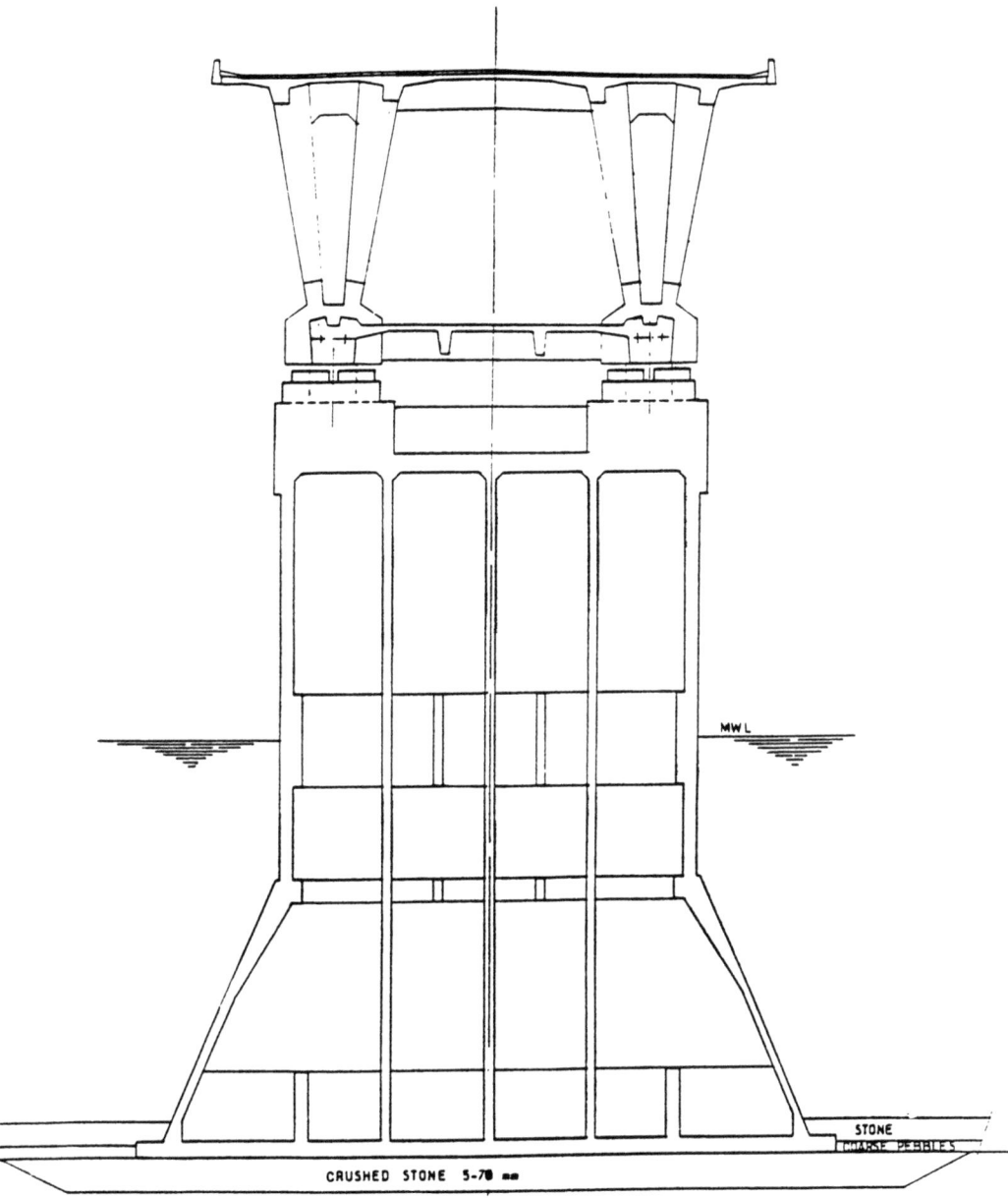

Figure 7

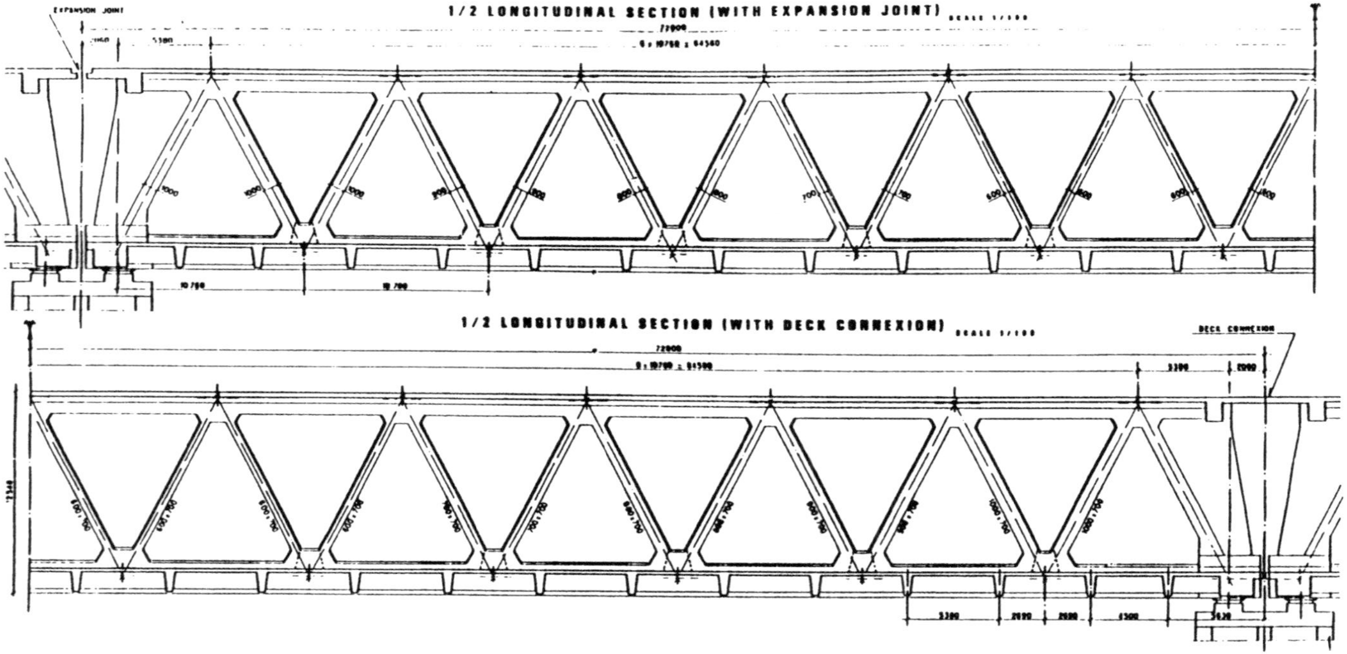
1/2 LONGITUDINAL SECTION (WITH EXPANSION JOINT)
EXPANSION JOINT
1/2 LONGITUDINAL SECTION (WITH DECK CONNEXION)
DECK CONNEXION

DECK ERECTION

Figure 8

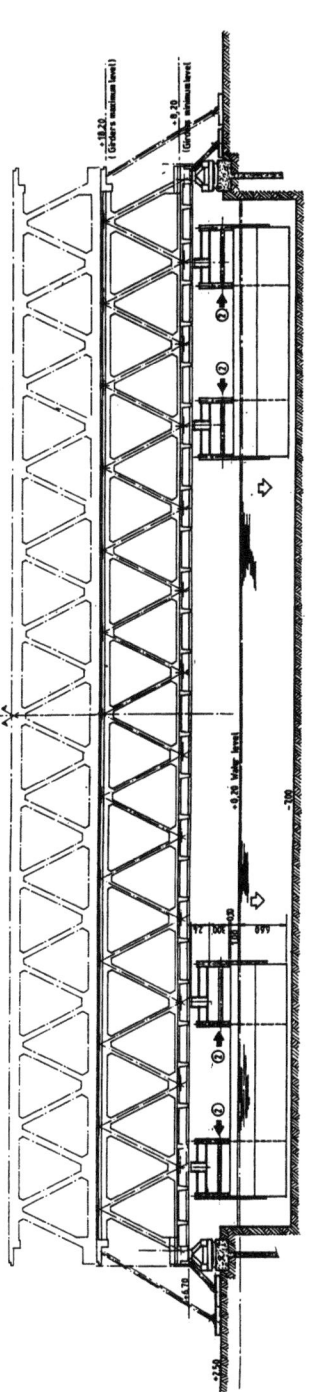

Figure 9

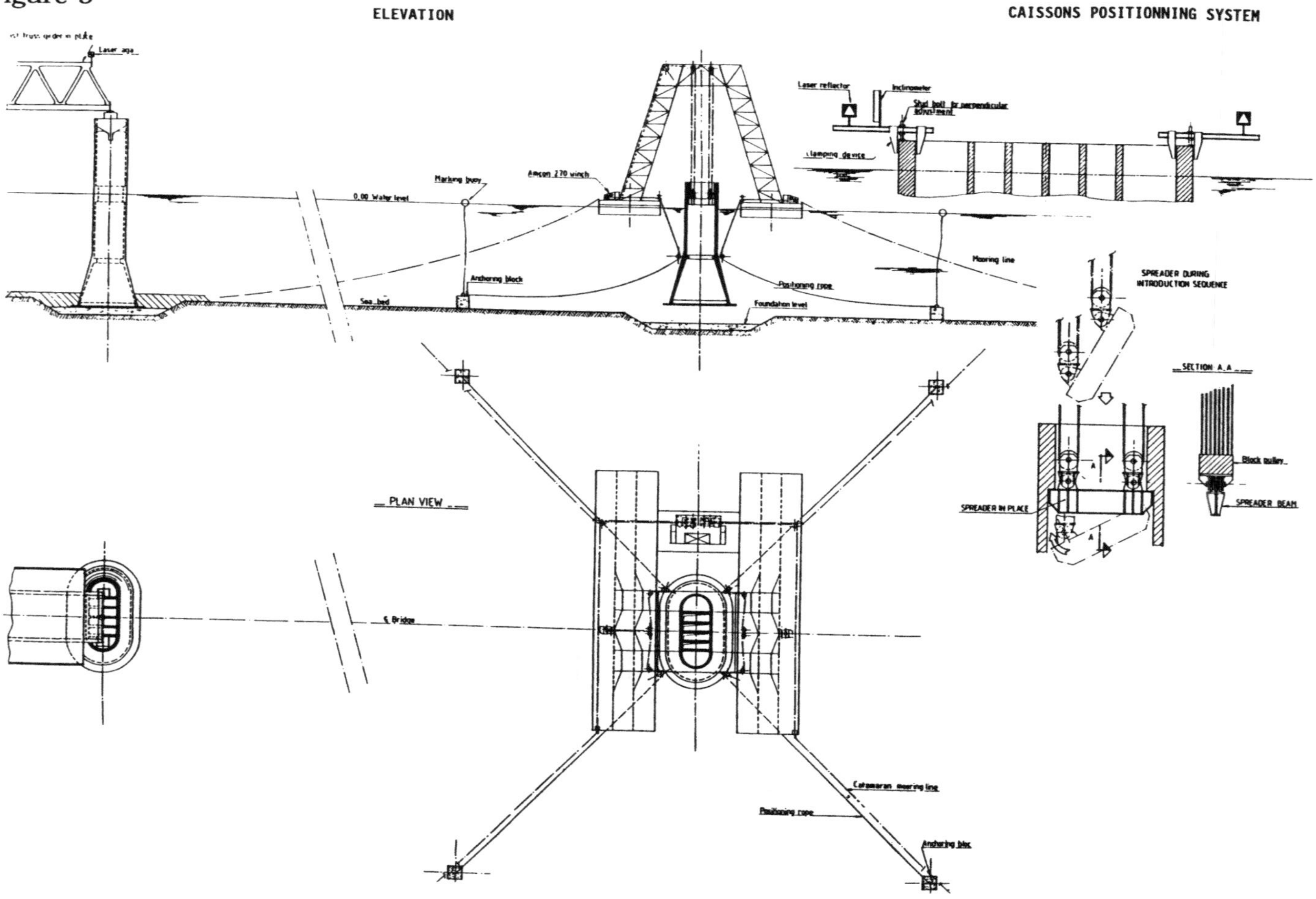
ELEVATION
CAISSONS POSITIONNING SYSTEM
Laser reflector
Inclinometer
Clamping device
Marking buoy
Amcon 270 winch
0.00 Water level
Anchoring block
Sea bed
Mooring line
Positioning rope
Foundation level
SPREADER DURING INTRODUCTION SEQUENCE
SECTION A.A
Block pulley
SPREADER BEAM
SPREADER IN PLACE
PLAN VIEW
Catamaran mooring line
Positioning rope
Anchoring bloc

railway tracks could thus be superposed, the train running through the box girder, the height of which should be increased.

We have also seen how inadequate the compressed steel sheet of a steel box girder is: it will thus be replaced by concrete which, despite its greater weight, easily takes longitudinal compression and transversal bending loads. Lastly, solid webs will be replaced by triangulation whose elements are less affected by instability problems. The increase in height of the box girder, to comply with functional requirements, will require longer spans, which, in this case, results in savings on piers and foundations, given the high-quality of the ground. These options, in the logic of our previous analysis, actually make the composite solution just as competitive as concrete box girders.

We can carry on with this analysis and contemplate replacing steel triangulation by concrete triangulation suited to the requirements of this technique, V-shaped webs, which also reduce transverse bending of the upper slab. What can be expected ?

Both steel and concrete are properly used. Given the strength ratio of 12 and the density ratio of 3, concrete triangulation will be about four times heavier that its steel counterpart.

The price of steel (around 18 000 F/ton) will tend to be decisive. Conversely, the price of concrete, which has to be worked more, is likely to increase (it will be about 4 000 FF/ton). The price ratio in favor of concrete will be slightly above 4. The outcome is thus uncertain and, given the great number of parameters, a detailed study will be necessary to show an advantage of 5 to 6% in favor of concrete with characteristic compressive strength of 50 MPa.

Since, as stated above, triangulation of webs allows concrete strength to be used to its best advantage and element sections are large, it is logical to go further and examine the gain achieved using high strength concretes. In this case, we have chosen a 30% increase, i.e. concrete with characteristic strength of 65 MPa.

The advantage of the concrete solution is thus increased to approx. 23%, i.e. savings of 16 to 17%. Despite the fact that we already note a reduction in the theoretical strength gain due to construction technological constraints, the gain remains considerable. In this particular case, concrete with characteristic strength of 80 MPa would have improved performance even further. Beyond this value, the technological constraints would have limited the increase in competitiveness.

It should be noted that, in usual conditions, the technological constraints will make strengths above 65-70 MPa useless for construction of bridge decks.

This analysis must be repeated for structures which do not have the same function.

These findings do not condemn very high strength concretes. They merely show that our present technologies, even the advanced ones like concrete triangulations, must progress if we want to use them to their best advantage.

3 Conclusion

1. High strength concretes can improve competitiveness of prestressed concrete structures.

The above analysis, quickly developed here, shows a particularly significant case.

2. The extra cost of high strength concretes, ranging from 100 to 150 FF/m^3, is often offset by the savings on pouring and reduced risk of poor workmanship.

Improved durability of works, unquestionable whatever the use of highstrength concrete, is achieved practically at no cost.

23 HIGH PERFORMANCE CONCRETE IN TUNNEL LININGS

J-M. BROCHERIEUX
SPIE Batignolles, France

1 Introduction

Generally, tunnels are important structures, often in severe aggressive environments and they need high strength and high durability.

We have to consider the actual evolution of tunnelling using mechanical boring machines combined with a prefabricated lining in concrete. So, we have to design an economical concrete suitable for an industrial process and allowing optimal prefabrication.

To achieve these performances (strength, durability, prefabrication) we have to make an optimal mix concrete with the following principles:

- reduced w/c ratio,
- minimum cement content,
- workability suitable for prefabrication with a high range water reducer,
- eventualy add an admixture like silica fume + high range water reducer.

We now present two actual examples showing how the required performance was achieved.

2 TGV Atlantique: Villejust runway tunnels, Paris

The twin runway tunnels south of Paris with their precast lining, required the casting and the erection of 50,670 concrete segments (Fig. 1 and 2) and thus the placing of 110,000 m³ of grade 30 MPa concrete of which the average compressive strength was 57 MPa with a standard deviation of 4.2 MPa which corresponds to a grade 50 MPa concrete.

The mix was designed by Béton Labo Services for a stiff consistency (10-30 mm slump) to be demoulded 9 hours after

High Performance Concrete: From material to structure. Edited by Yves Malier. © 1992 Taylor & Francis.
Published by Taylor & Francis, 2 Park Square, Milton Park, Abingdon, Oxon, OX14 4RN. ISBN 0 419 17600 4.

casting. To reach a minimum strength of 10 MPa so early necessitated a heat treatment, which was tested by SATM in their Chambéry laboratory.

The concrete leaving the batching plant at ordinary temperature was heated up to 40° C in a 1.5 m³ skip by electrodes immersed in it. The mix in kg/m³ was:

- Portland cement from Cantin plant	CPA 55	350
- Siliceous sand from CSS Etampes	0/0.3 mm	185
- River sand from CSS Villeneuve	0/4 mm	585
- Gravel from CSS Villeneuve :	4/12.5 mm	476
	12.5/22 mm	700
- Free water		154
- High range water reducing admixture :		
- Durciplast (Chryso)		4.2

The water/cement ratio, (154 + 0.70.4.2) / 350 = 0.45, ensured an excellent durability to these reinforced concrete segments.

Testing carried out by the designer showed that the long term strength of the mix was lessened by the heat treatment although it was rather moderate and although it was carried before placing concrete:

- untreated concrete, 20° 7/28/90 day strength= 44.2/58.7/ 62.8 MPa
- treated concrete, 40°C 7/28/90 day strength   = 39.4/49.2/ 55.5 MPa

In January/February 1988 10 experimental segments were cast on a programme sponsored by the Joint Venture : Spie-Batingolles, SOGEA, Franki. Four segments with various cement contents (from 300, 350 to 400 kg/m³) and without heat treatment and six segments with the addition of silica fume in the ratio SF/Portland cement from 0.06 to 0.08 ; silica fume from the Pechiney plant of Laudun (Gard) was used. In Table 1 are recorded the mix proportions of the six segments, from which 4 were cored at 28 days, and 2 others were used to test the effect of heat treatment (5M untreated, 6M treated).

In Table 1 the 28 day results are recorded. Table 2 gives the total porosity measured on the part of the core nearest to the formed surface of the segment (the intrados): it was very low but, with or without silica fume, of the same order of magnitude. Microstructural analysis and water penetration test would have been necessary to distinguish them.

Hand finishing of the extrados was much easier with the SF mixes, but a number of bug holes appeared and with the limited number of segments cast, it was not possible to find a remedy to this defect.

Table 1 : TGV Atlantique, Villejust tunnels. Results of the testing of 2 reference segments 1T and 2T and four MS concrete segments 1M (S/C = 0.078), 2M (S/C = 0.082), 5M and 6M (S/C = 0.061).

	C kg/m³	S kg/m³	W l/m³	Adj C+S	W/C+S	Fc MPa	R28 MPa	Rc MPa	R90 MPa
1T	347		150	1.07	0.43		57.6	50.2	70.3
2T	397		158	1.2	0.40		67	66.4	72.5
5M	326	20	148	1.4	0.43	61.5	77.3		86.1
1M	320	25	148	1.13	0.43	62.4	62.6	64.5	75.0
2M	365	30	162	1.11	0.41	65.9	63.7	71.0	75.3
6M	326	20	150	1.40	0.43		68.3		71.5

R28 : compressive strength of 16x32 cm control cylinders
R_C : compressive strength of specimens of length/diameter ratio 2.0 cut from 150 mm diameter cores
W : effective water content, total water on saturated surface dry aggregates
F_C : compressive strength calculated with the formula of Féret-Larrard with R cement = 61 MPa, deducted from the results on reference concretes 1T, 4T, 7T, 2T with the aggregate efficiency coefficient G = 0.55.

As well as for the companion segments, heat treated SF segments had long term strengths lessened, and in the present programme the difference was probably increased because the untreated segment was cast with a 13° C mix and so its 28 day cylinder strength was exceptionally high : 77.3 MPa. The 6M segment cast with concrete preheated to 40° C in the skip reached only 68.3 MPa at 28 days.

Table 2 : TGV Atlantique, Villejust Tunnels, Porosity test results

Specimen	Porosity (volume) (%)		Apparent dry density t/m³
1 T	10.3 -----		2.48
	| ----- 9.8		
2 T	9.35 -----		2.51
1 M	9.1 ------		2.41
	| ----- 9.0		
2 M	8.9 ------		2.40
3 H	9.7		2.46

From this programme, it was concluded that, particularly in aggressive soils, the SF concrete would permit lower costs of the precast lining, more than would be expected considering the weight reduction of the 60 MPa segments. To achieve this, it would be necessary that the minimum cement content specified by the present rules be lowered taking into account a W/C + k S ratio instead of a W/C ratio ignoring the addition of silica fume and particularly when W/C + kS < 0.45. With such a water/cement ratio, the minimum cement content for aggressive exposures required by ENV 206, is only 300 kg/m³ of Portland cement compared with 385 kg/m³ with the present French codes.

3 Conclusion

What can we deduce from the reported research and test results ?

a) For precast linings of grade 40 to 70 MPa, silica fume is noted mandatory: its cost is high in France, in consequence it is very seldom used and batching plants are not equipped to use this addition with the best efficiency. So it is less costly in the Paris area to produce a grade 60 MPa with a high strength Portland cement (CPA HP) and a high range water reducing admixture only.

b) In China, using very high cement contents (600 kg/m³ and more) and powerful water reducing admixtures, grade 51 to 60 MPa concretes are currently used for cast in situ railway bridge decks with the aggregates found in the vicinity.

c) However, for the strength grade considered high strength but not very high strength concretes (i.e; 80 to 100 MPa), in the French terminology of de Larrard HP not VHP concretes, durability criteria would give preference to MS mixes when chemical aggression is to be feared (sea water, sulphates) or alkali-silica reaction. With an S/C ratio less than 0.15 ; it seems that the decrease of the pH of pore water related to the partial consumption of calcium hydroxide, should not be feared from the reinforcement protection against corrosion point of view, as the denseness of the cement paste is greatly increased.

d) In order that innovation be not stopped in France, it appears necessary that the additions of pozzolans such as fly ash or silica fume, or even limestone fillers be no longer ignored by the codes, but that their cementitious effect be taken into account. These complements to the basic Portland cement content are practically a must for very high strength mixes in order to control the heat development during the first days of hardening. For HP mixes these additions are also useful particularly in aggressive environments.

4 The Channel Tunnel

4.1. Description
The Channel Tunnel comprises 3 parallel tubes (Fig. 1)

- two running tunnels, inside diameter = 7.60 m.
- one service tunnel, internal diameter = 4.80 m.

The service tunnel is connected to the running tunnels every 375 m by way of a pair of pedestrian cross passages. Those tunnels have an internal diameter of 3.30 m. Morever, the running tunnels are connected directly by 2.00 m internal diameter pressure relief ducts. Their role is purely aerodynamic. The total length is 51 km, comprising:

- 4 km under land on the French side,
- 10 km under land on the UK side,
- 37 km under the sea.

Three TBM's (Tunnel Boring Machines) on the French side and 3 TBM's on the UK side worked their way 40 m under the sea bed from their respective sides, at Sangatte (near Calais) and at Dover, in order to meet mid-way in 1990.

Figure 1a: Cross passage

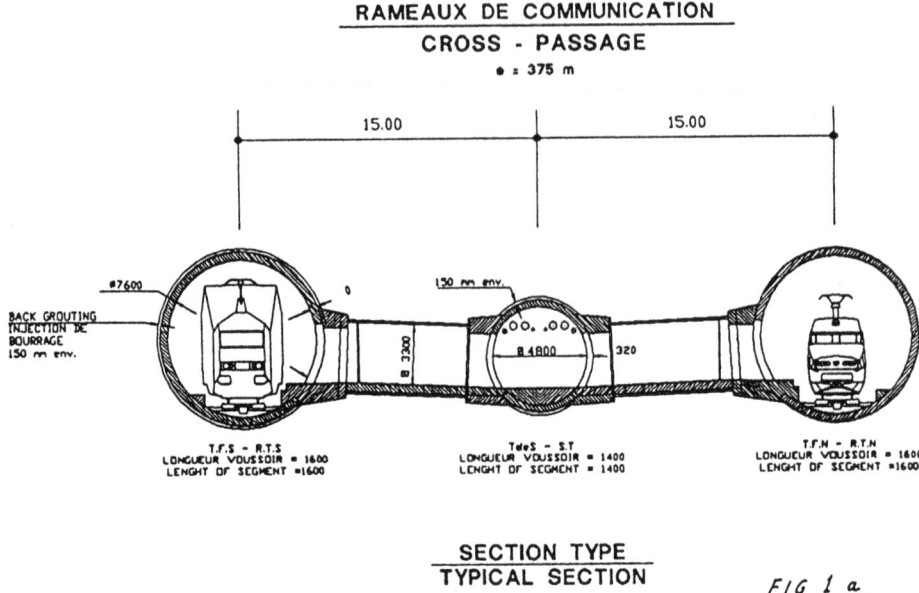

RAMEAUX DE COMMUNICATION
CROSS - PASSAGE
● : 375 m

SECTION TYPE
TYPICAL SECTION

FIG 1 a

Figure 1b: Pressure relief duct

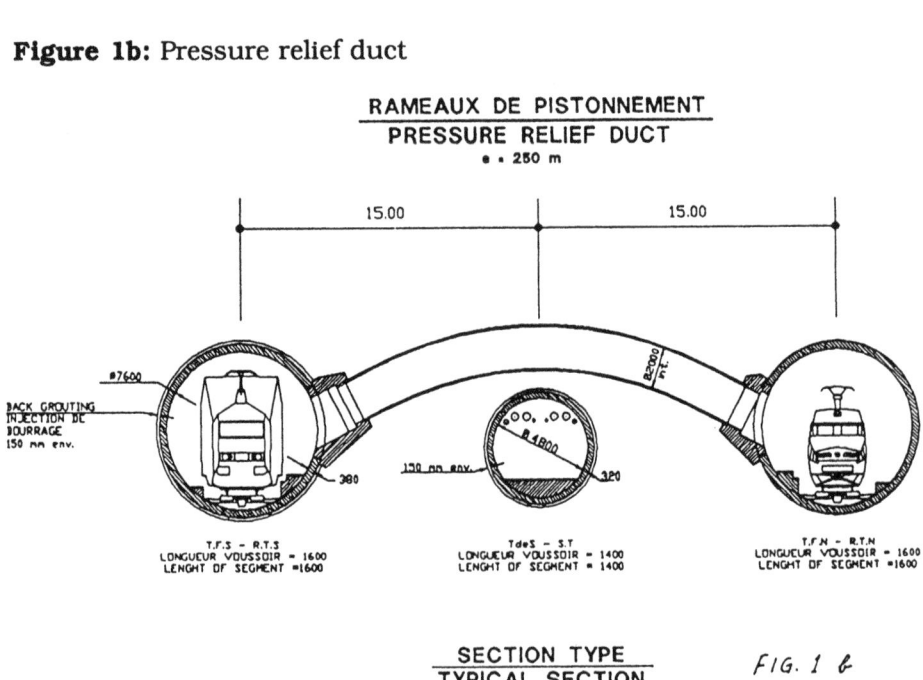

RAMEAUX DE PISTONNEMENT
PRESSURE RELIEF DUCT
● . 250 m

SECTION TYPE
TYPICAL SECTION

FIG. 1 b

352

4.2 Site Data (Fig. 2)

The location of the tunnels 40 m under the sea bed makes use of geological strata which are favourable for the tunnelling operations. The UK drives are mostly in fairly impermeable Chalk Marl. Their TBM's are open face. On the French side near the coast more permeable chalk formations require the use of a closed system (TBM's shield and provision for compressed air techniques).

The different geological environments and tunnelling methods lead to the difference in the approach to the tunnel lining systems on the French and UK sides.

On the French side where the tunnels are partly in permeable ground the linings have to be watertight. In addition to the pressure exerted by the ground, the lining has to support the head of water, which approaches 100 m at mid channel.

4.3 Loading on tunnel lining

In view of the ground geology, the total hydraulic head has been allowed for. In addition, the soil pressure has been calculated taking account of creep effects. Suitable load allowances are provided for at locations where the proximity of so called Gault Clay layers (underlying the Chalk Marl) is sufficiently close to cause significant uplift forces.

4.4 Constitution of lining on the French side

The tunnel lining is made of precast reinforced concrete segments (Figs. 3 and 4).

Each ring is made of five segments plus one key. Radial joints are articulated, and vertical joints are flat. Watertightness is ensured by means of an elastomer joint seal placed in a groove that completely surrounds each segment. The segments are picked up by the back-up train of the TBM and a conveyor takes the segments to within the reach of the TBM erector arm. The principal dimensions of the segments are shown in the Table 3 (French side and English side).

4.5 Quantities

There are 226 000 segments on the French side, a mounting to a total of 500,000 m^3 of concrete. Most of this concrete is a 45 MPa required concrete at 28 days.

The amount of cast in situ concrete is 250,000 m^3.

4.6 Compressive strength

The capacity of the prefabrication works depends on the necessary time to strip the casting segments. To be in accordance with the tunnel works, we must produce a segment every minute or half-minute. The prefabrication works includes:

- two lines of prefabrication of service tunnel segments with 45 moving molds on each line,
- three lines of prefabrication of running tunnel segments with 45 moving molds on each line.

Figure 2: Channel Tunnel

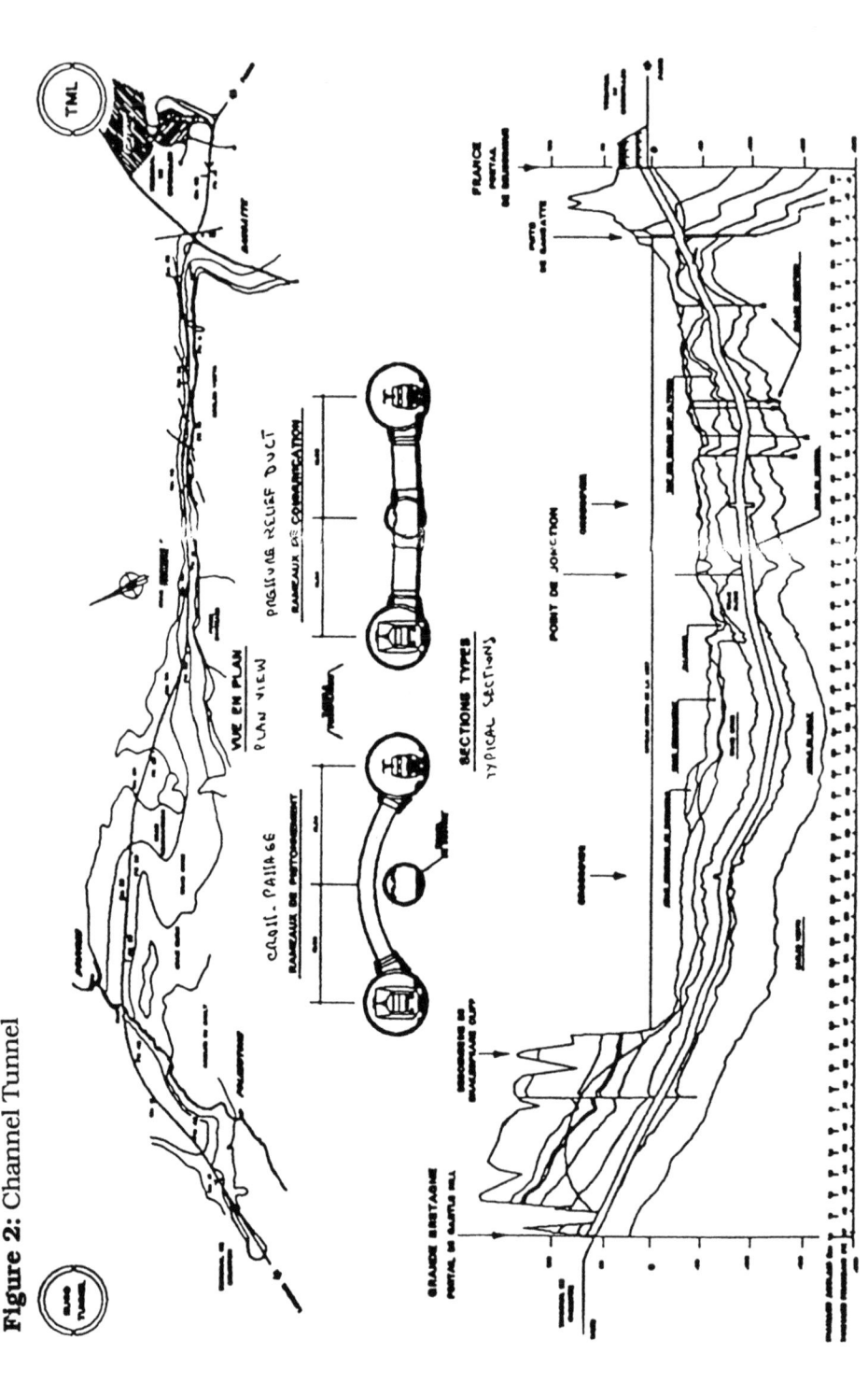

Figure 3: Composition of a typical ring

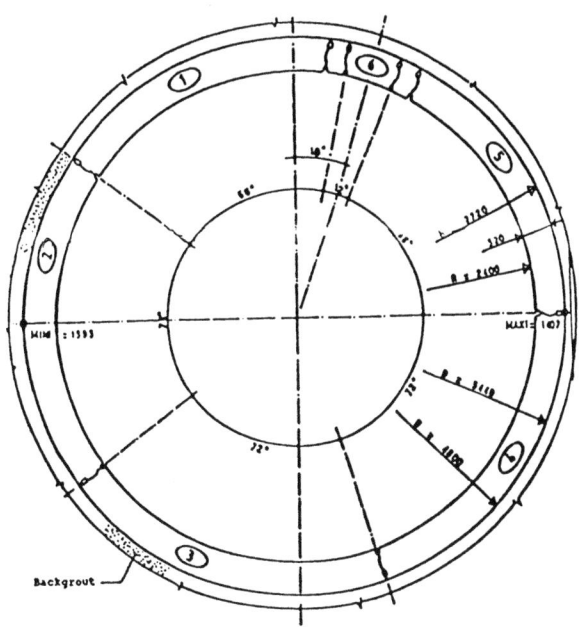

Figure 4:

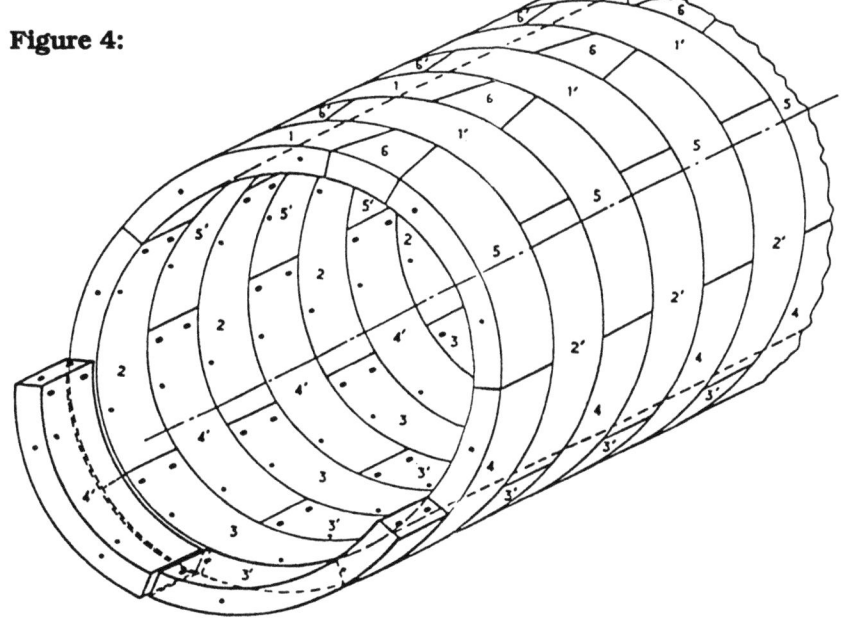

Table 3 : Channel Tunnel dimensions

	Running tunnels (French)	Service tunnels (French)	Running tunnels (English)	Service tunnels (English)
Internal diameter	7.60 m	4.80 m		
Segment thickness	0.40 m	0.32 m	0.27 to 0.36 m undersea	0.22 to 0.41 m
Ring composition	5 segments + 1 key	5 segments + 1 key	8 segments + 1 key	6 segments + 1 key
Length of a ring	1.60 m	1.40 m	1.50 m	1.50 m
Weight of a segment	8 t	3.5 t	9 t max	

4.7 Heat curing

The segments move forward at a rate dictated by the rate of concreting and they are normally 6 to 8 hours old when they exit the tunnel. The curing temperature is adjusted so as to give the concrete a strength of at least 15 MPa (preferably 20 MPa) in order to minimise corner and edge spalling at the most exposed parts during mold stripping. Heat curing allows us to get early very high compression strength but the ultimate strength is reduced. This concrete has to allow rapid stripping with a high strength at 8 hours, and its mix is important to get optimal capacity of the prefabrication works.

4.8 Concrete mix

Apart from the temperature of the concrete, the evolution of the strength depends of the type and content of cement and w/c ratio. The cement used is a CPA 55 PM from Dannes and Lumbres. The main requirements are:

- strength of AFNOR mortar

2 days :	25 MPa	s = 1.6 MPa
28 days :	56 MPa	s = 2 MPa
- C_3A = 7.5 %		s = 0.5 %

- heat of hydration (Langavant) at 12 hours = 167 j/g, s = 10.7 j/g
- cement content : 400 kg/m^3
  W/C ratio < 0.35

Fig. 5 indicates the evolution of strength in relation to the maturing of the concrete, first on 11 x 22 AFNOR samples with normal curing (20° C), secondly on samples in heating curing

356

Figure 5: Strength evolution - Control sheet

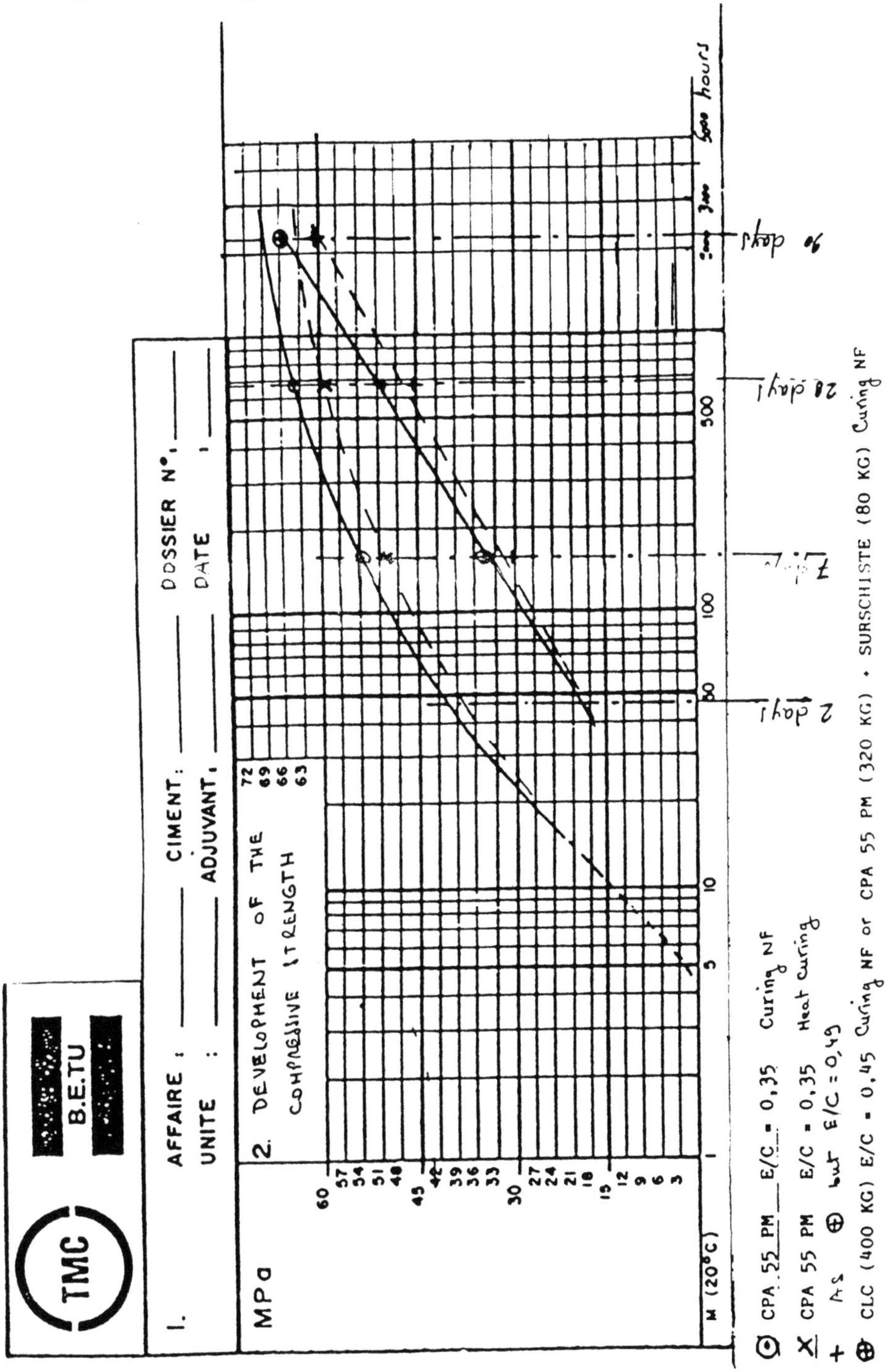

storage. We compared the evolution of the strength of four concrete mixes to cast in situ with :

Cement : CLC 400 kg/m³ or CPA 328 kg/m³ + fly ash 80 kg/m³ supplied by SURCHISTE of the Charbonnage De France group. W/C = 0.45 and 0.487.

In these cases, we have strength on AFNOR cylinders with a normal curing (20° C).

4.9 Workability

The segments are cast in molds horizontally, like in Villejust. This needs a very stiff concrete with slump equal to zero. Then workability is measured with the Vebe test, 30 and 70 s.
We used two high range water reducing admixtures:

- Durciplast from Chryso
- Sikament HR 401 from Sika.

The content is about 6 l/m³ .
The concrete for casting in situ is, by contrast, very plastic with a slump of 120 mm, up to 2 1/2 hours after mixing.

4.10 Strength results

Concrete compressive strength tests were carried out at the prefabrication laboratory. Fig. 6 shows the compression test results obtained during the first production year (1988) on standard and heat cured 110 x 220 AFNOR cylinders. All concrete produced during 1988 had to meet the class 55 requirements. The evolution of strength between 28 and 90 days shows that the strength on heat cured cylinders reachs 55 MPa at 90 days.
In Fig. 5, we have the same evolution at 90 days with a composite cement (CPA + Fly ash) with W/C = 0.45. We get this rapid evolution of the strength with high pozzolannic activity fly ash.
In Fig. 7, we have transferred the TML results concrete strength, a Norwegian Construction Publication. The offshore platforms built in Norway were made with OPC cement and they get the same evolution with the CPA 55 PH used in TML. The platform built in Scotland was made with a mix of OPC + fly ash. The increase in strength between 28 and 90 days is 22 % compared to 28 % with the Surchiste fly ash.
In Fig. 8, we compare the strength of different platforms in the North Sea with the TML French and English side and we get a correlation between 150 x 300 cylinders + 150 mm cubes and 100 m cubes in Norway Normalisation.

4.11 Segments of same thickness: a choice

The thickness of the concrete segments is 0.40 m for the running tunnel and 0.32 m for the service tunnel, irrespective of ground

Figure 6:

PROGRAMME JMN.BAS 31-01-1989
PRESENTATION GRAPHIQUE DES RESULTATS DE CONTROLE BETON PREFA
GRAPHICAL PRESENTATION OF CONCRETE CONTROLE TESTS AT THE SEGMENT PREFAB.YARD
RESISTANCE CARACTERISTIQUE= 45 KPA
CHARACTERISTIC STRENGTH= 45 KPA
RESULTATS POUR LA PERIODE JAN.-DEC.1988

RESULTS FOR THE PERIOD JAN.-DEC.1988

ANNEE	MOIS NR	FE28(MPA) NON TRAITE	ECART TYPE	RES.CARACT.(FC28 EN MPA) +1.3=ECART TYPE
YEAR	MONTH NR	FE28(MPA) NO HEATING	STAND.DEV.	CHARACT.STRENGTH (FC28 IN MPA) +1.3=STAND.DEV.
88	1	65.7	9.3	57.09
88	2	60.5	4.6	50.98
88	3	61.8	6.4	53.32
88	4	63.9	4.9	51.37
88	5	65.5	4.09	50.317
88	6	61.4	4.48	50.824
88	7	61.9	4.67	50.941
88	8	63.9	4.91	51.383
88	9	62.5	3.1	49.03
88	10	61.6	1.9	47.47
88	11	60.5	4.6	50.98
88	12	65.0	4.41	50.733

Av. 63.15 MPa

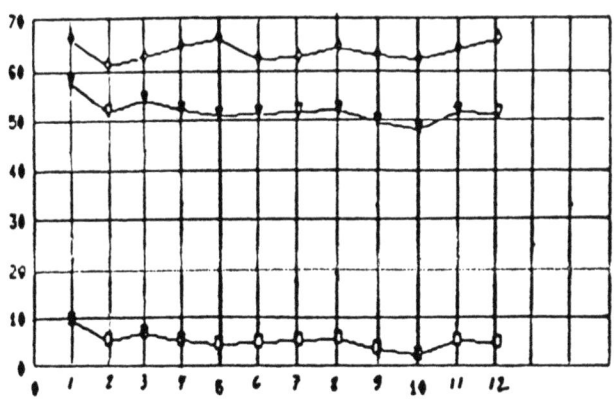

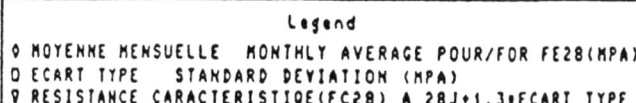

MOIS NR.   MONTH NR.

Legend
◇ MOYENNE MENSUELLE   MONTHLY AVERAGE POUR/FOR FE28(MPA)
□ ECART TYPE   STANDARD DEVIATION (MPA)
▽ RESISTANCE CARACTERISTIQE(FC28) A 28J+1.3=ECART TYPE

Figure 7: Age factors for different cements and platforms

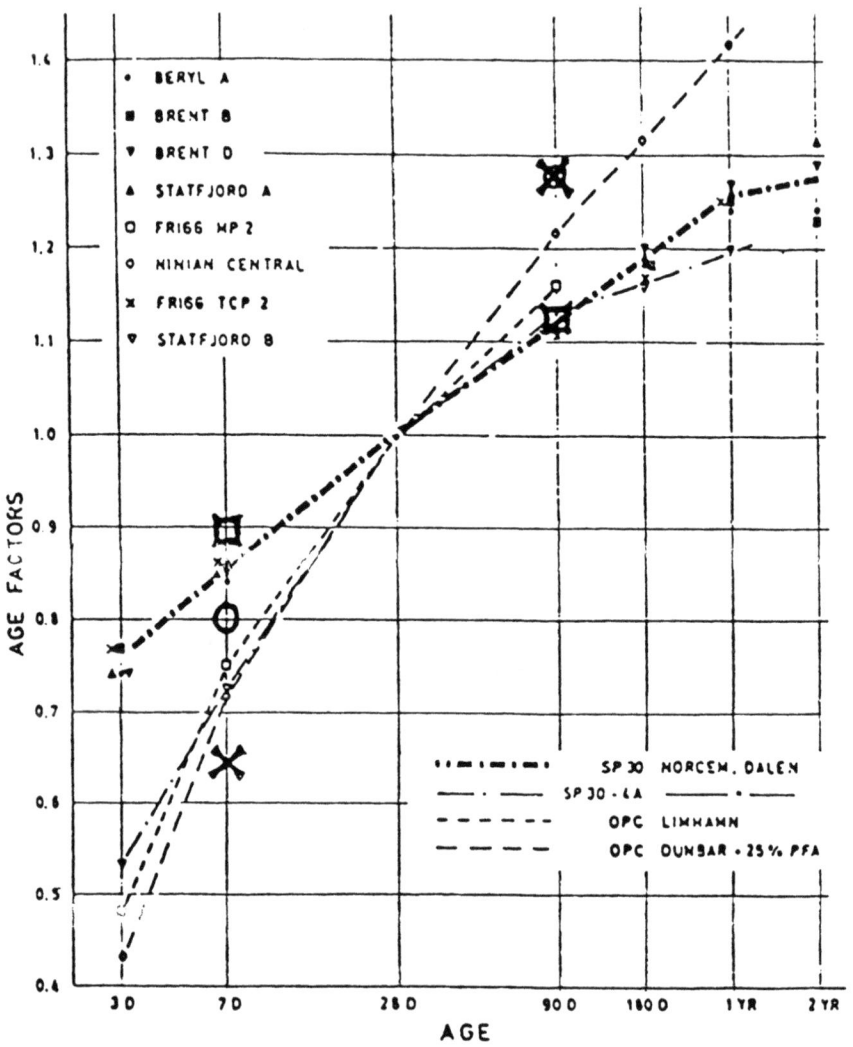

CLC 45 LUMBRES
CPA 55 PM + PFA (20 % SURSCHIST)
CPA 55 PM

Figure 8: Development of concrete quality

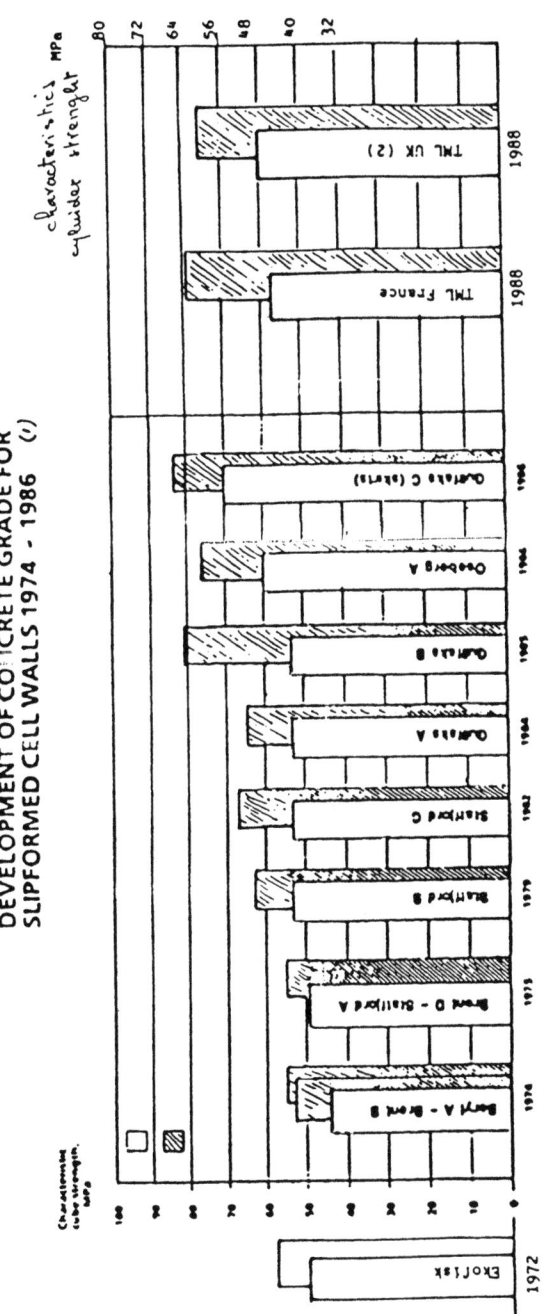

DEVELOPMENT OF CONCRETE GRADE FOR
SLIPFORMED CELL WALLS 1974 - 1986   (1)

(1) Tests N on cubes 100 (EXOFISK on cubes 100 and cylinders 150 x 300)

(2) Tests UK on cubes 150 and cylinders 150 x 300

conditions. The thickness of segment is related to two basic data :

- the internal diameter of the tunnels: 7.60 m for the running tunnels and 4.80 m for the service tunnel. Of course they depend on the dimensions of the trains and of the power lines,
- the external diameter as bored by the TBM. From the choice of this diameter and so, the space available for the precast lining, the economical consequences are of the greatest importance for tunnels of such length, as between Sangatte's well and the meeting point of the French and UK sections is 18 km.

Many factors advocate segments of minimum thickness:

a) decrease the quantity of excavated materials and its disposal area,
b) the cost of the TBM is not proportional to its diameter, it rises faster,
c) for a handling capacity of say 8 tonnes (transport from the casting plant and erection by the TBM), an increase in the thickness of the segments leads necessarily to a reduction of its width and so to a greater number of circumferential joints and of erection operations. Circumferential bolted joints are costly and the allotted time for erecting and bolting of a ring is on the critical path,
d) the cement and aggregate content of a segment is proportional to its thickness. A thicker segment of lower compressive strength would have the same unit cement content, as it is durability and not strength which in the present case fixes the cement content (Fascicule 65 of the French Ministry of Transport requires a minimum cement content of $700 / D^{0.20}$, D in mm being the nominal maximum size of aggregate). The only saving would be on the admixture,
e) the load on the lining is proportional to its external diameter.

4.12 Grade 55 MPa concrete v. grade 28 MPa concrete, a comparison

With grade 28 MPa and 8 tonne segments, the number of segments would have doubled as the number of rings. The quantity of excavated materials would have been multiplied by 1.20 and the length of the rubber gaskets by 1.87. With the number of rings the time necessary for their erection and bolting would have doubled, and cast iron segment lining would have been cheaper.

4.13 Consequences of using 45 MPa segments?

On the 1.8 km of the tunnels for which a grade 55 MPa is deemed necessary, there are the following options:

- use cast iron segments,
- use grade 45 MPa concrete but with a segment thickness of 0.50 m for the railway tunnel segments (55/45. 0.40 = 0.50) and

0.40 m for the service tunnel segments. However, as has been said before, it means using this thickness on 18 km of tunnels,
- leave the thickness unchanged but use more steel reinforcement.

This last alternative would have been chosen if a grade 55 concrete was not produced: the extracost on the reinforcement was compared to the extracost resulting from an increase in cement and admixture content of the mix. The corresponding extracost was 35 F per runway tunnel segment compared to ten times more for supplementary reinforcement.

Meanwhile, having produced now more than 50,000 segments, the increase of concrete strength resulting from the durability requirement of a maximum W/C of 0.35 was sufficient to reach the 55 MPa grade without any increase in cement or admixture content.

4.14 Higher strength concrete segments, are they possible and economical ?
Higher grades of Portland cement than those considered at the design stage of the mix are currently produced in France (CPA HP). With such cements a modern precasting plant would be able to produce 65 MPa concrete and so, to deliver runway tunnel segments of less than 0.35 m thickness.

However, for the Channel Tunnel at the design stage, only well known and well tested technologies were considered, the economic effect of untried and possibly unsuccessful innovations was not to be met. In Fig. 8 showing recent progress in concrete strength on several major project is shown, the 55 MPa grade is still amongst the higher.

4.15 High performance concrete segments, UK section
For the UK section, the boring method used did not require a constant thickness of the precast lining, which varies from 0.27 to 0.54 m for the runway tunnel segments. Therefore the Isle of Grain precasting plant is designed to produce only one 65 MPa concrete grade as controlled on 150 mm cubes. Comparative tests carried out in the plant's laboratory have shown that it corresponds to a compressive strength of 52 MPa on cylinders with a length/diameter ratio of 2.0, which is lower than the conventional ratio of the ISO table.

Average compressive strengths for the UK segment concretes are :

- at 28 days: 75 MPa on cubes and 59 MPa on cylinders,
- at 90 days: 95 MPa on cubes and 77 MPa on cylinders.

The increase in strength from 28 to 90 days is almost 30 %, which is a little more than the increase of our Portland/Fly ash mix used for cast-in-situ concrete (see Fig.7). But the UK mix uses 310 kg/m³ of Ordinary Portland Cement plus 30 kg/m³ of fly ashes, but the ashes complying to BS 3892 Part 1, so have less than 12.5 % retained on the 45 µm sieve, so they are much finer ashes.

5 Conclusion

For practical and economic reasons, high strength concrete is produced for the Channel Tunnel segments. For the French section 90 % of the lining was designed for a characteristic specified strength of 45 MPa and on only 10% of the lining was a 55 MPa strength specified (in the deepest section). Durability requirements (minimum Portland cement content, maximum W/C) resulted in an average compressive strength of the segments of about 63 MPa and the same trend is shown by the UK segment production.

24 HIGH PERFORMANCE CONCRETE IN NUCLEAR POWER PLANTS

J-L. COSTAZ
Electricité de France - SEPTEN, France

1 The main construction work for a PWR nuclear power plant

The French NPP construction program is characterized by the standardization of all its components, including the civil works. This leads to the definition of standardized plant series (fig. 1) :

- 900 MW electric
- 1300 MW electric
- 1400 MW electric.

We shall be paying particular attention to the last PWR class whose lead plant is under construction at Chooz in the Ardennes region (fig. 2) and outlining the differences with other series where appropriate.

A nuclear plant comprises several independent buildings (fig. 3) with the various basements only interlinked in special cases of high seismic loadings. From the structural view point, four types of structure are distinguished :

The cylindrically-shaped **Reactor building** has to protect the reactor against a variety of externally-generated hazards and must also contain radioactive products in the event of a break in a high-pressure pipe conveying the water of the reactor primary system or the steam of the turbine secondary system. This means it constitutes the "reactor containment" and must be capable of withstanding an internal pressure of about 4 atmospheres of an air-steam mixture (0.4 MPa gauge).

Worldwide, this type of containment has taken several forms:

- cylindrical or spherical metal containment
- cylindrical reinforced concrete containment
- cylindrical prestressed concrete containment.

High Performance Concrete: From material to structure. Edited by Yves Malier. © 1992 Taylor & Francis.
Published by Taylor & Francis, 2 Park Square, Milton Park, Abingdon, Oxon, OX14 4RN. ISBN 0 419 17600 4.

Figure 1: Nuclear power plants in France

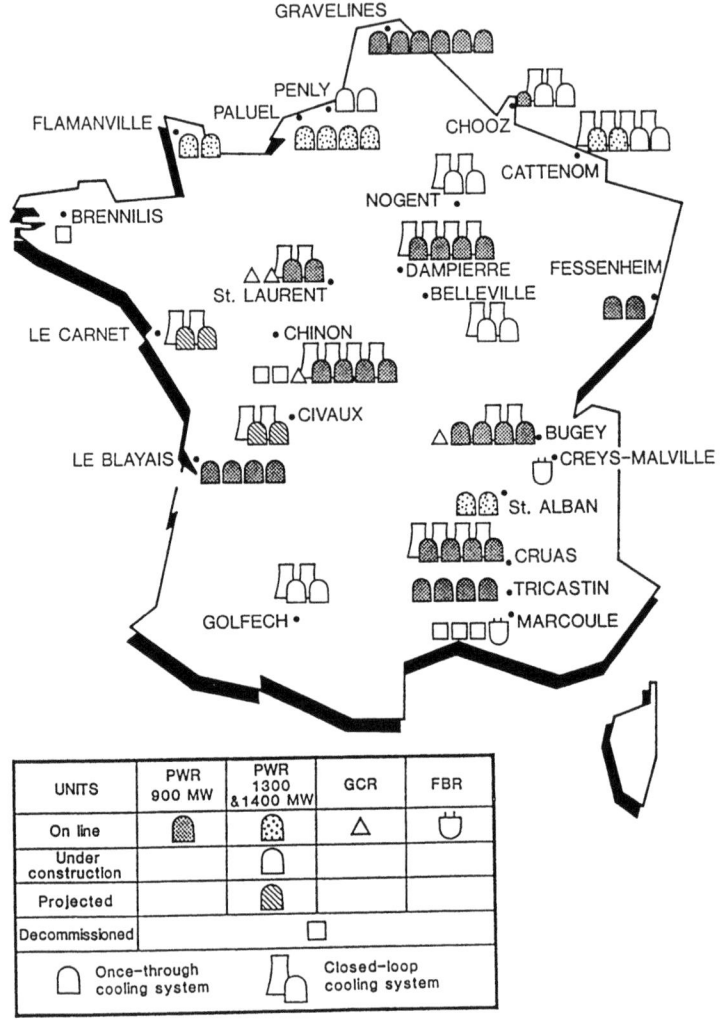

UNITS	PWR 900 MW	PWR 1300 &1400 MW	GCR	FBR
On line			△	
Under construction				
Projected				
Decommissioned				

Once-through cooling system   Closed-loop cooling system

Figure 2

Figure 3: Masses of building (PWR 1400 MW N4)

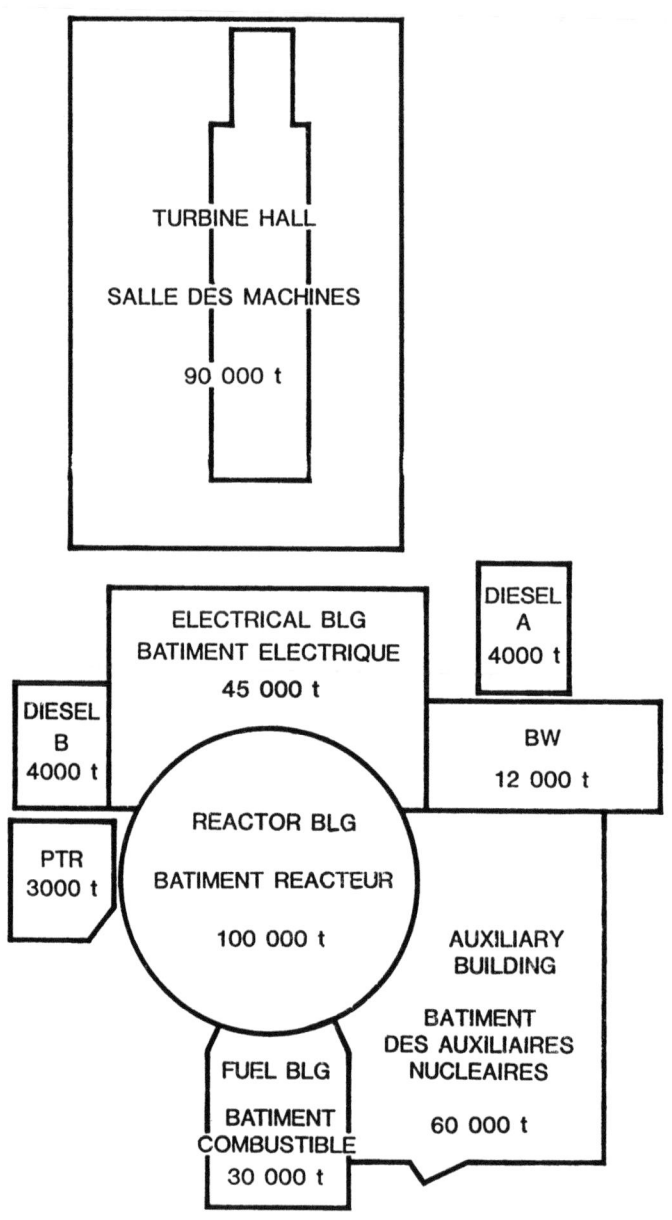

Figure 4

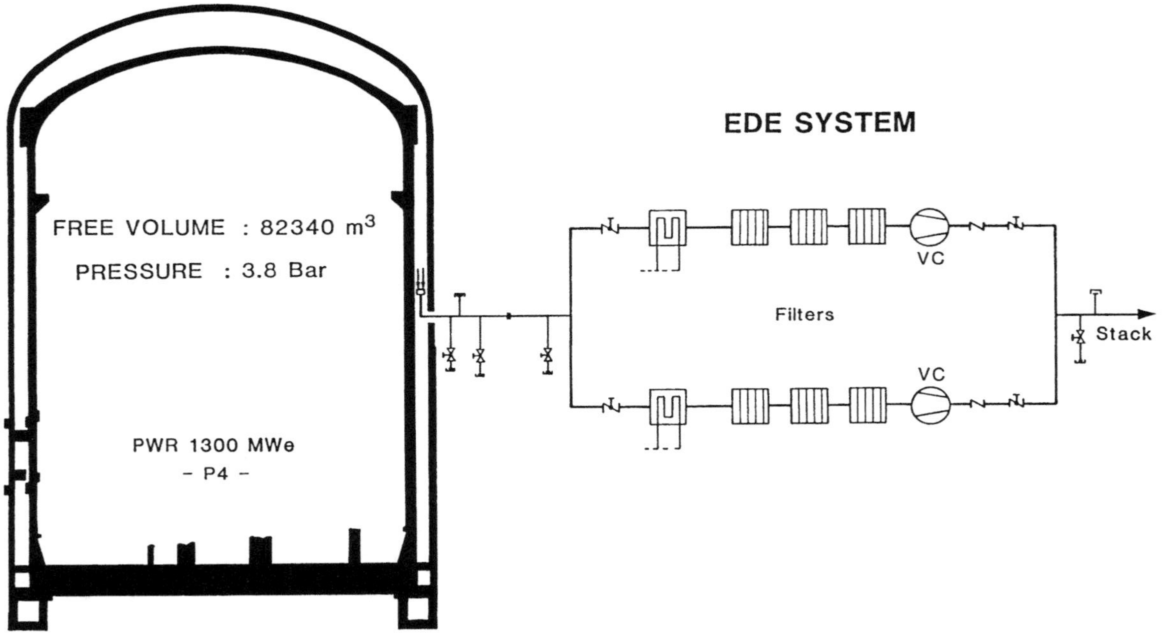

EDE SYSTEM

FREE VOLUME : 82340 m³

PRESSURE : 3.8 Bar

PWR 1300 MWe
– P4 –

Filters

VC

VC

Stack

In France, the containments are all made of prestressed concrete, which provides a quality and safety level higher than those encountered with reinforced concrete containments, especially in earthquake behavior - the prestressing forces must simultaneously oppose the tensile loads caused both by accident pressure and by earthquake-induced overturning moment.

Containment of radioactive products differs between each standardized plant series. For the 34 units of 900 MW class, the prestressed concrete containment is internally lined with a 6 mm thick steel liner anchored to the concrete. For the 22 units of 1300 MW or 1400 MW class, the unlined prestressed concrete containment itself provides a first barrier ; leaks through the concrete are recovered in a space maintained at slight negative pressure before filtration and release to the atmosphere (fig. 4). This space consists of a second containment made of reinforced concrete and completely surrounding the first.

The other **main nuclear island buildings** surround the reactor building are:

- the fuel building
- the nuclear auxiliaries building
- the electrical and safeguard auxiliaries building.

They do not have to withstand internal pressure but are subject to externally-generated hazards (plane crashes, explosions) and earthquakes. Hence these reinforced concrete buildings are stiffened by a nomber of vertical walls which, with the base mat and floors, form highly rigid structures.

The **turbine hall** is not protected from externally-generated hazards. Assurance only has to be provided that it will not collapse in the event of an earthquake. This means it is a standard industrial building. Various types of structures are encountered in different countries:

- USA: structural steelwork
- West-Germany: reinforced concrete
- France: reinforced concrete up to the turbine generator set level and structural steelwork above.

The concrete structure is a conventional beam and column structure.

In this building there is a special structure : the turbine generator set pedestal which must exhibit high rigidity despite its plane dimensions and its spring-mounted supports (fig. 5).

The **hydraulic structures** for turbine cooling consist of large tunnels which are made of traditional reinforced concrete or are of the "Bonna" prefabricated type.

Figure 5: Turbine generator pedestal 1400 MW N4 class

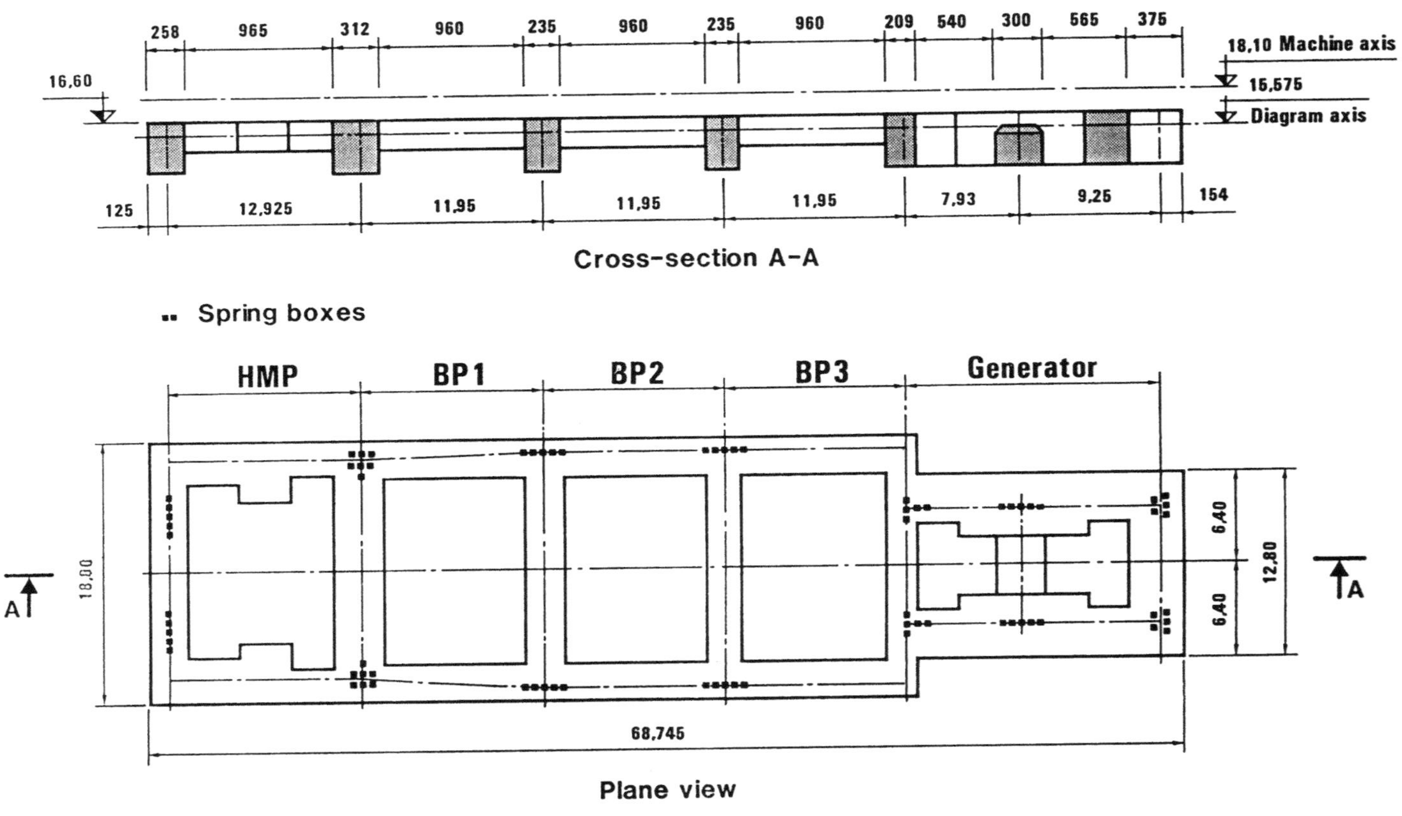

Cross-section A–A

.. Spring boxes

Plane view

When required by the site (absence of water in sufficient quantities), cooling takes place in high, natural-draft air cooling towers. We shall return to this subject in sections 2.2.

In summary, the main constrcution work of a 1400 MW unit (Chooz) is characterized by the following quantities :

- 200,000 m^3 of concrete
- 300,000 m^2 of formwork
-   15,000 t of re-bar
-     1,500 t of prestressing tendons
-     1,000 t of embedded parts
-     6,000 t of structural steelwork
- 2 million working hours.

2 Benefits of high performance concrete (HPC)

Given the diversity of the structures described above, the benefits of using HPC in place of traditional concrete must be examined on a case by case basis. At first sight, they appear slight for the turbine hall, although the reduction in beam weight could be to the advantage of a prefabricated solution. But the turbine hall is not on the critical path.

An exception is the turbine pedestal whose volume could be quite significantly reduced with HPC provided prestressign is used ; the current high re-bar content would make a reinforced concrete solution impractical. Study of such a structure is included in our 1990-1992 program. Although the problem of vibration isolation can be resolved by spring-mounted suspension, unknowns still persist concerning reduction of material damping and the maintenance with time of prestressing forces in a relatively high-temperature environment subjected to sustained vibrations.

For walled and floored nuclear island buildings, the current concrete thicknesses (0.60 - 1m) are dictated by resistance to externally-generated hzards or by biological shielding from radiation.

The mass of the concrete is very useful for shocks and the benefit of HPC is not obvious, except if fibers are added; in which case project economy could suffer.

This leaves the two structures specific to a nuclear power plant :

- the reactor building
- the atmosphere cooling tower.

We shall now examine these in detail.

2.1. Reactor building

Figure 6 gives the main dimensions of the 900, 1300 and 1400 MW containments.

Figure 6: Trends in french nuclear program reactor containments

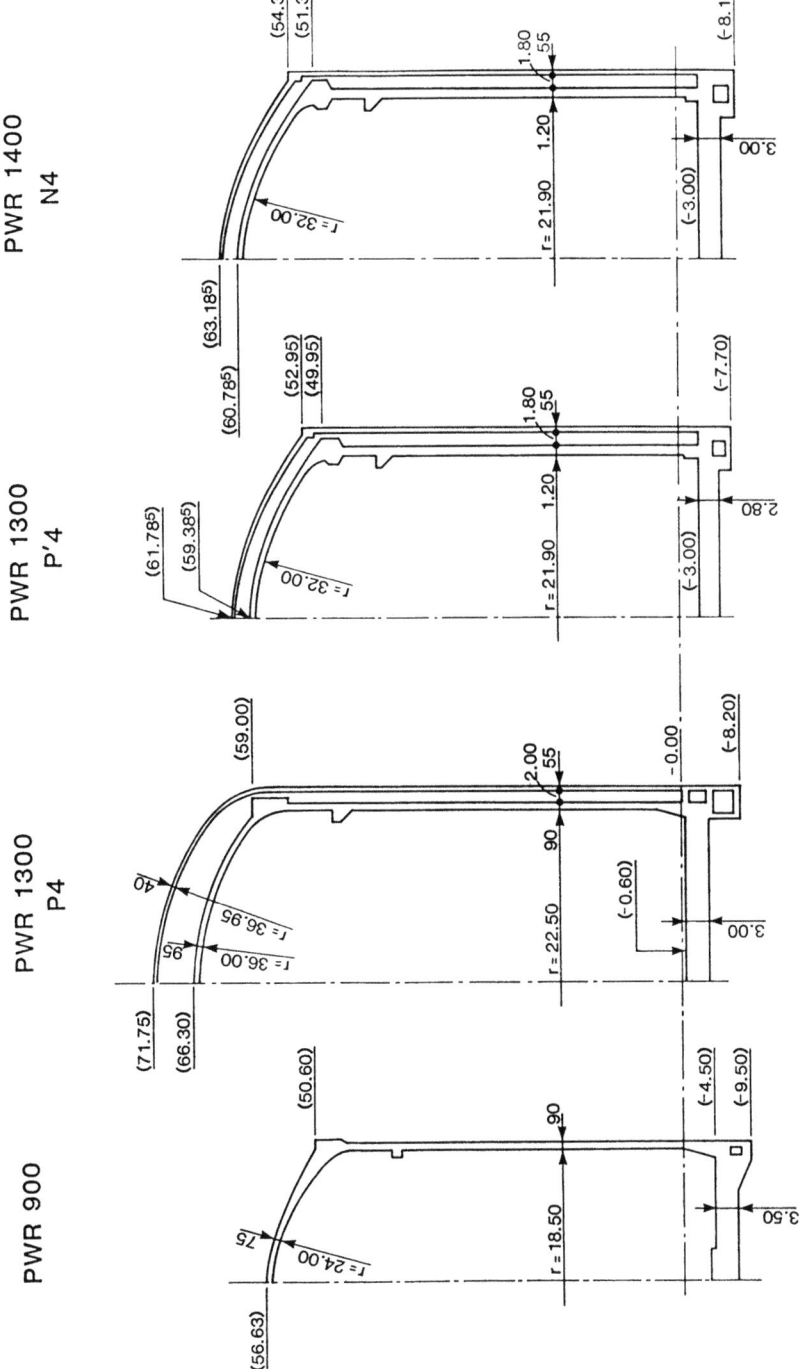

For twin-walled containments (1300 and 1400 MW), prestressed concrete must perform a combined strength/leaktightnesse function. With regard to leaktightness, G. Ithurralde's paper shows the adavantages of HPC which considerably reduces the cracking of early-age concrete. As regards strength, the thickness is currently set for technolgical reasons of prestressing duct housing. At constant average thickness (1.20 m), a 60 MPa HP concrete would allow discontinuation of crack-generating extra thicknesses :

- omission of gusset between base mat and cylinder,
- omission of reinforcement around the diameter 8 m equipment access point,
- reduction of the thickness of the ring between cylinder and dome.

The reduction in average thickness and improved utilization of HPC calls for design changes :

- hemispherical rather than torospherical dome,
- new prestressing layout (helical) and possibly external prestressing,
- new prestressing techniques to reduce friction (greased strands inside polyethylene ducts).

These innovations will be analyzed for the future plant series to be started around the year 2000 and referred to as PWR 2000. For current projects, the use of HPC would allow local shape improvement, better airtightness and reduction of concrete creep.

The Civaux containment study described in the paper by G. Ithurralde led to a formula ideally suited to the problem but also highly innovative and expensive. After a few industrial adaptations, this concrete will be full scale tested on a Civaux 1 standard structure with a view to use on the Civaux 2 containment.

2.2 Cooling towers

2.2.1 Introduction to role and principle of cooling towers
In view of the Carnot principle, thermal machines such as the turbines of fossil-fired (coal, fuel) or nuclear electricity-generating plants need, in order to operate, a heat source producing steam and a heat sink. The conversion of the thermal energy contained in the steam into mechanical then electrical energy at the generator outlet has an overall efficiency of between 30 and 40 % depending on the plant type. The remaining thermal energy is released into the environment, either directly into a river or the sea (once-through) or through an atmospheric cooling tower (closed-loop). It must be remembered that a plant unit of 900-1400 MW electric uses a

cooling water flow of 40-60 m³/s which it heats up by about ten degrees C.

Coastal sites do not generally raise problems for "once-through" facilities. On the other hand, riverside sites often have limited flow - this is why the closed loop solution with a cooling tower is highly satisfactory for avoiding excessive heating of river water.

A cooling tower basically consists of an air circuit and a water circuit leading to a heat exchange vessel in which the air and water are brought into contact. By a stack effect, i.e. by natural draft, the cooling tower shell provides air circulation. When, as is the case with the wet cooling towers at EDF thermal and nuclear plants, the air and water are placed in direct contact in the heat exchange vessel, the hot air escaping from the shell is saturated with steam : mixing with the colder outside air produces by steam condensation a plume whose structure is similar to a natural cloud.

2.2.2 Characteristics of current cooling towers (Table 1 and Fig. 7) Atmospheric cooling towers are generally walls with a shape similar to a hyperbolic torus, resting on a large number of columns, diagonals or other supports linked by an annular footing which transmits loads to the ground.

Among the variety of cooling towers, those referred to as "air/water counterflow" feature a structure supporting the heat exchange vessel inside the tower consisting of gantries and beams.

Table 1: Comparison of characteristics of the french power plants' cooling towers

	PLANT			TOWER					
	SITE	Unit	Power	Height	Diameter		Throat thickness	Reinforcement	
					Throat	Lintel		Layers	% Throat section
NUCLEAR	BUGEY	4-5 (1975-77)	900 MW/2	127 m	61 m	95 m	18 cm	2	V:0,13 H:0,34
	DAMPIERRE	1-2-3-4 (1977/79)	900 MW	165 m	77 m	122 m	21 cm	2	V:0,38 H:0,38
	CATTENOM	1-2-3-4 (1983-86)	1300 MW	165 m	84 m	177 m	21 cm	2	V:0,36 H:0,36
	GOLFECH	1 (1986)	1300 MW	179 m	83 m	122 m	20 cm	2	V:0,50 H:0,50
	CHOOZ	1 (1986)	1400 MW	173 m	85 m	138 m	25 cm	2	V:0,50 H:0,50
FOSSIL FUEL	PONT SUR SAMBRE	2 (1961-62)	250 MW	93 m	43 m	63 m	10 cm	1	V:0,17 H:0,19

Figure 7: Cooling towers evolution

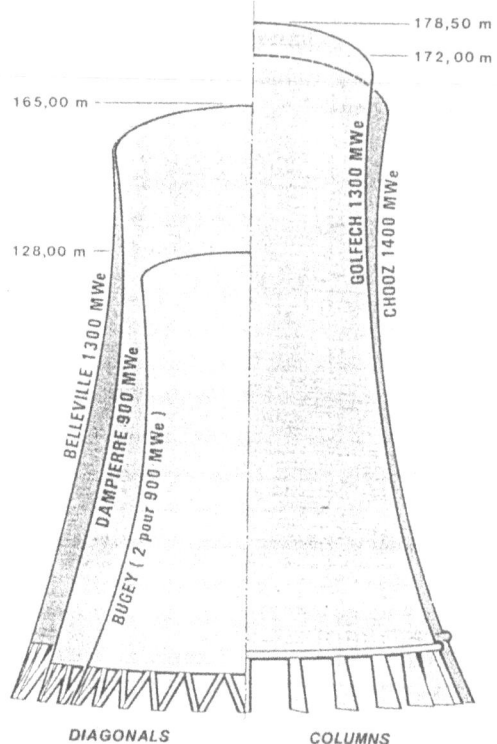

These structures whose height can reach 180 m and with a diameter at the base of about 140 m have a thickness of only 20 cm in the straight section of the shell which is, relatively speaking, much thinner than an egg-shell. They are subjected to a variety of loadings:

- weight
- differential settlements
- wind
- thermal effects:
. gradient across concrete thickness due to inner air/outlet air difference
. seasonal temperature variations
. sunshine (facing shadow/facing sun)
. concrete shrinkage.

These major loadings are combined in accordance with the rules of the Technical Specifications so as to preclude any risk of excessive cracking and buckling failure.

Cooling towers are also subjected to physical/chemical and biochemical hazards (ice, run-off water, algae, etc).

Since these structures have a direct impact on plant availability, special requirements on the shell concrete and on the re-bars must be met to limit the risks of corrosion on the steel reinforcement.

Shell distortion is regularly monitored, by topographic monitoring and visual inspection, for any variation with time. Monitoring of the 20 shells already built for NPP's has shown that they have all exhibited some defects, these defects (mainly cracks) are preferentially located in the top and bottom sectors of the shell (i.e. in the thickest parts).

2.2.3 Why use HPC ?

One of the factors enabling improvement of cooling tower durability is the concrete composition which provides :

- good resistance to corrosive water
- good compactness
- good resistance to freeze/thaw cycles
- minimum porosity
- low shrinkage and low cracking.

HPC is apparently the one material which meets all these requirements, hence the idea of analysing a concrete with silica fume to improve compactness.

Some tests have already been run on 350 kg/m^3 - cement concrete with added 35 kg silica fume. The results were highly satisfactory.

2.2.4 Study of Civaux cooling tower

A study is now underway at the Civaux site. The objective is twofold:

- formulation of a high-performance concrete containing silica fume using site materials and a cement produced close to the site,
- allow for resistance increases in the dimensional design.

The main sections in which concrete thickness and re-bar could be reduced are the shell base, its support, the crown and the internal structures. The expected gain should offset the additionnal cost of m^3 concrete. The aim of the study is to build a longer-life cooling tower without extra cost.

3 Conclusion

High performance concrete has not yet been used in any NPP in the world. The major reasons are probably the reduction of all the nuclear programs and the need of experiences of this new material in non nuclear structures. Nevertheless, we think HPC ought to be used in the future nuclear plants because of its following main qualities:

- water, air and steam tightness
- high strength
- good behaviour with corrosive water
- low creep and shrinkage
- acceptable price.

25 FULL SCALE USE OF HIGH PERFORMANCE CONCRETE IN BUILDING AND PUBLIC WORKS

G. CADORET*and P. RICHARD
Groupe Bouygues, Direction Scientifique, Paris, France

1 Introduction

In the climate of stiff international competition now characteristic of large-scale projects, the durability of the structure has become a key factor in a contractor's ability to remain competitive. Durability, which is most often expressed as a minimum service life (Denmark Great Belt, Channel Tunnel, Morocco's Hassan II Mosque), ensures that investments will pay off.

Thus, the need to offer objective guarantees of the durability of structures leads the Contractor to develop his know-how in the fields of project design, construction and operation.The design quality of a project is reflected by how well that project satisfies the needs expressed by the client, taking into account a given environment and incorporating the appropriate construction techniques and technology.

The construction quality the other hand, is essentially based on mastery of those techniques, site organization, quality control of materials and quality assurance records.

Service life quality is the result of having accounted for control and maintenance procedures during the project design phase.

Therefore, in conjunction with his services, the Contractor must inevitably develop its skills in the areas of:

Design : Analysis of new structures in order to take advantage of the properties of high performance concrete, in particular.
Construction : Development of new construction techniques involving equipment and construction methods, as well as the type of materials used.
Service Life : Computer modelling of the aging of structures and the evolution of deteriorations.

Moreover, as the geographical location of construction sites is varied by definition, optimum use must be made of local materials.

High Performance Concrete: From material to structure. Edited by Yves Malier. © 1992 Taylor & Francis.
Published by Taylor & Francis, 2 Park Square, Milton Park, Abingdon, Oxon, OX14 4RN. ISBN 0 419 17600 4.

In addition, technical progress must overcome the sometimes contradictory requirements of national codes and standards.

Finally, taking an approach comparable to our approach to the durability of structures, we must verify that the concrete itself complies with the three criteria of rheology, strength and durability.

1.1 Rheology

The rheology (or workability) of concrete placed on site must comply with the requirements dictated by the type of elements to be built, the construction methods and the time needed to carry out the works. Otherwise, incidents or even accidents will inevitably occur.

Rheology is therefore a critical characteristic, whose value must be determined through a thorough analysis, taking into account the construction methods and production requirements. The analysis must also reflect the local conditions (materials and climate) under which construction will take place.

1.2 Mechanical characteristics

Strength is the physical magnitude most often associated with concrete. Codes prescribe strength values for different concrete ages, and the contractor must ensure that such values can be attained under the probable production conditions.

It is appropriate to note here that site constraints due to the organization of the works are more often than not a key factor in selecting the materials and equipment to be used. Among these constraints, the most frequent involve formwork stripping, striking of falsework, handling, storage, post-tensioning.

If the works are to proceed on schedule, such requirements must be quantified and detailed analysis (taking into account the natural variations in supply and climate) must permit an assessment of how well the proposed concrete satisfies the project requirements. It is worth emphasizing here that concrete strength requirements apply to specific structures or parts of a structure.

For this purpose we use computer modelling of the hydration process in order to define the development of strength as well as temperature distribution throughout the structure or part of structure.Three categories of data are entered into the model:

Concrete parameters

- relation strength/maturity
- heat generation
- initial temperature of concrete

Environmental parameters

- wind velocity
- temperature
- relative humidity

Construction method parameters

- concrete cross section
- formwork insulation

Through this approach :

- the jobsite can verify the accuracy of the production hypothesis (cycles and phasing)
- the design department can base its calculations on concrete parameters which better reflect the actual condition of the structure.

1.3 Durability
On site, durability is the result of the concrete maturing process. The condition of concrete upon removal of the formwork (strength, degree of hydration, temperature gradient, etc.) determines its ability to withstand the aggressive environments to which it may be subjected.

Since 1983, the Bouygues Group has conducted a research program devoted to high performance concrete. In this paper, we present the data collected on our building and public works projects.

2 High strength concrete

Table 1 shows the main projects on which high performance concrete has been used, the aims we had defined and the quantities produced, with the main results. The Table does not include on-site, but experimental-scale applications of this material. This comment refers namely to trial use of shotcrete made with silica fume, for tunnel linings or retaining walls.

2.1 A86 Highway Bridge, Paris
This project marked the first time that silica fume was used in an actual project. It involved pump-placing of 70 m^3 of concrete for the hollow abutment slab of a highway bridge. The purposes of this undertaking were the following :

- development of equipment suitable for the production of concrete containing silica fume (homogeneity, production capacity),

Table 1

| PROJECT | DATE | OBJECT | CONCRETE COMPOSITION PER (M3) | | | | STRENGTH (MPa) | | | | QUANTITIES PRODUCED |
| | | | CEMENT (KG) | SILICA FUME (KG) | WATER/ CEMENT | SLUMP (CM) | 28 DAYS | | 90 DAYS | | (M3) |
							AVERAGE	COMPRESSIVE	AVERAGE	COMPRESSIVE	
PLM A 86	1984	DEVELOPPEMENT OF EQUIPEMENT	CPA HP 400	25	0,47	5	75	70	/	/	70
GREAT ARCH LA DEFENSE	1985	LOWER DECK (WORKABILITY-STRENGTH)	CPA HP 425	/	0,39	20 - 25	60	55	67	62	15 000
	1989	UPPER DECK (DITTO-PUMPABILITY)	425	30	0,40	22 - 25	65	60	73	68	15 000
RE ISLAND BRIDGE	1987	SEGMENT PRECASTING	CPA 55 400	30	0,40	15	68	60	/	/	35 000
SYLANS VIADUCT	1986	PRECAST DECK X ELEMENTS	CPA 55 400	/	0,38	20 - 23	69	61	/	/	2 000
	1988	PARAPET DURABILITY	375	35	0,38	15	75	/	/	/	/
PERTUISET BRIDGE	1988	DECK (B 60)	CPA HP 400	30	0,38	14	75	70	/	/	1 000
JOIGNY BRIDGE	1988	DECK (B 60)	CPA HP 450	/	0,36	20	78	68	90	80	1 100
HASSAN II MOSQUE (MAROCCO)	1988	MINARET CYCLE DURABILITY OF IMMERSED STRUCTURES	CPJ 45 375	25	0,45	15	37	37	/	/	5 000 (W/SILICA FUME) 1 000 (W/O SILICA FUME)
	1989	STRENGTHENING OF FOUNDATIONS + SUPERSTRUCTURES	CPA HP 425	30	0,39	23	96	92	105	100	400
KWUNG TONG BY PASS	1990	SEGMENT PRECASTING	OPC (BS 12)	/	0,38	18	70	65	/	/	12 000 (PLANED : 65 000)

- monitoring of the rheology of fresh concrete,
- measurement of the advantages afforded by the addition of silica fume and plasticizer.

Measurements taken by the Laboratoire Régional de l'Est Parisien in charge of construction quality control are provided in Table 2.

Homogeneity of fresh concrete

This homogeneity was measured by analysis of fresh concrete samples with results as shown in Table 2.

Comment : The variances between the design mix and the composition of control samples mainly pertain to the proportion of fines. This can be attributed to the variable fines content of the sand used. Once this adjustment is made, the control samples remain well within the usual mix proportions.

The results of concrete compressive strength tests shown on Figures 1 and 2 indicate that changing the composition (silica fume, plasticizer) and the mixing cycle allowed a strength gain of about 35 MPa at 28 days (concrete produced on site commonly has compressive strength of about 40 MPa). As for workability, slump tests and flow measurements demonstrate that these parameters can be controlled even for low-slump (5 cm) concretes.

2.2 La Grande Arche Paris-la-Défense Figure 3

This project is exceptional both by virtue of its size and of the construction methods used. Regarding the concrete, it is worthy of note that of all the bidders on the project, only Bouygues proposed cast-in-place erection of the main girders of the upper deck, on falsework, placing the concrete by pump. Specifications for the project called for a characteristic strength of 50 MPa at 28 days for the concrete of the main girders, the connections with the wings (side buildings) and the pier caps.

An analysis of production, transport and placing conditions indicate that a concrete with a flowing consistency, able to maintain its slump value for at least 60 minutes, was in order. Studies of concrete composition that we conducted led to design of two different mixes, suitable for the concreting of the lower and upper decks.

Generally, the concrete process is considered to involve three phases: batching and mixing, transportation and acceptance of mixes prior to pumping, and concrete placing. These phases were followed up by two additional procedures, one involving the organization of concrete controls, and the other involving curing techniques and the monitoring of concrete maturity in the structure.

Table 3 shows the concrete mixes used for the lower and upper decks of La Grande Arche.

Table 2

	Mix Design	CONTROL (LREP)	
		Measurement n° 1	Measurement n° 2
Cement + Silica Fume	424	444	436
Sand	720	682	709
Coarse Aggregate	1 050	1 092	1 035
Total water	185	173	190

Figure 1

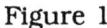

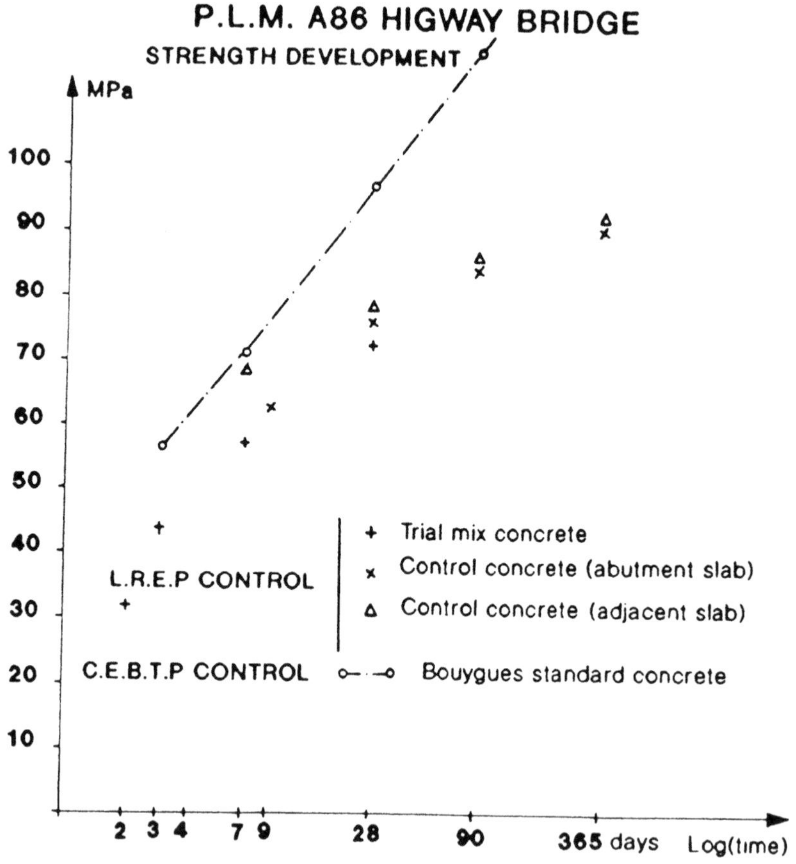

P.L.M. A86 HIGWAY BRIDGE
STRENGTH DEVELOPMENT

+ Trial mix concrete
x Control concrete (abutment slab)
△ Control concrete (adjacent slab)
o—·—o Bouygues standard concrete

L.R.E.P CONTROL

C.E.B.T.P CONTROL

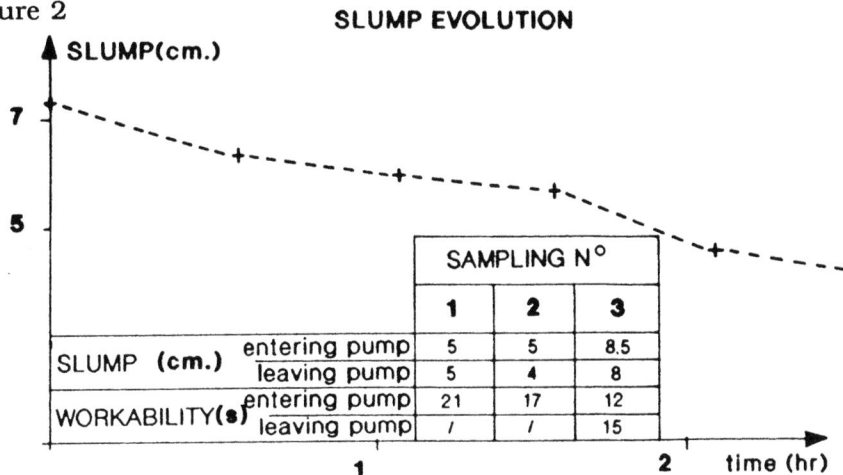

Figure 2

SLUMP EVOLUTION

		SAMPLING N°		
		1	**2**	**3**
SLUMP **(cm.)**	entering pump	5	5	8.5
	leaving pump	5	4	8
WORKABILITY(s)	entering pump	21	17	12
	leaving pump	/	/	15

Table 3

	LOWER DECK WITHOUT SILICA FUME	UPPER DECK WITH SILICA FUME
CEMENT (CPAHP)	425	425
Silica Fume	-	30
Fine Sand	90	50
Sand	640	655
Gravel 5/20	1 044	1 033
Water	165 (175)	170 (190)
Retarder	17	10
Plasticizer	0 - 3 kg	0 - 3 kg
SLUMP	20 - 25	22 - 25
Max. Variation in water content $(1/m^3)$	10	20

Figure 3　La Grande Arche

Comment: In the Table, figures in parentheses represent the maximum amount of water authorized in our instructions to the batching plant. Note that adding silica fume results in a doubling of the permissible tolerance (from 10 to 20 liters per cubic meter of concrete).

Figures 4 and 5 summarize the results obtained in the main girders of the lower and upper decks. Data collected on concrete containing silica fume are presented month by month in Table 4. A statistical summary of results is provided by the curve in Fig. 6.

For the upper deck, concrete was placed by pump at an average rate of 40 m3/hr. The concrete travelled along the path shown in Fig. 7. During placing, the concrete was pumped at a pressure of about 60 bars.

In order to post-tension the girder sections cast in place by cantilevers starting from each wing building, concrete strength had to have reached 36 MPa. Strength development was monitored in situ by recording the concrete maturity coefficient (see Fig. 8). Therefore, the 36-hour cycle was kept up under conditions of complete safety for all sections of the upper deck.

Table 4

MONTH (1987)	7 days	28 days	90 days
MAY	58,7	66,4	69,8
JUNE	53,0	59,6	64,4
JULY	53,5	61,2	68,9
AUGUST	50,0	61,7	68,4
SEPTEMBER	44,4	58,0	65,0
OCTOBER	54,7	55,5	73,3
NOVEMBER	55,7	67,0	75,5
DECEMBER	49,5	59,7	66,5
Numer of tests Standard Deviat.	48 6,6	61 5,0	60 5,0
Caracteristic strength	46,9	59,0	65,9

Figure 4   GREAT ARCH AT LA DEFENSE

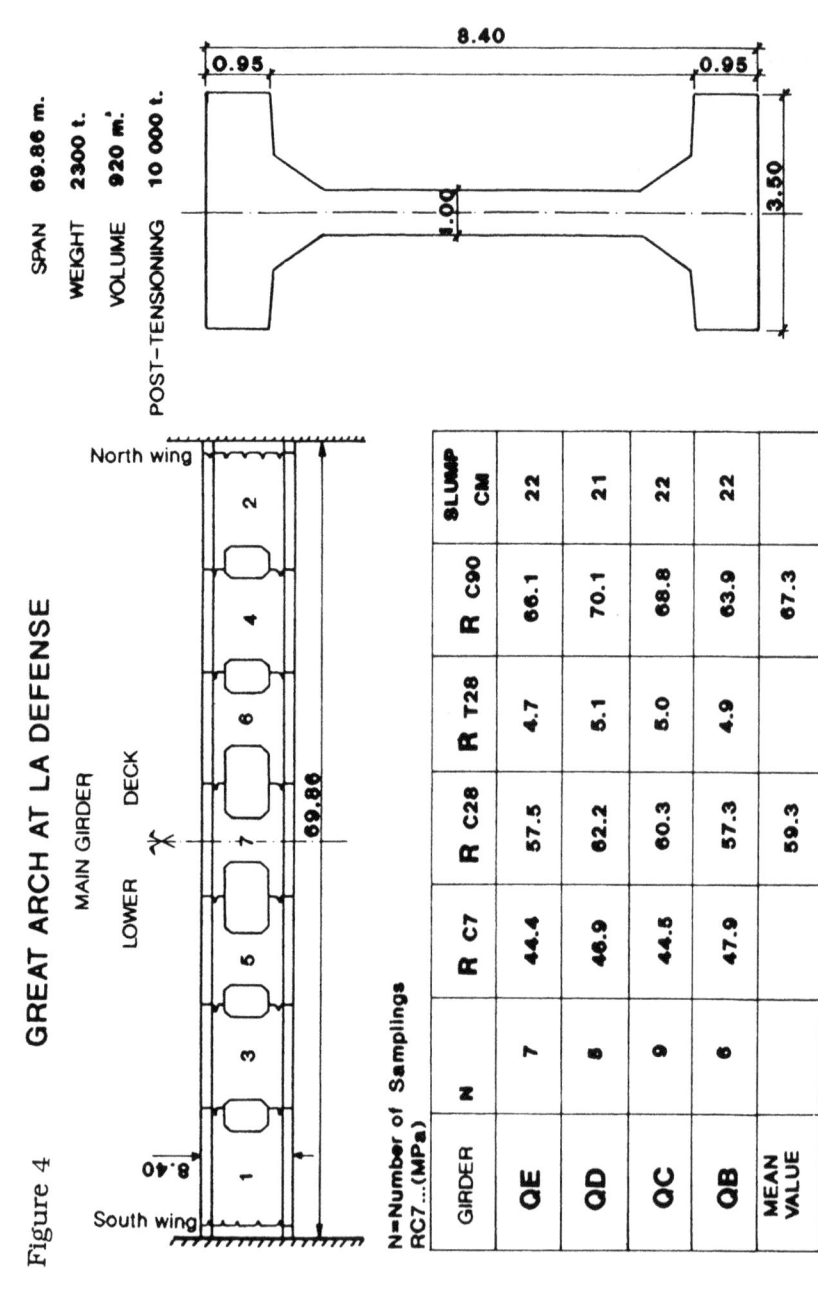

SPAN	69.86 m.
WEIGHT	2300 t.
VOLUME	920 m³
POST-TENSIONING	10 000 t.

N=Number of Samplings
RC7...(MPa)

GIRDER	N	R C7	R C28	R T28	R C90	SLUMP CM
QE	7	44.4	57.5	4.7	66.1	22
QD	5	46.9	62.2	5.1	70.1	21
QC	9	44.5	60.3	5.0	68.8	22
QB	6	47.9	57.3	4.9	63.9	22
MEAN VALUE			59.3		67.3	

Figure 5 GREAT ARCH AT LA DEFENSE

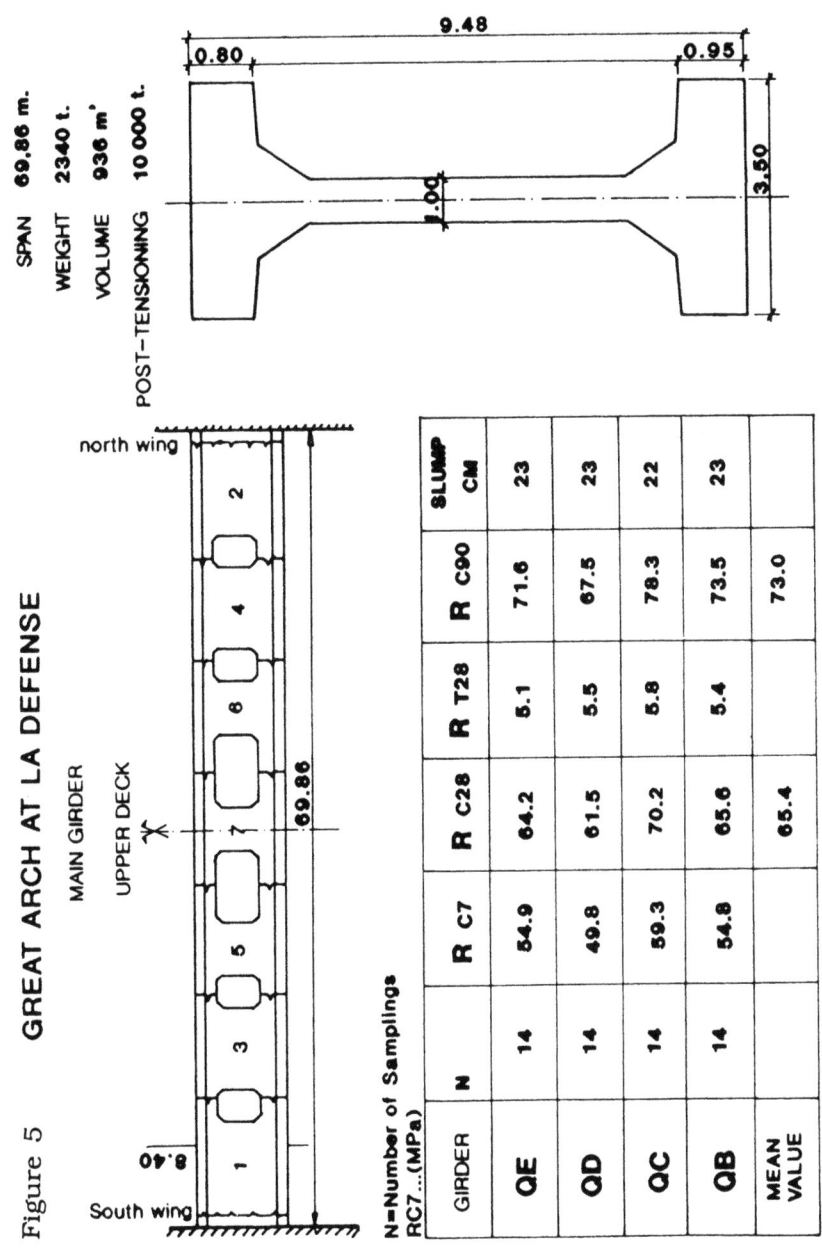

MAIN GIRDER

UPPER DECK

SPAN 69.86 m.
WEIGHT 2340 t.
VOLUME 936 m³
POST-TENSIONING 10 000 t.

N=Number of Samplings
RC7...(MPa)

GIRDER	N	R C7	R C28	R T28	R C90	SLUMP CM
QE	14	54.9	64.2	5.1	71.6	23
QD	14	49.8	61.5	5.5	67.5	23
QC	14	59.3	70.2	5.8	78.3	22
QB	14	54.8	65.6	5.4	73.5	23
MEAN VALUE			65.4		73.0	

Figure 6   GREAT ARCH AT LA DEFENSE

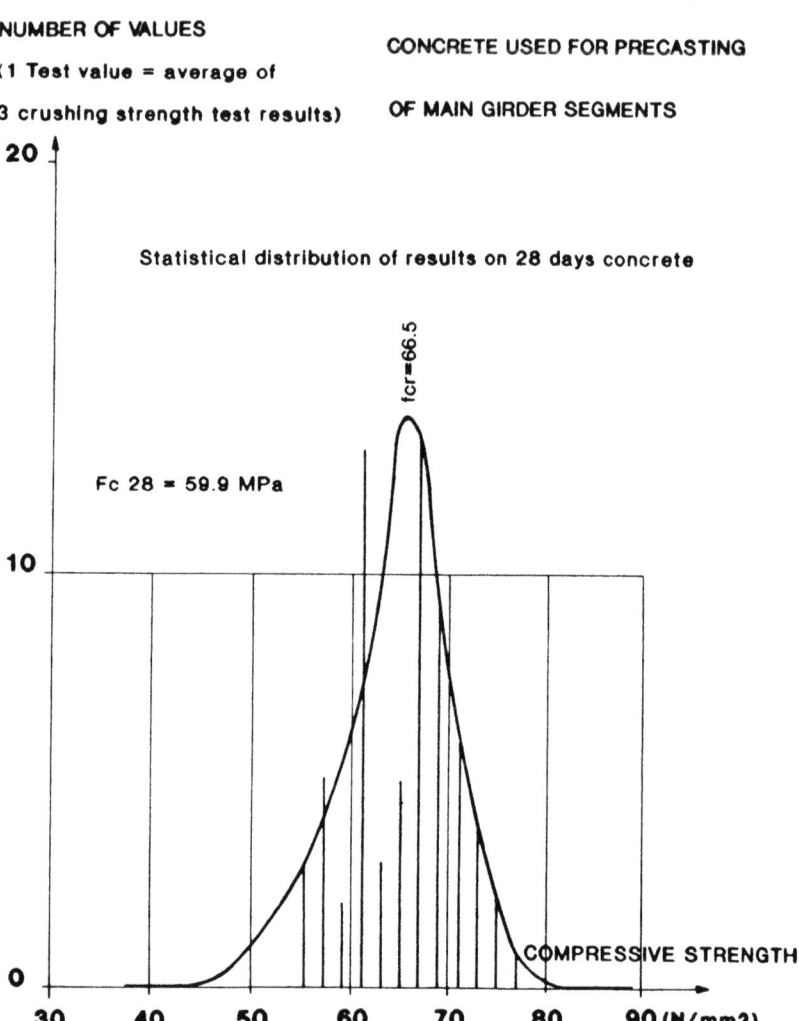

NUMBER OF VALUES

(1 Test value = average of

3 crushing strength test results)

CONCRETE USED FOR PRECASTING

OF MAIN GIRDER SEGMENTS

Statistical distribution of results on 28 days concrete

fcr=66.5

Fc 28 = 59.9 MPa

COMPRESSIVE STRENGTH

(N/mm2)

Figure 7

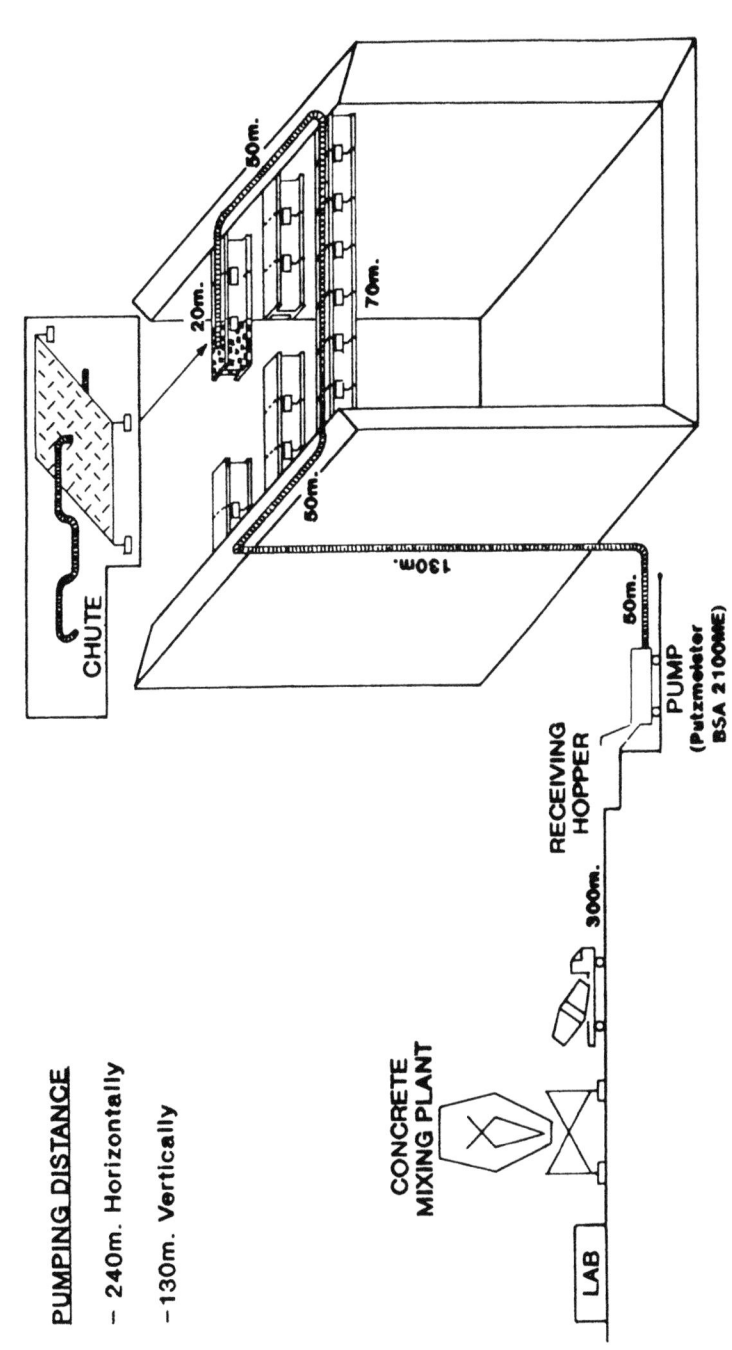

PUMPING DISTANCE

– 240m. Horizontally

– 130m. Vertically

CHUTE

20m.

50m.

70m.

50m.

130m.

50m.

PUMP
(Petzmeter
BSA 2100ME)

RECEIVING
HOPPER

300m.

CONCRETE
MIXING PLANT

LAB

Fig. 8   **GREAT ARCH AT LA DEFENSE**

CONCRETE WITH SILICA FUME (Rc=50MPa)

MATURITY CURVE FOR EARLY-AGE CONCRETE

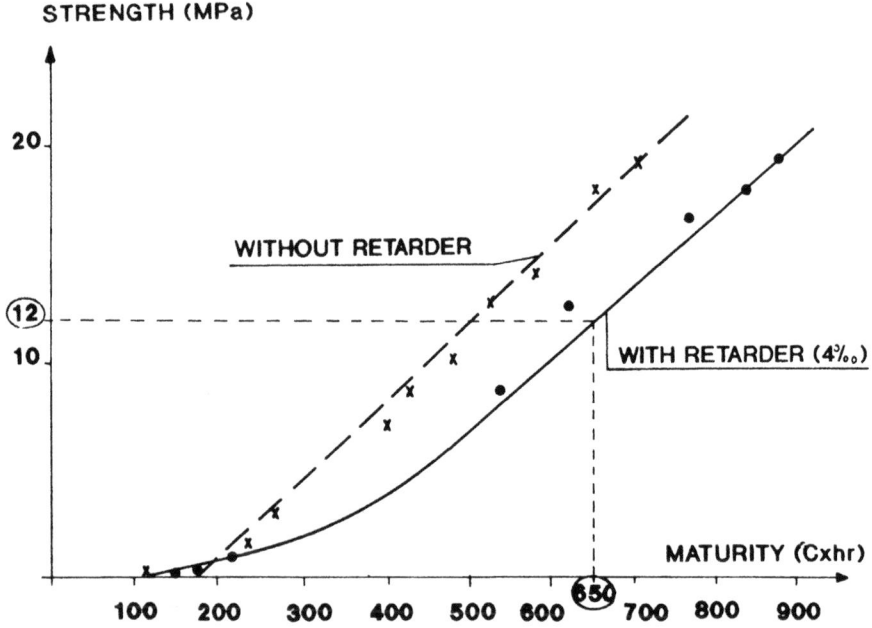

2.3 The Ré Island bridge, France   Fig. 9.

For this project, various concrete mixes incorporating silica fume admixtures were used in the construction of the deck and piles and the underwater concreting ("hydrocrete") of the disks which serve to connect the four pile heads.

The organization of segment precasting is summarized in Table 5.

The composition of the concrete used was as follows :

- Rapid-hardening Ordinary Portland Cement   400 kg
- Water   160 l
- Sand 0/3 m   805 kg
- Gravel 6/20 mm   1 195 kg
- Silica fume   30 kg
- Plasticizer (Melment)   15.05 kg

Slump : 15 cm

Figure 9   **The Ré Island bridge, France**

The 28-day strength for all concrete produced (798 samples) corresponds to an average strength of 67.7 MPa measured on test cylinders. Characteristic strength is 59.5 MPa (standard deviation 6.3 MPa). Results are broken down month by month and shown in Table 6. The statistical summary of results is illustrated by the curve in Figure 10.

As was the case for the Grande Arche project, concrete strength upon removal of formwork was verified through measurement of maturity. Experimental curves of maturity have been established based on tests conducted at the site laboratory of the Engineer for

Table 5

	Number of Segments	Concrete Volume	Quantity of steel	Quantity of cement (t) / d
Productivity	7/day	300 m3	37 t	120 t
Maximum	8/day	340 m3	45 t	136 t
Total	798	33140 m3	4 300 t	13 250 t

Table 6

MONTH (1987)	Strength tests Results (MPa)					28 days strength cement
	N	S 7d	S 28d	Stand. Deviat	F'C28	
MARCH	3	42,6	53,5	6,5	50,8	58
APRIL	33	50,8	59,5	5,2	52,7	59
MAY	61	52,2	63,4	5,9	55,7	55,8
JUNE	98	58,4	68,9	4,2	63,4	56,7
JULY	129	58,8	66,5	6,5	58,1	58
AUGUST	140	55,9	66,4	5,7	59,0	57,7
SEPTEMBER	154	56,1	67,3	4,9	60,9	56,3
OCTOBER	114	60,2	72,8	4,1	67,5	54,9
NOVEMBER	66	61,6	72,5	5,2	65,7	56,9
TOTAL	798	56,9	67,7	6,3	59,5	57,1

N = Number of tests
FC = Caracteristic strength

the project. In accordance with the Engineer's requirements, a maturity envelope was defined using a safety factor of 1.10. Figure 11 shows the relation between development of concrete strength and its maturity.

It is most interesting to note that use of this method enabled us to determine with great accuracy and without risk the proper formwork removal time. Thus, rather that the 15 hours taken into account during the design phase, formwork removal times were cut down to 10 hours for a period in excess of 4 months.

A final favorable repercussion of the above procedure was that control samples needed to be made only for purposes of periodic verification, in order to ensure that the maturity envelope defined in the course of design did indeed encompass the results obtained on site.

Figure 10   **RE ISLAND BRIDGE**

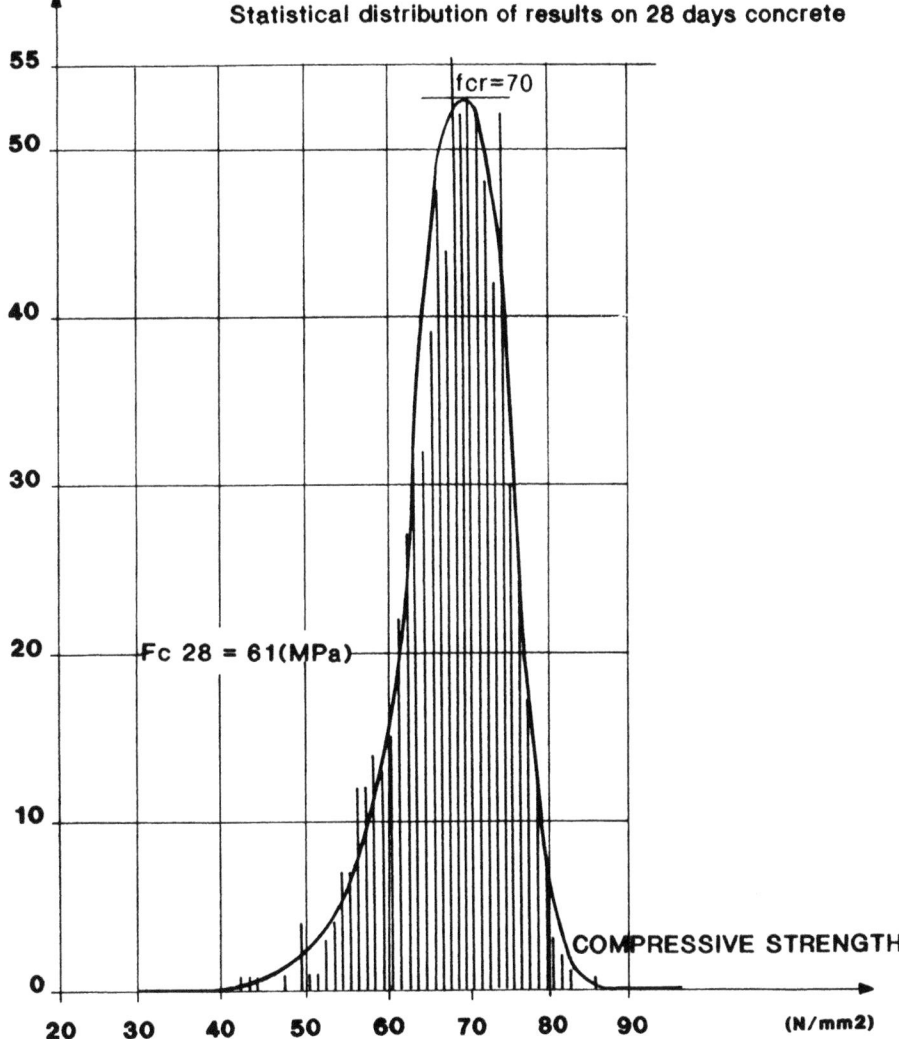

NUMBER OF VALUES

(1 Test value=average of
3 crushing strength test results)

CONCRETE USED FOR
PRECAST SEGMENTS

Statistical distribution of results on 28 days concrete

fcr=70

Fc 28 = 61(MPa)

COMPRESSIVE STRENGTH

(N/mm2)

RE ISLAND BRIDGE

Fig. 11

MATURITY CURVES / B40 A CONCRETE STRENGTH

SPECIMENS FROM SAMPLINGS OF
THE FOLLOWING CONCRETE:

o—o V27-1W (le 03.04.87)
•—• V27-1E (le 07.04.87)
■—■ V27-2E (le 10.04.87)
x—x V25-7E (le 21.05.87)

— — maturity/strength curve
(deduced from the lower envelope.
rising the coefficient:1.10)

maturity coefficient
(°Cxhr)

envelope curve
(coefficient 1.10)

25 MPa
22.5 MPa
20 MPa
12 MPa

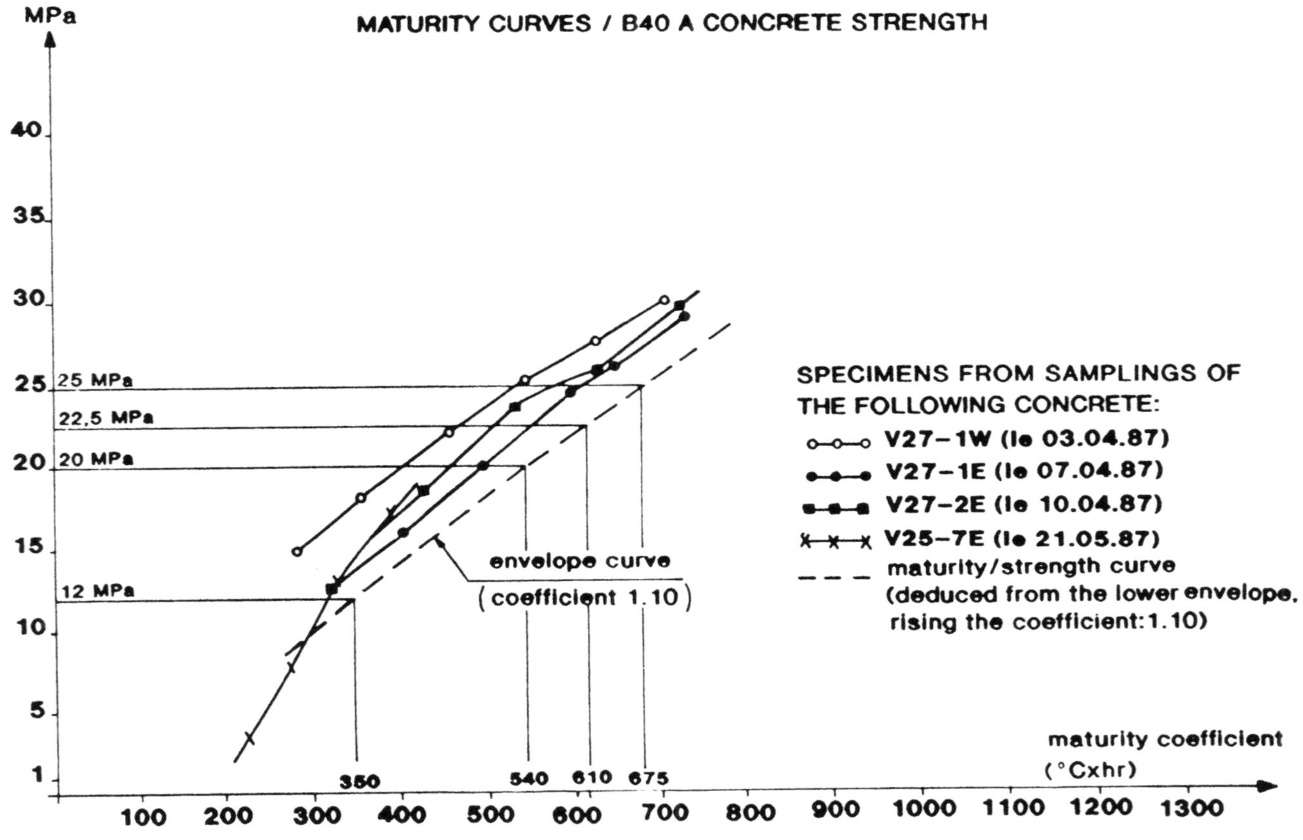

2.4 Perthuiset bridge, France Figure 12

This is a cable-stayed bridge with a 132-m span crossing the Loire River and constructed by sequential cantilevering. The 33 segments are cast in place and the required characteristic strength of concrete is 60 MPa at 28 days.

Figure 12

Composition of the concrete placed by pumping on site is as follows :

- Sand 0/5 mm	800 kg
- Gravel 5/10 mm	280 kg
- Gravel 10/20 mm	800 kg
- Ordinary Portland Cement (rapid hardening)	400 kg
- Water	140 l
- Silica fume	30 kg
- Melment (highrange water reducer)	16 kg

Strengths achieved are the following :

- 33 Mpa at 16 hours
- 60 MPa at 7 days
- about 80 MPa at 28 days.

On the occasion of this construction, we conducted experimental research works in cooperation with the Laboratoire Central des Ponts et Chaussées, in order to determine more precisely the laws governing the behavior of high performance concrete.

The research works that we had carried out previously had shown that the magnitude of creep and shrinkage as well as their rates were quite different from those deduced from application of the regulation.

As regards the geometry control of a structure under construction, it is necessary to know precisely the evolution in time of concrete creep and shrinkage in order for these characteristics to be taken into account in a design program incorporating the actual schedule of operations.

In the case of the above-mentioned bridge, Figure 13 shows the geometry variations when different laws of creep and the effect of the actual construction schedule are taken into account.

Figure 13

PERTUISET BRIDGE

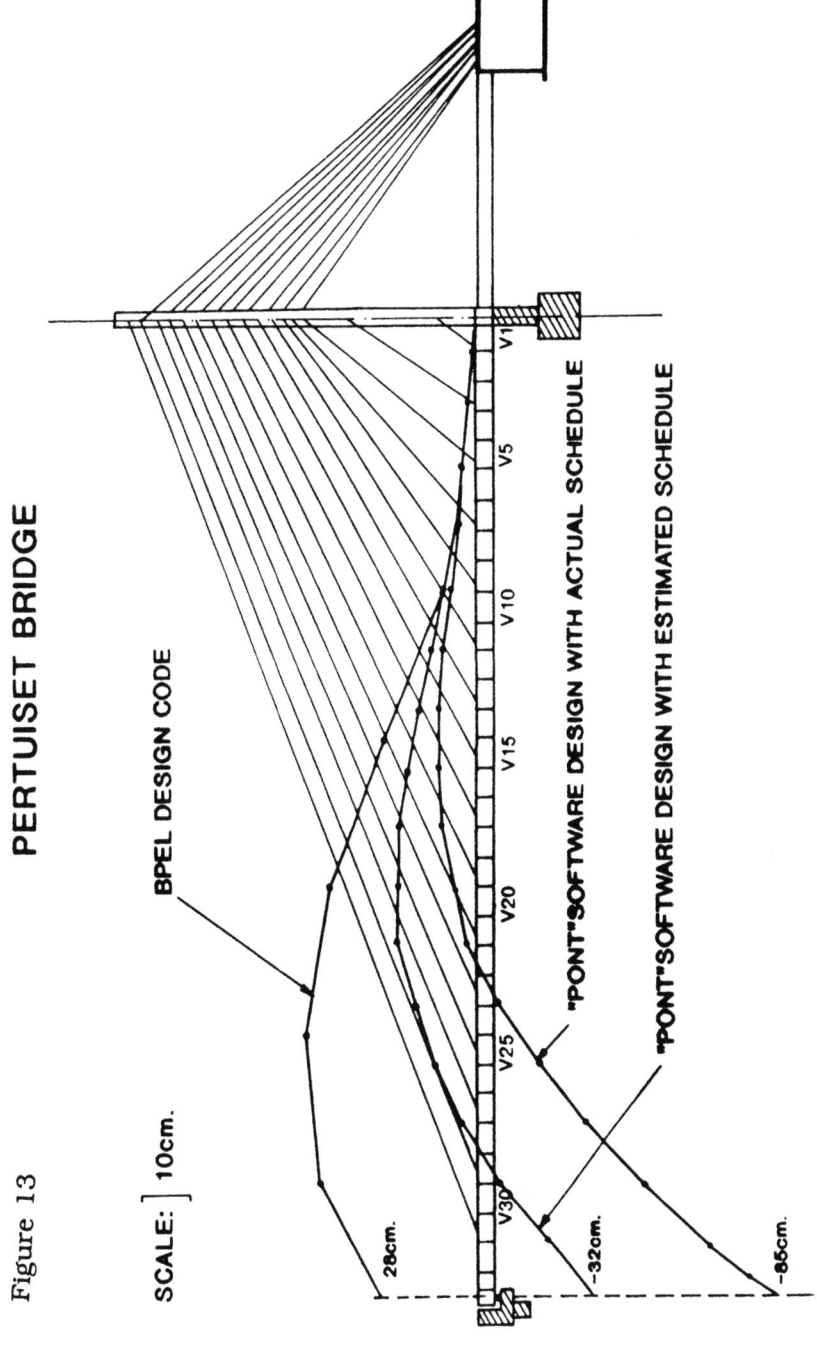

SCALE: ⌐ 10cm.

BPEL DESIGN CODE

"PONT"SOFTWARE DESIGN WITH ACTUAL SCHEDULE

"PONT"SOFTWARE DESIGN WITH ESTIMATED SCHEDULE

V1
V5
V10
V15
V20
V25
V30

28cm.
−320cm.
−85cm.

2.5 Sylans viaduct, France Figures 14 and 15
Concrete used for this structure does not contain silica fume
(except for construction of parapets). However, it is a high-strength
concrete placed on an industrial scale, that is why it is presented
here.

The structure is composed of 5,500 precast elements, each
weighing 0.8 ton. Production is presented in Table 7.

16 X-shaped elements for one segment were cast every day.
Strength at formwork removal was to be at least 25 MPa. That is
why, after a minimum curing period of 12 hours, this strength was
controlled by continuous recording of the concrete temperature.

Composition :

- Ordinary Portland Cement	400 kg
- Water	150 l
- Sand 0/5 mm	800 kg
- Gravel 0/5 mm	389 kg
- Gravel 10/15 mm	740 kg
- Melment (20% of dry matter)	
(16 kg - mix n° 1, 4% of the cement mass)	
(12 kg - mix n° 2, 3% of the cement mass)	

Slump = 20 cm and 23 cm

Figure 14

400

Figure 15 **Sylans viaduct**

Table   8 shows results at 28 days. Characteristic strengths are
60.3 MPa (standard deviation of 6.3 Mpa) for mix n° 1 and 58.5 Mpa
(standard deviation of 5.6 MPa) for mix n° 2. Figure n° 17 shows the
statistical distribution for mix n° 1.

Table 7

	Number of Segments	Quantity of Concrete	Quantity of reinforcement steel	Quantity of cement (t) / d
Productivity	16/day	5 m3	1,44 t	2 t
Total	5 500 u	1 720 m3	495 t	700 t

Figure 16

SYLANS VIADUCTS

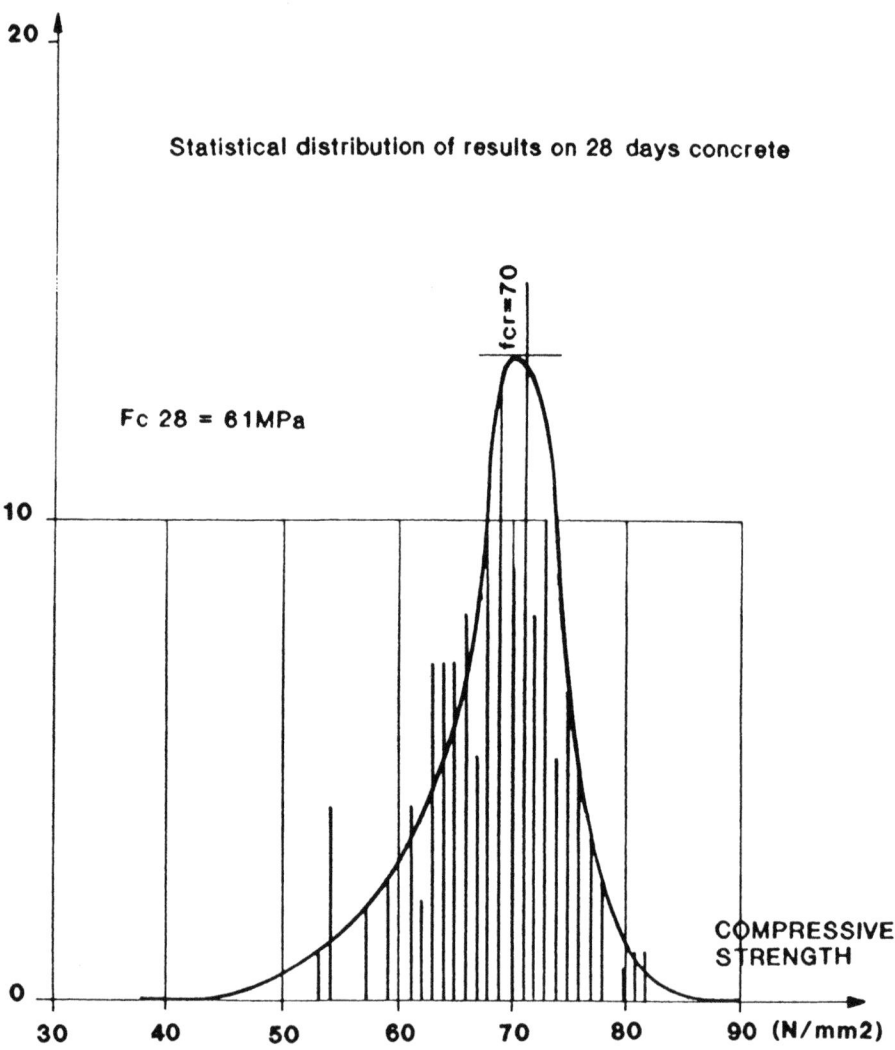

NUMBER OF VALUES

(1test value=average of

3 crushing strength test results)

CONCRETE USED FOR

PRECAST "X" ELEMENTS

Statistical distribution of results on 28 days concrete

fcr=70

Fc 28 = 61MPa

20

10

0

COMPRESSIVE
STRENGTH

30   40   50   60   70   80   90 (N/mm2)

Table 8.

MONTH (1987)	Strength Tests results			MPa)	28 days Strength on cement
	Number of tests	S 28d	Standard Deviat.	Fc 28	
January February	21	65,7	8,3	54,9	58,2
March	22	69,9	6,5	61,4	58,1
April	17	69,7	6,1	61,8	57,9
May	16	70,2	5,3	63,3	57,9
June	22	66,6	6,3	58,5	57,5
July	20	70,1	5,6	61,1	57,4
August	20	67,4	5,6	61,1	57,4
September	9	67,4	3,5	62,8	57,1
MIX 1	147	68,5	6,3	60,3	57,7
October	17	65,9	4,5	60,1	57,9
November	13	64,2	6,7	55,5	58,5
Décember	17	66,8	5,7	59,4	58,8
MIX 2	47	65,8	5,6	58,5	58,4

FC = Caracteristic strength

2.6 Joigny bridge Figures 17 and 18
This structure, carried out within the framework of the National Project on high performance concrete, has a high performance concrete deck (1000 m^3), with concrete produced using two batching units of local ready-mix concrete suppliers.

Figure 17 **Joigny bridge**

Figure 18

Composition of the concrete designed by the Contractor using the concrete mixing plant was as follows:

- Cement (Ordinary Portland Cement)   451 kg
- Ultra-fine sand   105 kg
- Sand 0/5 mm   649 kg
- Gravel 6/20 mm   1030 kg
- Total water   165 l
- High range water reducer   11.25 l
- Retarder   4.5 kg

The various batches were subject to consistency and weight controls the results of which are shown in the following diagrams :

- proportioning of materials : Figure 19
- slump measurement : Figure 20

Besides, the detailed analysis of actual production conditions (actual strength of cement - quantity of water) shows that production faithfully reproduces the results of site and laboratory tests on trials mixes (Figure 21).

Figure 19   **JOIGNY BRIDGE**

BATCHING OF RAW MATERIALS

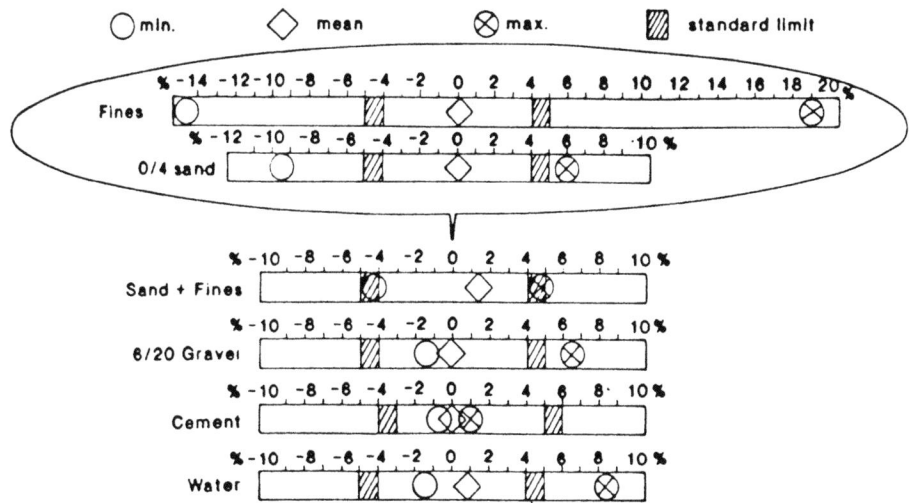

Figure 20

SLUMP TEST

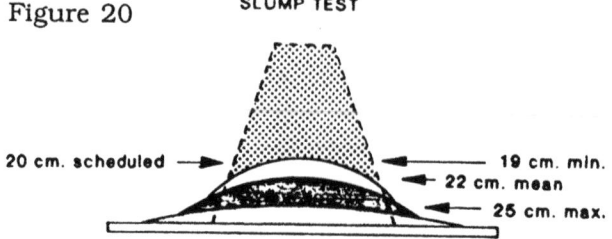

20 cm. scheduled →
← 19 cm. min.
← 22 cm. mean
← 25 cm. max.

Figure 21

PRODUCTION / SUITABILITY COMPARISON TABLE

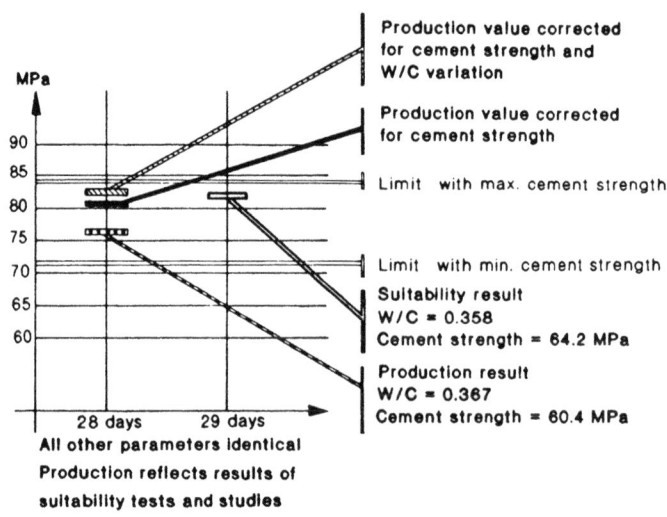

MPa

Production value corrected
for cement strength and
W/C variation

Production value corrected
for cement strength

Limit with max. cement strength

Limit with min. cement strength

Suitability result
W/C = 0.358
Cement strength = 64.2 MPa

Production result
W/C = 0.367
Cement strength = 60.4 MPa

28 days   29 days
All other parameters identical
Production reflects results of
suitability tests and studies

STRENGTH V. TIME

Compression test on 16 / 32 cm. sample cylinders

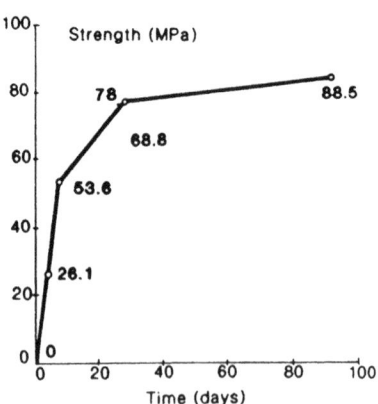

2.7 The great Hassan II Mosque, Morroco Figure 22

Increasing the height of the Minaret of the Great Hassan II Mosque by 25 meters entails the following main structural changes:

- strengthening of walls and columns,
- modification in the bearing sections of piers and mullions located in the openings,
- underpinning of foundations in order to increase their bearing capacity.

Given the ageing of the concrete elements forming the structure and the loads applied, concrete used for underpinning had to have as little impact as possible on the existing structure (compatibility of relative deformations). In this context, we chose to design a high-strength concrete whose behavior with respect to long-term deformations (shrinkage and creep) is compatible with deadline requirements of construction phases.

For this structure, characteristics of high performance concrete used are thus the rates of establishment of deformation and in particular, the high initial rates of the relative deformations of shrinkage or creep.

Figure 22

Concrete composition

For this structure, the need to minimize differential relative deformation between structural parts in concrete of different ages and characteristics, has led us to place high performance concrete.

With respect to water and silica fume contents as well as the type of cement, the composition of this concrete was similar to that of a concrete for which we had studied creep and shrinkage. So we concentrated our efforts on a range of concretes already known, whose creep and shrinkage behavior modeling allowed us to check if methodologies and phasing defined for strengthening works were relevant.

The composition selected for the concrete was as follows :

- Cement (Ordinary Portland Cement)	425 kg
- Silica fume	40 kg
- Aggregate 5/15 mm	1045 kg
- Calcarenite sand 0/5 mm	755 kg
- Total water	165 l
- High-range water reducer (2000 PF)	13.8 l
- Retarder (Melretard)	2 l

Due to the high density of reinforcement (400 kg/m^3) and low concrete placing rates (10 m^3/h), the slump was 24 to 22 cm and had to be maintained for at least one hour.

Concrete characteristics

During the execution of the works, samples of the concrete produced were taken in order to determine and check its characteristics of shrinkage, thermal expansion, heat generation and mechanical strength. In addition, through additional financing provided by the national research Project on high strength concrete, instrumentation and additional tests on creep have been undertaken. These research works, presented in the paper by Cadoret and Courtel, will be published when completed.

Mechanical strength

It was controlled by 19 samples taken during production of the 400 m^3 of high strength concrete. Compressive strength tests were performed in France (LCPC and CEBTP) on series of 3 specimens (16 x 32), ground in order to reduce scattering generally noted in the usual preparation and crushing procedures for concrete specimens. The compressive strength results to date are:

- at 28 days :   95 MPa on average
                (minimum : 85 ; maximum : 102)

2.8 Kwung Tong By pass, Hong Kong   Figures 23 - 25
It is a highway bridge, 3.7 km long, comprising two decks consisting of precast segments. Segments are precast on site, using the mixing plant of a local ready-mix concrete manufacturer.

Figure 23

Figure 24

Figure 25

The main characteristics of the segment production which started in September 1989, are as follows:

- daily production rate of one segment per precasting cell
- early age strength requirements (21 MPa on test cubes at 10 hours), without resorting to heat treatment external to the concrete itself.

Concrete composition
This mix results both from our production requirements and the specific local conditions for selection of the materials as well as their quantity.

- Cement (OPC)	500 kg
- Fines stone	766 kg
- Crushed gravel 5/14 mm	345 kg
- Crushed gravel 14/20 mm	575 kg
- High-range water reducer	4 l
- Retarder	0.1 l
- Total water	190 l

Mechanical characteristics

150 samples have been taken on the 600 m3 already produced, providing the following compressive strength results measured on test cubes (15 x 15 cm):

- at 7 days   65 MPa
- at 28 days   80 MPa

If we apply the relation defined by L'Hermite to equate strength measured on test cube and cylinder, the following values are obtained :

* compressive strength at 7 days on cylinders   56 MPa
* compressive strength at 28 days on cylinders   70 MPa

3 Conclusions

High performance concrete is not a new material. In 1949, Eugène Freyssinet had already used a special excess water removal technique, after casting, in order to increase compaction of segments cast in the precasting yard and said: 'one hour after casting, our concretes, the total thickness of which could be reduced to 12 centimeters in some cases, had comparative strength of more than 50 MPa, with ultimate load reaching 100 MPa after a few days.

The sale of admixtures such as high-range water reducers, has allowed the industrial production of concretes whose rheology can be adapted to construction of structures containing high density reinforcement. In addition, mixing plants now have batching and automation systems, allowing reproducibility and satisfactory control of concrete production.

These technological considerations combined with savings in operation (durability, net surface area, reduction in supports etc..), production (workability, form removal time) or design (reduction in quantities) have led to full-scale uses.

For seven years, our Construction Division has been building structures that use and capitalize upon the advantages of **high performance concretes**. We think that this material has reached full development and should henceforth be considered by all participants in construction **as a new opportunity for them of designing, constructing and operating**.

*   Gaël CADORET is now head of Material Division of the Technical Research Center of CIMENTS FRANCAIS Group.

26 THE DEVELOPMENT OF HIGH PERFORMANCE CONCRETE IN NORTH AMERICA

P.-C. AITCIN and P. LAPLANTE
Université de Sherbrooke, Québec, Canada

1 Introduction

In North America, the development of high performance concrete is closely associated with the construction of columns and shear walls in high-rise buildings. To the best of our knowledge, very few high performance concretes have been used on a large scale for the construction of other structural elements.

Around 1970, some designers started using concretes with compressive strengths higher than 20 to 30 MPa for the construction of columns in high-rise buildings. At that time, the concrete and steel markets in the construction industry were very well-defined: the horizontal market was reserved for concrete; the vertical for steel. It was unthinkable that one day concrete could replace steel in skyscraper construction. Nevertheless, this happened within less than 15 years.

Given the technological context and mind-set of the day, only truly pioneering spirits would have dared to attempt to double the compressive strength of the concrete used in columns overnight. Barely 15 years later, the compressive strengths of these first high-performance concretes look very modest. One must keep in mind that portland cements then lacked today's performance, that they were less finely ground, and that the water reducers then available were much less effective than the high-range water reducers presently used. These lignosulfonate-based water reducers were limited in dosage by secondary drawbacks (excessive retardation and excessive entrainment of large bubbles). These drawbacks resulted from impurities from the sugars and surfactants present in the wood. Moreover, one must remember that the concrete industry in North America was just embarking on the use of fly ash and slag, which had already gained popularity in Europe.

It was in this context that, in 1975, the 262-meter Water Tower Place was built in Chicago. The concrete used for the construction of the columns of the first stories had a compressive

High Performance Concrete: From material to structure. Edited by Yves Malier. © 1992 Taylor & Francis.
Published by Taylor & Francis, 2 Park Square, Milton Park, Abingdon, Oxon, OX14 4RN. ISBN 0 419 17600 4.

strength of 60 MPa. The main structural characteristics of this building and the main characteristics of the concrete used to build the columns supporting the highest load will be presented in detail in order to show what could be done at that time by the pioneers of high-performance concrete.

In the second part of this paper, the characteristics of three modern high-rise buildings built in North America with high-performance concrete will be given in order to illustrate the latest technological developments that have been stunning the concrete industry.

Of course, such technological developments could not happen without one or more technological breakthroughs in the domain of the material itself. As mentioned above, the concrete producers trying to make high-performance concrete at the beginning of the seventies were facing a difficult challenge with the materials available. It was practically impossible to lower the water/cement ratio of the concrete below 0.40 and still provide workability acceptable to the contractor.

With the development of superplasticizers, synthetic water reducers able to disperse cement grains much more efficiently without any secondary effects for dosages ten times that of lignosulfonate, concretes with water/cement ratios between 0.20 and 0.30 became possible.

The marketing of silica fume, an industrial by-product from the fabrication of silicon metals and ferrosilicon alloys, helped concrete producers to produce low water/cement ratio mixtures. Nowadays, field concretes having water/cement ratios between 0.20 and 0.30 a with 200-mm slump one hour after mixing can be delivered.

Of course, such concretes do not contain not enough water to hydrate completely all the cement particles, which puts into doubt drastically lowering the water/cement ratio. An analysis of the source of compressive strength in concrete reveals, however, that the concrete industry waited more than 100 years to rediscover the essence of Feret's work and his famous formula: the compressive strength of concrete is closely linked to the compactness of its microstructure.

The principal objective when fabricating high performance concrete is to obtain the most compact microstructure possible when solidification begins. This means minimally lowering the volume of water not used to hydrate cement grains during the early first step of hydration in order to have cement particles as close as possible to each other. Ideally, of course, one must find the means to provide later on the water molecules necessary to completely hydrate all the cement particles in order to take advantage of the binding properties of the hydrated calcium silicates that could be developed from the anhydrous cement grains. Pushing this idea to its extreme, H.H. Bache successfully made a laboratory microconcrete with a water/cement ratio of 0.16 and a compressive strength of 280 MPa, using, of course, very strong aggregates.

In fact, when trying to decrease the water/cement ratio of a high performance concrete, it rapidly becomes evident that the factor limiting compressive strength is no longer the intrinsic strength of the hydrated cement paste or the bond between the hydrated cement paste and the aggregate. Unlike normal strength concrete, the intrinsic compressive strength of the aggregate used ultimately limits the concrete's strength.

In order to show this technological evolution that has so drastically changed the concrete industry in North America, we have selected four case studies in which high performance concrete was used:

Water Tower Place built in 1975 in Chicago;
the Nova Scotia Plaza built in 1987-88 in Toronto, Canada;
the Two Union Square Building, built in 1988-89 in Seattle, Washington; and
One Wacker Place,under construction in Chicago.

2 Water Tower Place

Water Tower Place, built in Chicago in 1975, represent the most spectacular construction of the pioneer period. During this time, concrete compressive strength inched upwards from 20 to 30 to 60 MPa.

To make a 60-MPa concrete, Material Services Company had to call on all the technological resources available at the time. The best cement on the market was selected by trial and error. While it didn't provide the highest compressive strength on 2-in cubes, it did yield the best rheological behavior in the presence of a water reducer. Similarly, the best fly ash available in Chicago, in terms of strength and rheology, also had to be found. The same applied for determining the lignosulfate with the higher water/cement ratio reduction but the fewest secondary effects. This concrete had to have a minimum slump of 100 mm after one hour, so that the placing crew could be able to place it as easily as a normal-strength concrete.

In Table 1, the composition and the main characteristics of the concrete used to build the columns of the first stories of Water Tower Place are given, (similar to the composition of that used to build River Plaza). It represented the cutting edge of technology in 1975. Material Services Company also had to develop a highly effective quality-control program in order to win the confidence of designers and owners who were quite sceptical.

One particularly attractive aspect of high performance concrete for designers and contractors was that by simply varying concrete compressive strength, prefabricated forms could be used for the

Photo 1: Water Tower Place (Chicago)

Photo 2: Compressive strengths of the concrete used in the columns

columns on every floor. This prefabricated form system was very economical in terms of manpower, since the same form cross-section could be used.

3 The Scotia Plaza

While Toronto's Scotia Plaza (1987-88) wasn't Canada's first high-rise building with high performance concrete, it was the country's first with a design compressive strength of 70 MPa. It also ranks as the world's first high-rise building with a high performance concrete containing blast furnace slag. Moreover, it was the first field high performance concrete to yield extensive field data throughout the construction: 142 loads of high performance concrete were carefully controlled. In Table 2, the composition and the mechanical characteristics of the concrete used to build Scotia Plaza are reported.

Finally, the concrete delivered to the site had an average compressive strength of 93.4 MPa with a standard deviation of 6.5 MPa, which corresponds to a design strength of 85.1 MPa, much higher than the designer expected.

Such a premium of compressive strength was given only because there was the Bay Adelaïde building on the designers table in Toronto designed at 85 MPa. In fact, it was found that such compressive strengths had to be reached in order to make concrete more economical than steel in building a new high-rise building in downtown Toronto.

The development of a 93.4-MPa average compressive strength high performance concrete was done in a dry batch plant. It was necessary to develop a mixing technique to make very low water/cement ratio directly in the drums of the transit mixers.

The delivery of such a concrete in summer necessitated the development of a quite original delivering technique in which the concrete was cooled using liquid nitrogen. It was also necessary to develop a very particular quality-control system in order to optimize the ascending speed of the jumping forms. This system was based on pull-out tests made with special metal anchors. The contractor had five pull-out tests, plus a correlation curve giving the compressive strength as a function of pull-out strength. The contractor was responsible for asking to have the anchor pulled out when he thought that the concrete had reached the minimum strength specified in order to allow stripping. The average of the 3 last pull-out tests were used to calculate the theoretical compressive strength of the high performance concrete. With time, the contractor learned when to carry out the pull-out tests, a remarkable feat when one realizes that concreting for the Nova Scotia Plaza was done while external temperatures varied between -20°C and + 30°C.

Table 1: River Plaza, Chicago 1974
(same concrete as for Water Tower Place)

	kg/m³
Water	195
Cement	505
Fly Ash	60
Coarse Aggregate	1 030
Fine Aggregate	630
Water Reducer	975 cm³

Age (d)	MPa
7	50.3
28	64.8
56	72.4
91	78.6

Table 2: Scotia Plaza, Toronto 1988

	kg/m³
Water	145
Cement (US Type I)	315
Slag	137
Silica Fume	36
Coarse Aggregate	1 130
Fine Aggregate	745
Superplasticizer	5.9 L
Water Reducer	900 cm³

Age (d)	MPa
7	66.9
28	83.4
56	89.0
91	93.4

4 Two Union Square Building

Two Union Square Building, a 58-storey skycraper in Seattle, Washington, presents a very innovative design. Built in 1988-1989, it stands as the world's first structure with high performance concrete averaging 120 MPa and using tube-confined concrete. Confining high performance concrete helped alleviate high-rise oscillations due to wind. Some New York high-rise buildings built with steel frames sway so much that workers in the top third of the buildings have to leave work due to motion sickness on windy days. People in the steel industry claim, however, than the problem lies more with the workers than with the steel.

Even though a compressive strength of 90 MPa was sufficient for the structural compressive load, the designer specified 120 MPa so that the columns would be rigid enough so that occupants on the top floor would enjoy the same comfort as those on the ground floor during storms.

The composition of a similar concrete used to build Two Union Square in Seattle is shown in Table 3. It was made in the laboratory by the author.

Table 3: Two Union Square, Seattle(USA)
Personal results

	kg/m^3
Water	130
Cement (type I/II)	513
Silica Fume	43
Coarse Aggregate	1 080
Fine Aggregate	685
Superplasticizer	15.7 L/m^3

Age (d)	MPa
2	64.8
7	90.7
28	119
91	145

Such concretes were possible because high-quality materials were available in Seattle. Both very efficient portland cement and aggregates are available there. The cement used is ASTM Type I/II portland cement, which meets the standards for Type I and II cements. It has a very low alkali content, less than 0.4% in terms of Na$_2$O equivalent. The aggregates were from a glacial gravel pit. Aggregate particles were eroded from a very hard syenite rock. In spite of the fact that very high compressive strength has been reached with these aggregates, meticulous petrological examination revealed that some were slightly altered, leading us to conclude that 150 MPa field concrete could be reached with natural aggregates.

Using the same material as the concrete producer, the author was able to make a 152-MPa laboratory concrete measured on 100 x 200-mm specimens with faced ends.

Normal strength concrete used in the building and the high-performance concrete used in the columns were placed at night in order to avoid the daily traffic jams. Almost 1000 cubic yards of concrete were placed nightly.

The four steel pipes in which the 120-MPa concrete was confined measured ten feet in diameter. They were made of 3/8" steel sheeting with 12-inch Nelson studs welded every foot in all directions along the inside. The steel pipes were fabricated in Korea and transported by barge across the Pacific Ocean for delivery to the site.

5 One Wacker Place

One Wacker Place, under construction in Chicago aims at beating a number of world records. Its finished height of 290 m will make it the world's tallest concrete structure. Columns in the first thirteen stories will use a 80-MPa concrete, gradually dropping to 40 MPa in the upper storey. The slab concrete will be a record-breaking 60 MPa. The slabs will be supported by post-tensioned beams with a run of 5.8 m to 10 m and spaced at 2.8 to 3 m.

The designer calculated that this design using high-performance concrete will be 25 m shorter than a similar steel structure with the same number of storey. Reductions in total building height translate into savings in facing the building, in all service accesses (including elevators shafts), and in maintenance and cleaning costs.

Some 100 000 cubic meters of pumped high performance concrete will go into this building. Two sets of flying forms will be used to build a full storey per week.

6 The future of high performance concrete in North America

It is clear that high performance concrete will be used increasingly for high-rise buildings throughout the United States, forcing concrete producers to learn how to make concrete whose water/cement ratio will continue to drop.

Cement producers will have to develop specific cements to facilitate the fabrication of high-performance concrete with a satisfactory rheological behavior for one hour after mixing: specifically, cement that yields concrete with a 200-mm slump one hour after mixing.

High performance concrete will also incorporate less cement and superplasticizer, as these two key ingredients become more efficient. On the other hand, it will have ever-higher fly ash or slag content.

The two main limiting factors in the years to come will be aggregate strength and the resistance of designers. At present, making high performance concrete over 150 MPa with natural aggregates or crushed rocks that are currently used in North America to make regular concrete seems beyond our grasp. Undoubtedly, someone will clear this technological hurdle only to face another looming just beyond the horizon.

Finally, the concept of high-strength concrete will gradually come to mean high performance in the wider sense of the term. Applications will be more specific, especially in terms of their longer durability, abrasion resistance, higher rigidity, higher tension resistance, lower creep characteristics, or the rapidity with which final creep is reached. In such specific applications, the high level of compressive strength will be seen as value added.

27 EXPERIMENTATION : THE KEY TO A RESEARCH-DEVELOPMENT STRATEGY IN CIVIL ENGINEERING

Y. MALIER
ENS Cachan, France

Over the decades up until the beginning of the 1980s, a separation between **fundamental research**, **applied research**, and **development**, was commonly acknowledged by all those in charge of research, both public and private. For a few years now it has become clear in most industrialised countries that, if there is still a distinction, it no longer amounts to a separation.

On the contrary, it is easy to demonstrate, using precise examples drawn from the various industrial sectors (aeronautics, food industry, production engineering, metrology, civil engineering, etc.), that it is from the existing continuous chain between these three factors that the most recent innovations have been derived - the other determining factors being concerned with cost, the engineer's talent, and mere chance.

The **specific features** of civil engineering, whether **technical** (a strong prototype industry), **economic** (nature of the market), **contractual** (guarantee of quality and safety of the structures), **geographical** (cost of local resources), or **human** (an industry with much manpower) and **sociological** (construction consists in creating the national heritage rather than consumer goods) imply that the chain, to be effective, should comprise an additional link: experimentation.

1 The part played by experimental structures in "new ways for concrete"

At the heart of the research and development relationship, experimental structures play an essential part, since they are the only real way :

1. to determine the **specified models** for engineers rigorously, using research results,

High Performance Concrete: From material to structure. Edited by Yves Malier. © 1992 Taylor & Francis.
Published by Taylor & Francis, 2 Park Square, Milton Park, Abingdon, Oxon, OX14 4RN. ISBN 0 419 17600 4.

2. to improve and/or demonstrate the **feasibility of new materials** and/or new placing methods, formerly little known outside research laboratories,

3. to evaluate the contribution of these new materials to **the behaviour of the structure** (construction, service, durability),

4. to open up **economic prospects** and to apply the conclusions to the means of optimisation,

5. to contribute to **determining new research directions**, by making them partly end-user orientated, which is the best guarantee of good valorisation of research in the future,

6. to analyse rigorously ways to improve **working conditions** on our sites,

7. and, provided that the **reproducible character** of the experimentation is demonstrated, to convince the various partners in the building process (owners, contractors, design offices, firms, ... and control organisation).

2 The conditions for effective experimentation

Without any claim to being exhaustive, we think we can draw from our experience a few essential conditions.

1. A convinced owner

In this respect we wish to say once again what an essential and outstanding part the "Direction des Routes" (Roads Administration) and the "EDF" (French Electricity Board) have played in the National Project.

2. A design office and contractors open to innovation.

3. Precise and limited objectives

... Owing to the fact that the many technical specific mentioned in the introduction are better suited, as far as innovation is concerned, to three large steps perfectly mastered than to the one big leap envisaged by the researcher, who sometimes underestimates some of these aspects.

4. A reproducible structure representative of a technically and/or economically important market

Thus, rather than aiming for the exceptional, our leitmotiv in the field of experimentation is "a structure of ordinary dimensions on an ordinary site".

5. Thorough instrumentation
Thorough, but also realistic, from the researcher's point of view, of course, but equally from the point of view of compliance with the contractors' methods and schedules for the project?

6. Broad diffusion of results and lessons learnt.

7. A trusting relationship between all the partners (owners, contractors, design offices, firms, laboratories, etc).

28 THE JOIGNY BRIDGE: AN EXPERIMENTAL HIGH PERFORMANCE CONCRETE BRIDGE

Y. MALIER, ENS Cachan, France,
D. BRAZILLIER, DDE Saône & Loire, France
S. ROI, Dalla Verra, France

After extensive research and development in French laboratories, the French Ministry of Public Works and the National Project on New Concretes team agreed to build an experimental bridge using high performance concrete. The organizations wanted to demonstrate the feasibility of building a typical prestressed bridge with high performance concrete, using unsophisticated means and materials that could be found throughout France.

When choosing the nature and location of the bridge, several prerequisites were followed:

- The bridge should be representative of the current bridge construction and should not be unusual in any way.
- The bridge should not be located near large towns or industrial areas with convenient facilities.
- The 28-day strength of the concrete should be 60 MPa. Generally, French bridges are built with 35 to 40 MPa concretes.
- The bridge should be built using local cement and aggregates. To check their capabilities, local ready-mixed plants should furnish the concrete. The use of silica fume should be excluded since it is not necessary to obtain the specified strength.
- The bridge should be designed according to the French building code dealing with prestressed concrete structures. All the possibilities offered by the enhanced strength of the concrete should be exploited.
- The bridge should have monitoring instruments built in to follow its behavior and to verify the validity of the conceptual approach. (This condition led to the specifying of cast-in-place construction methods in order to avoid interferences with the instrument caused by temporary stress distributions from construction).

High Performance Concrete: From material to structure. Edited by Yves Malier. © 1992 Taylor & Francis.
Published by Taylor & Francis, 2 Park Square, Milton Park, Abingdon, Oxon, OX14 4RN. ISBN 0 419 17600 4.

1 The bridge

The bridge was built crossing the river Yonne near the town of Joigny, approximately 150 km southeast of Paris. Aesthetical and economical considerations led to the classical design of a balanced, continuous three-span bridge, with span lengths of 34.00, 46.00 and 34.00 m, a height of 2.20 m and an overall width of 15.80 m (Photo 1 and Table 1).

Table 1

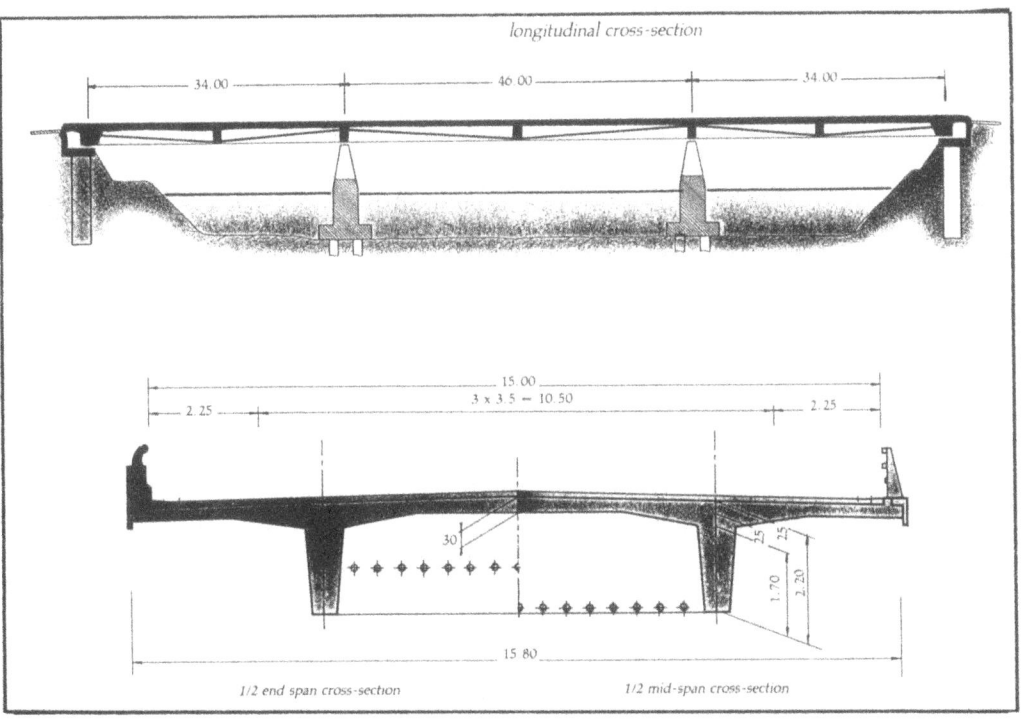

The bridge's double-tee cross-section is made of two main beams with trapezoidal cross-sections and upper slab. The distance between the beam axes is 8.00 m. Their minimal width, at the bottom chord, is 0.50 m.

Photo 1: The bridge of Joigny

Photo 2: The bridge's 13 external tendons allowed simple and accurate prestressing force measurements

2 Prestressing

The structure is prestressed longitudinally by 13 external tendons, each made of 27 15 mm strands (Photo 2). External tendons offered three advantages:

- The width of the webs could be reduced to a minimal value determined chiefly by normal stress considerations. A lighter structure was thus obtained.
- The tendons may be easily replaced if necessary. This is a plus because the 60 Ma is more durable than ordinary concrete and may have a longer life span than the high-tensile steel.
- In the case of this experimental structure, the layout allowed simpler and more accurate prestressing force measurements in the tendons.

3 Design stress

The bridge was designed according to the French codes BPEL (Béton Précontraint aux Etats Limites [Limit-States Design of Prestressed Concrete]) and BAEL (Béton Armé aux Etats Limites [Limit-States Design of Reinforced Concrete]). These codes have been upgraded to incorporate 60 MPa concretes since they previously dealt only with concrete strengths up to 40 MPa.

The use of these codes, which include different safety factors, led to an actual maximum compressive stress of 30 MPa in the lower fiber to the central span's mid-section during the last stages of prestressing.

4 High-strength vs. ordinary concrete

It should be emphasized that comparisons carried out during the preliminary design of the bridge showed that the concrete quantities could be reduced from about 1 395 m^3 when using ordinary 35 MPa concrete to 985 m^3 with 60 MPa high-strength concrete (Table 2). This 30 percent reduction in concrete volume led to a 24 percent load reduction on the pier, abutments, and foundations.

The decreased dead weight also induced savings by reducing the number of prestressing strands. Since the height-to-span ratios were not identical in the two solutions, the steel savings were not as large as they could have been.

Table 2: comparison 60 MPa / 35 MPa

	Solution 60 MPa	Solution 35 MPa
Height/span ratio	1/20.9	1/18.4
Concrete	985 m3	1 395 m
Longitudinal tendons	47 tonnes	52 tonnes
replaceable cables	yes	no
reinforcement	160 tonnes	157 tonnes
Total weight of bridge	3 200 tonnes	4 230 tonnes

5 Concrete design

Laboratory test were run to define a mix design allowing the production of ready-mixed concrete with:

- A 28-day mean strength of about 70 MPa with a minimal standard deviation.
- The ability to be transported 30 km on a boat from the concrete plants and delivered fresh to the construction site.
- The ability to be pumped through 120 m long pipes.
- A high workability and a sufficient setting time.

The concrete batch constituents shown in Table 3 also included a water-reducing admixture and retarder.

Table 3: Batch constituents

River Yonne sand	0 - 4 mm
Corrective sand	0.080 - 0.13 mm
Coarse aggregate	6 - 20 mm
HS PC	450 kg/m^3
Water	161 L/m^3

6 Fresh concrete

As the concrete plant was producing concrete for the bridge, tests were run on every batch. The water-cement ratio remained between 0.36 and 0.38. The entrained air contents were within 0.5 and 1 percent. The slumps, measured at the site, were over 200 mm for more than 2 hours.

7 Compressive strength

The concrete strength was measured according to French standards using 160 x 320 mm test cylinders cast in metallic molds. The average compressive strengths are shown in Table 4.

Table 4: Average compressive strengths, MPa (160 x 320 mm references cylinders)

3 day	26.1
7 day	53.6
28 day	78.0
57 day (150 mm diam in situ cores)	86.1
1 year	102.0

At 28 days, the minimum and maximum strength values were 65.5 and 91.7 MPa and the standard deviation reached 6.8 MPa. The French construction code (Fascicule 65) required that:

- The mean strength be at least equal to the characteristic strength plus 1.3 times the standard deviation.
- The minimum strength value be larger than the characteristic strength minus 3 MPa.

These 28 day strength requirements were easily met. Comparative tests were performed on different types of samples (Table 5).

Table 5: Comparative sample test, MPa

160x320 mm reference cylinder	160x320 mm ground bearing faces	110x220 mm cylinders	100 m^3 cube	150x300 mm in-situ core	100x200 mm in-situ core	70x40 mm in-situ core
71.9	71.2	67.7	83.9	76.7	84.2	75.5

8 Tensile strength

The tensile strength was measured on cylinders with the "Brasilian" splitting test. The average tensile strength was 5.1 MPa on 28-day samples.

9 Concrete placement

The contract required that the concrete be placed in one continuous phase. To assure continuous placement, two independent ready-mixed plants were to deliver the same concrete. As planned, the 1 000 m^3 of concrete was placed without incident by two pumps in 24 hours.

10 Instrumentation

Monitoring instruments were built into the bridge to verify the assumptions in the calculations, to check its long-term behavior, and to assess its durability compared to ordinary bridges. The experiment is being run over several years by the "Laboratoire Central des Ponts et Chaussées (Central Laboratory for Roads and Bridges)", which is headed by the Ministry of Public Works. The three goals of the experiment are to measure :

Thermal evolution. The evolution was measured during the setting phase at mid-central span and in the massive end blocks. The maximum temperatures measured were 73 C in the middle of the end blocks, 57 C in the webs, and 32 C in the uper slab.

A special finite element analysis program was run to check these values. To calibrate the calculations, a quasi-adiabatic test was carried out to get the temperature variations of a concrete specimen. The results were in good agreement with field measurements.

Prestressing forces and deformations. The time evolution of the tensile forces in the longitudinal tendons were measured. The first measured values were 5 100 kN per tendon near the anchorages

430

and 5 200 kN at mid-central span. The calculations predicted
5 100 kN and 4 900 kN, respectively. The differences are attributed
to an overestimate of friction at the deviator transverse beams.

The deformations of a cross section located near the middle of
the central span are continuously recorded. The rotations of the
same span are also measured and recorded. As of the writing of this
article, the measured deformations were 15 percent smaller than
those predicted. This discrepancy is attributed to an underestimate
of the Young's modulus.

Creep and shrinkage. The major part of the bridge experiment is
devoted to the study of shrinkage and creep of the concrete. Sixteen
tests are to be carried out during the next few years on various
specimens:

- Two shrinkage tests (with and without drying) and two creep
  test (with and without drying and loaded at 28 days) will be
  performed on specimens made on-site with field concrete.
- Four shrinkage tests and eight creep tests (loaded and unloaded
  at different ages) will be performed on laboratory specimens
  made with concrete using the same materials and mix
  proportions as the field concrete.

The results of these tests will be available in a few years.

11 Conclusions

The first French prestressed concrete bridge designed and built
with a 60 MPa characteristic strength and following the French
building codes was successfully completed in early 1989. The bridge
is going to be incorporated into one of the main French roads. This
first achievement of the National Project on New Concretes is to be
followed by further steps:

- In accordance with the French Road Direction wishes, the use of
  60 MPa will be extended to the construction of most French
  bridges in the coming years.
- Other experimental bridges implementing concrete with even
  higher strengths will be built in the near future.

29 EXPERIMENTAL MONITORING OF THE JOIGNY BRIDGE

I. SCHALLER, F. DE LARRARD, J-P. SUDRET,
P. ACKER and R. LE ROY
LCPC, Paris, France

High strength concretes (60 to 80 MPa) and very-high-strength concretes (80 to 100 MPa) are currently the subject of many researches. However, most of these researches are aimed at investigation of the materials rather than analysis of the structures built with the concretes. This is why a structure such as the Joigny bridge is of special interest of research in this area. It was accordingly decided, as part of the "Voies Nouvelles du matériau Béton" (New Ways for Concrete) National Project, that this bridge would be monitored experimentally, by the Laboratoire Central des Ponts et Chaussées and the Autun Regional Laboratory jointly.

1 Objectives of the experiment

The experiment had three main objectives :

- to investigate thermal effects in this concrete during setting, especially in massive sections;
- to check the behaviour of the bridge during its construction, especially in the various prestressing stages, and to compare it with the predictions yielded by the engineering calculations ;
- to track the medium-term performance of the bridge (for two years), with special emphasis on the effects of creep in the high-strength concrete structures.

2 Experimental plan

The instrumentation of the deck of the bridge was chosen on the basis of the three objectives mentioned above. A section of the deck 1.67 m from mid-span was instrumented with:

- 21 acoustic strain gauges to track the strains in the concrete over time (see Fig. 1); in this case, acoustic strain gauges were

High Performance Concrete: From material to structure. Edited by Yves Malier. © 1992 Taylor & Francis.
Published by Taylor & Francis, 2 Park Square, Milton Park, Abingdon, Oxon, OX14 4RN. ISBN 0 419 17600 4.

chosen in preference to resistance strain gauges because of their excellent long-term reliability;
- 29 platinum temperature probes (see Fig. 2), 21 of which were paired with the acoustic strain gauges to allow the temperature corrections necessary for proper interpretation of the results; the eight additional temperature probes in the core of one of the ribs and in the axis of symmetry of the section served to monitor the temperature rise of the concrete during hardening.

Figure 1 : Positions of acoustic strain gauges.

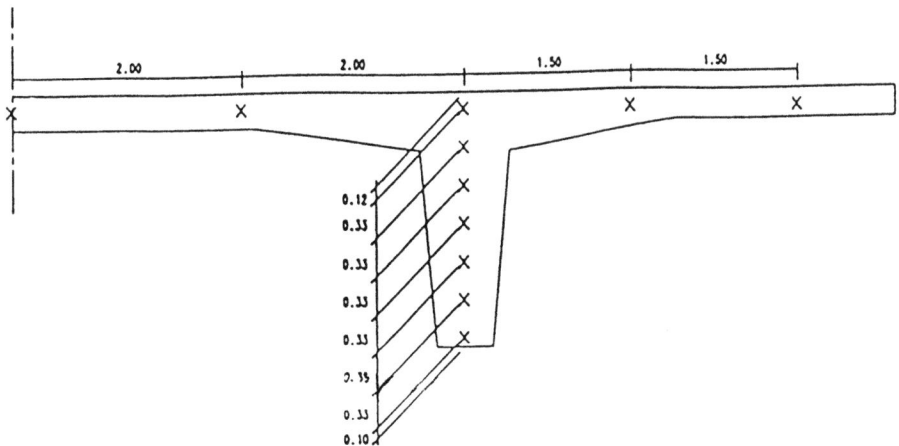

Figure 2 : Positions of temperature probes in the most heavily instrumented half-section.

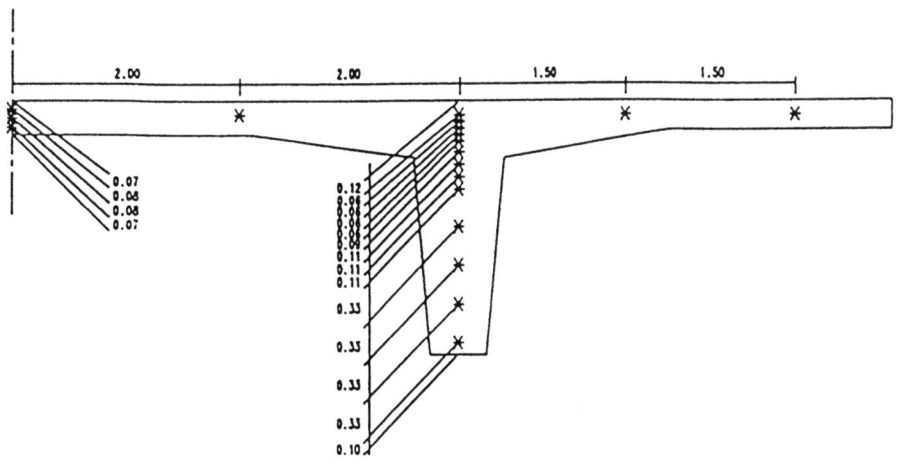

Three other temperature probes were implanted in the right-bank end, a very massive part, to measure temperatures during the setting of the concrete.

Two silica inclinometers were attached under the deck in the central span, at 4.40 and 6.50 m from the pier on the right-bank side (see Fig. 3), to track the rotations or curvatures resulting from instantaneous and deferred deformations of the deck. The positions of these two inclinometers were determined on the basis of the engineering calculations (the rotations must be large enough to be significant but not so large as to exceed the measurement span of the instruments).

An data-logging unit using teletransmission made it possible to vary the frequencies at which all of these measurements were made.

In addition, the longitudinal prestressing forces were measured periodically using a prestressing tendon vibration method. This method, developed at the LCPC, makes it possible to determine the tension of a cable from its measured frequency using the formula for vibrating strings, $f_n = nT/2l\mu s^{-1}$.This method, very rapid to apply, gives a rather precise indication of the tension of the prestressing tendons.

Concurrently with these measurements on the structure, shrinkage and creep tests were conducted at the LCPC on concrete taken from the site and on a concrete prepared in the laboratory from the same materials.

Finally, finite-element calculations (TEXO program) are planned to simulate the thermal effects occuring in the structure during the hardening of the concrete. In order to parameterize these

Figure 3 : Positions of clinometers.

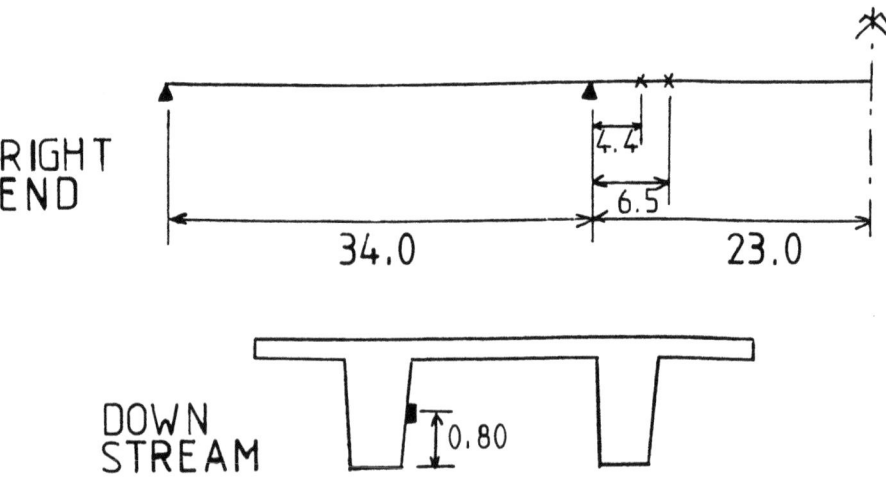

calculations, a quasi-adiabatic test was carried out at the site starting on the day or pouring. This test consists of measuring the temperature rise of sample of concrete in an insulated enclosure. The calculations, once validated by temperature measurements taken in the field, can identify the zones of the deck that are most stressed as the concrete hardens.

3 Shrinkage-creep tests

As mentioned above, the experimental plan comprised the measuring of shrinkage after mould release and creep at different loading levels and different ages. Two types of curing were applied to the specimens after mould release : storage with no exchange of moisture, and with drying at 50% R.H., both at 20±1°C, conditions bracketing the actual conditions to which the concrete is subjected in the structures.

3.1 Materials
The composition and strength values of the concrete are given in table 1. The cement content is explained by the high-strength values aimed at, which also made it necessary to use a large proportion of superplasticizer. A retarding agent was also used to prevent the concrete in place from setting before the casting of the deck was completed (this took about 24 hours). The aim was in fact to allow the concrete to adapt to the deformations of the formwork during the casting of the successive spans of the structure. Four 1000 mm x 160 mm samples were taken on site during the casting of the deck. Samples of the raw materials were also stored so that additional specimens could be prepared in the laboratory (2 batches) for early loading tests.

3.2 Tests
The specimens are cylinders 160 mm in diameter and 1 m long. The sections 250 mm from the ends were fitted with three inserts,

Table 1 : Composition of HSC (kg/m^3)

Aggregates	Coarse 5/20	1027
from the Yonne	Sand 0/4	648
Fine sand 0/1:		105
Portland Cement 'CPA HP' from Cormeilles:		450
Added water:		158
Superplasticizer(melment 40% dry content):		11.25
Retarder (melretard)		4.5
Slump		230 mm

attached in advance to the inside of the mould, at 120° intervals along the perimeter of the section. The measurements after mould removal were therefore made over a length of 500 mm in which the Navier-Bernouilli principle applies. The symmetrical placement of the insert also served to avoid spurious bending effects.

Metal rods are placed in the inserts and two plates bear against the upper ends of the rods (one at the lower section, the other at the upper section). A mechanical comparator is then used to measure the distance between the two plates, giving the exact mean shrinkage of the concrete over the 500 mm baseline.

The concrete is loaded using a flat "Freyssinet" -type jack on which the specimen rests. A bottle containing oil and nitrogen is first pressurized, then suddenly connected to the jack via a three-way valve. Loading is therefore accomplished in a few seconds, making it possible to obtain the early creep of the material. The system also ensures excellent maintenance of the applied load, because of the great compressibility of the nitrogen.

The experimental plan is given in Table 3. It was impossible to load the Joigny concrete very early, because setting was considerably delayed by the high proportion of retarder. For each age and each loading level, the tests were carried out on specimens with and without drying in an enclosure at $20 \pm 1°C$ and $50 \pm 10\%$ R.H. The specimens protected from drying were covered with an adhesive aluminium sheet.

Table 2 : Shrinkage Tests. $\in_\infty$ is the extrapolated long-term shrinkage after demolding.

N°	1	2	3	4	5	6
Batch	site	lab. (1s tbatch)	lab. (2n db)	site	lab. (1s tb.)	lab. (2n db)
Curing	sealed	S.	S.	Unsealed	U.	U.
$\in_\infty. 10^{-6}$	87	130	150	544	578	578
α	.43	.44	.41	.67	61	.56
$\beta(hours^\alpha)$	6.9	19.5	18.2	307	210	111

3.3 Analysis
3.3.1. Shrinkage (Fig. 4)
This is described by a function of the type:

$$\epsilon_s = \frac{\epsilon_\infty \, (t-t_0)^\alpha}{(t-t_0) + \beta}$$

where ϵ_∞ is shrinkage deformation, and t time. The values of parameters ϵ_∞, α and β used to smooth the curves are given in Table 2. With regard to autogenous shrinkage (measured on sealed specimens), the long-term value (mean value 120×10^{-6}) is comparable with the one measured on NSC elsewhere [1]. However, a certain part of the deformation, occurred before 3 days, has not been measured. So the total autogenous shrinkage, from the setting to the old age, is supposed to be somewhat higher than that of NSC.

The drying shrinkage is the difference between the total shrinkage of a drying specimen and the autogenous shrinkage of an identical specimen that has undergone no dessication. As for total shrinkage of drying specimen, and following the French building code "BPEL 83" [2], this type of specimen, made with NSC, would lead to the values $\epsilon_\infty = 482 \times 10^{-6}$. Then the drying shrinkage of this type of HSC ($\epsilon_\infty = 430 \times 10^{-6}$) appears to be of the same order than the one of NSC, with a comparable kinetic. As usual, the kinetic of autogenous shrinkage is faster, because the hydration is almost stabilised after some weeks, whereas the drying process continues during several years.

Table 3 : Creep and mechanical tests. *S : site, L1 : 1st laboratory batch. L2:2nd laboratory batch. **S : sealed specimens, U: unsealed specimens.

N°	1	2	3	4	5	6	7	8	9	10	11
Batch	S.	L2.	L2.	L1.	L1.	L1.	S.	L2.	L2.	L1.	L1.
Curing**	S.	S.	S.	S.	S.	S.	U.	U.	U.	U.	U.
f_c(MPA)	78	40	40	?	58	70	78	40	40	58	70
to (d.)	28	3	3	5	7	28	28	3	3	7	28
σ(MPA)	?	14.4	9.6	7.7	14.0	22.6	21.6	13.7	9.0	13.4	22.3
σ/f_c(%)	?	36	24	?	24	32	28	34	22	23	32
$\epsilon_{c\,r}10^6$	460	582	367	297	613	981	1910	1230	593	1294	1742
$K_{c\,r}$	?	1.78	1.68	1.69	1.92	1.99	3.89	3.95	2.90	4.2	3.44
β(d. 1/3)	17.0	3.6	4.5	12.2	12.9	18.6	32.5	10.9	10.9	32.7	23.2

Figure 4a : Shrinkage of Series A HSC (strains in millionth vs time in hours). The references of the specimens are reported in table 2.

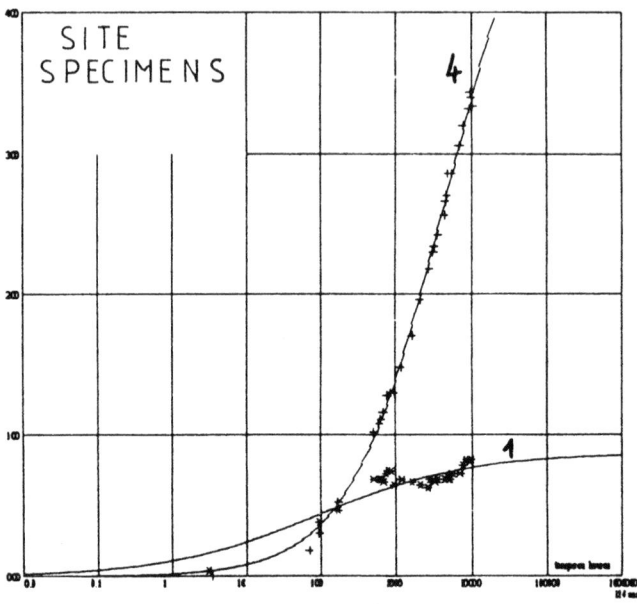

3.3.2. Creep (Fig. 5A and 5B)

The sum of elastic deformation, shrinkage deformation, and creep is measured on loaded sealed specimens. The creep term is smoothed by the following empirical expression:

$$\in_{cr} (t, t_0) = K_{cr} (t_0) (\sigma/E_{i28}) f(t - t_0)$$

where E_{cr} is the basic creep deformation, t time, t_0 the age of the concrete at loading, K_{cr} the creep coefficient, σ the stress applied at t_0, E_{i28} the modulus at 28 days, and $f(t - t_0)$ a kinetic function. The kinetic function is assumed to be

$$f(t - t_0) = \frac{(t - t_0)^{1/3}}{(t - t_0)^{1/3} + \beta}$$

Figure 4b : Shrinkage of Series A HSC (strains in millionth vs time in hours). The references of the specimens are reported in table 2.

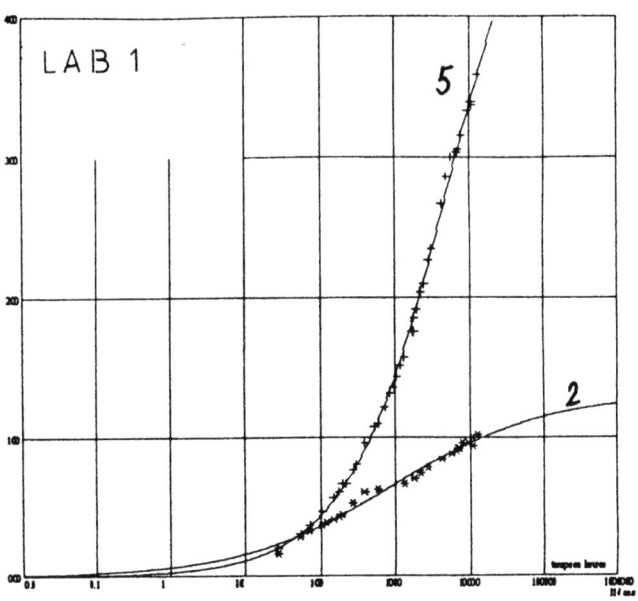

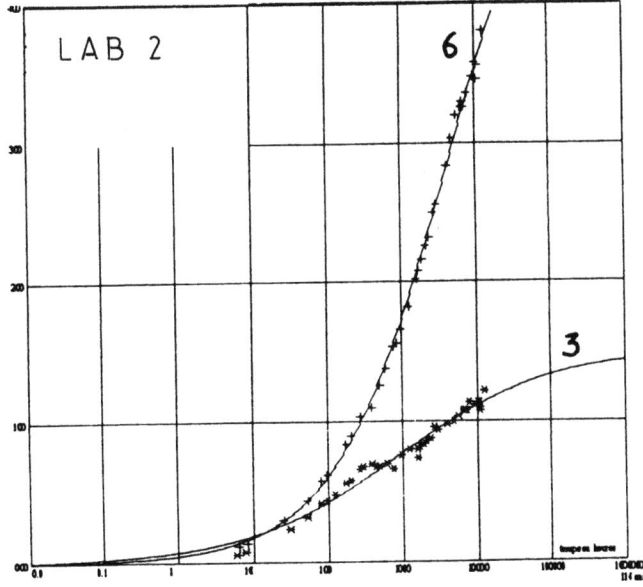

Figure 5a : Creep of Series A HSC (strains in millionth vs time in hours). The references of the specimens are reported in table 3.

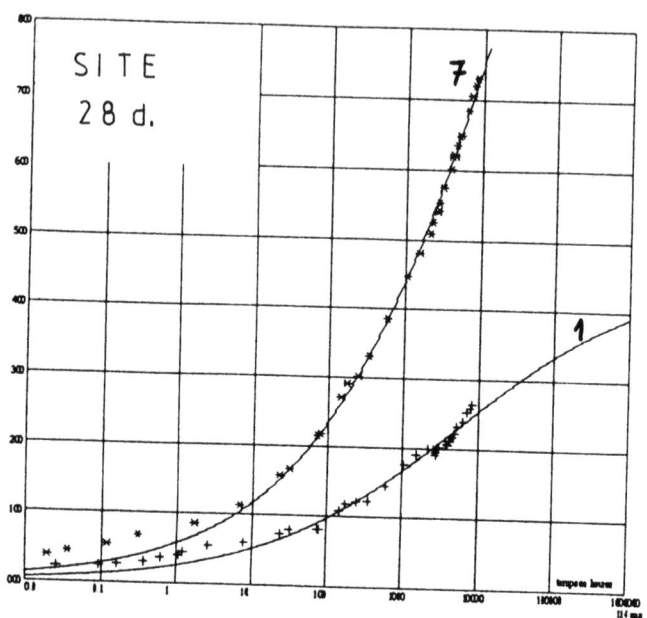

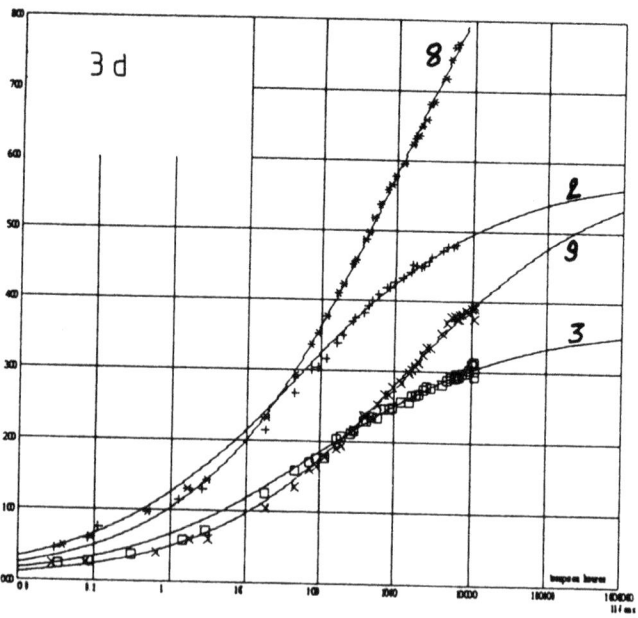

Figure 5b : Creep of series A HSC (strains in millionth vs time in hours). The references of the specimens are reported in table 3.

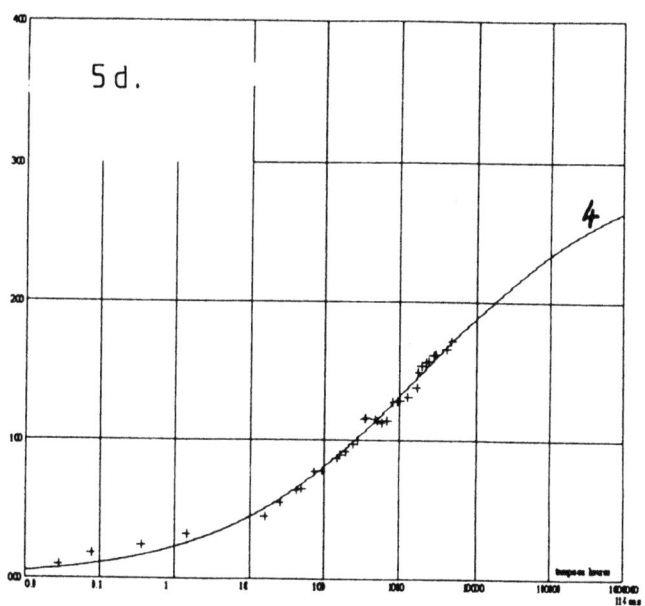

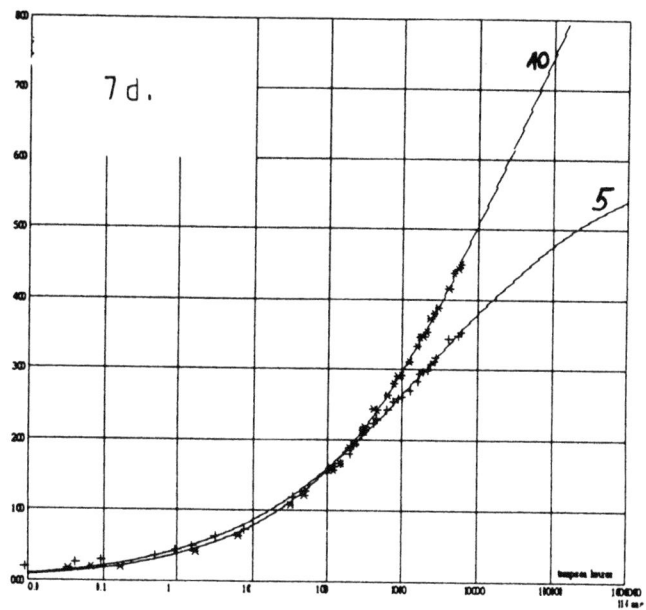

Figure 5c : Creep of Series A HSC (strains in millionth vs time in hours). The references of the specimens are reported in table 3.

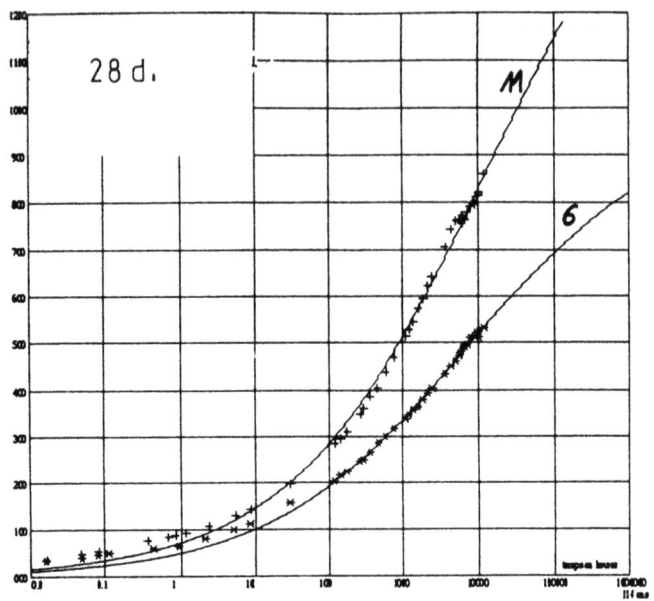

Table 4 : Creep of Normal Strength Concrete according to the BPEL 83 Code [10] using sealed (S) and unsealed (U) specimens

Curing	S.	S.	S.	S.	U.	U.	U.	U.
t_o (d.)	3	5	7	28	3	5	7	28
$K_{c r}$	1.04	1.03	1.02	.92	4.01	3.93	3.86	3.30
$\beta(d.^{1/3})$	10	10	10	10	10	10	10	10

The values of K_{cr} and β are given in Table 3. Compared with silica-fume HSC (SFHSC) [3,4], this HSC exhibits a very different behavior:

- while creep of SFHSC practically does not depend on drying (no drying creep), here the creep of drying HSC is about twice the basic creep;
- while the kinetics of creep are constant for SFHSC, here it depends on the age at loadings ; it is slower as the concrete is older (see Fig. 6). This is logical, if we consider that the rate of creep decreases when the hydration goes further;
- while the basic creep coefficient of SFHSC strongly decreases in the early days [3,4], there is no clear trend, at a first glance, for the present HSC. However, it is interesting to plot the creep coefficient against the applied stress, regardless of the age at loading (see Fig. 7). Tests at 3 days show that the creep is not linear (unlike, once again, SFHSC [3,4]). Furthermore, a linear relation appears between creep coefficient and stress. It is concluded that, for loadings after 3 days, there is a unique non-linear relationship between final deformation and applied stress;
- while SFHSC has a much lower creep than NSC (except when loaded at very early age), this HSC has a quite higher basic creep, and a total creep comparable with the one of NSC (see Table 3). The increase of basic creep may be attributable to the high superplasticizer content. Furthermore, the water/cement ratio is not low enough to induce a decrease of the hydrates volume and an internal selfdesiccation, which are the probable causes of low creep in very high-strength silica-fume concretes [5].

4 Results of monitoring after one year

An analysis of the measurements made during the construction of the bridge has been carried out.

4.1 Thermal effects

4.1.1 Results of measurements
A number of points can be made about thermal effects during the hardening of the concrete. The temperature evolution at the various temperature probes exhibits a stage of latency, a stage of temperature rise, and a stage of return to ambient temperature. Since the concrete in question was deliberately highly retarded, the latency period is very long (between 48 and 60 hours); the time of onset of the temperature rise depended on the age of the concrete, but there was some interaction between the first concrete poured, already starting to set, and later concrete, in which setting was triggered by the foregoing (temperature-activated hydration reaction). The maximum temperatures observed

443

Figure 6 : Evolution of the kinetic creep coefficient with time.

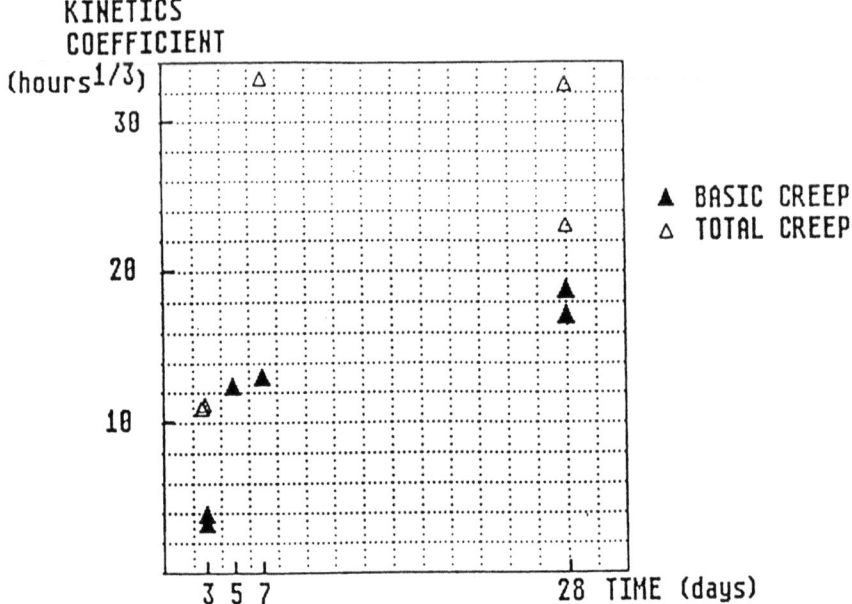

Figure 7 : Relation between creep coefficient and applied stress.

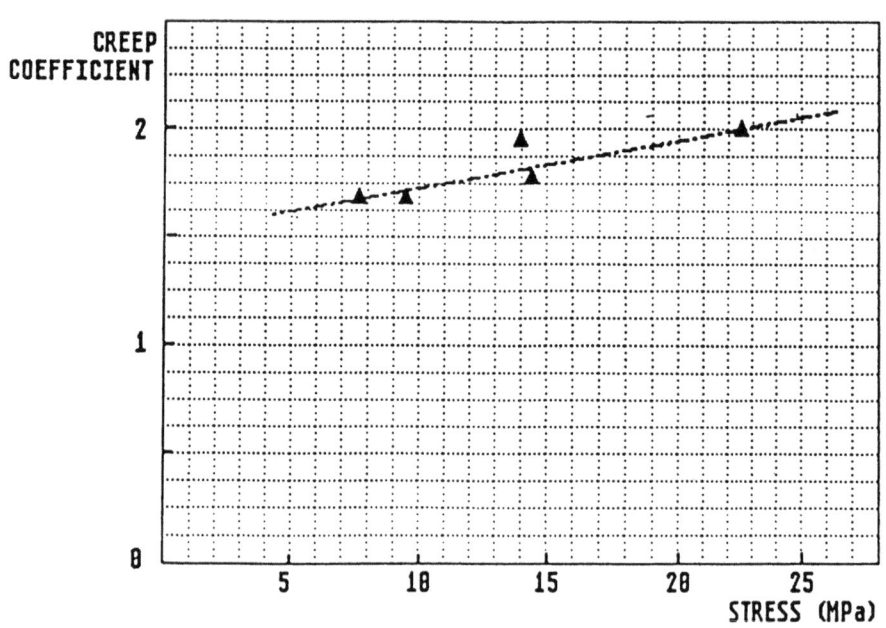

444

were 73°C in the core of the end mass; 57°C in the rib; and 32°C in the slab at the axis of symmetry of the deck section, as shown by Figs 8 and 9.

4.1.2 Numerical simulation

Finite elements calculations have been performed thanks to TEXO, a part of CESAR, the LCPC code. The principle of these calculations is presented herein. The heat equation is solved, taking into account a source term:

$$\delta Q = Q \exp\left(-\frac{E}{R\,T}\right) \delta t \quad (Arrhenius \; equation),$$

where Q is the heat already emitted by a unit volume of concrete;
E is the activation energy;
R is the constant of perfect gases;
T is the absolute temperature.

The function Q is determined by a semi-adiabatic test. A 160 x 320 mm sample of fresh concrete is placed in an insulated box, and the internal and external temperatures are recorded during a week. Heat exchanges between inner and outer volumes are a function of the form:

$$\delta Q = \alpha \; (\theta_e - \theta_i) + \beta \; (\theta_e - \theta_i)^2$$

coefficients α and β depending mainly on the calorimeter. So, the actual adiabatic curve can be calculated from the semi-adiabatic one (see Fig. 10).

For simulating the actual structure, one must determine the following parameters:

- initial temperature of concrete ($\theta_0 = 15°C$);
- exchange coefficients at the surface, with forms (2,8 W/m^2 °C) and without (6,1). Here, this value can be only approached, since it is very dependent of the wind speed;
- conductivity of concrete. Generally, this parameter is taken constant in the whole structure, neglecting the role of reinforcement. But, in this case, the end mass was heavily reinforced (300 kg/m^3), with numerous bars placed following the heat flux. Using the homogeneization theory, the conductivity of reinforced concrete can be bracketed by the two following limits:

445

Figure 8 : Evolution of temperature in right-bank end spacer.

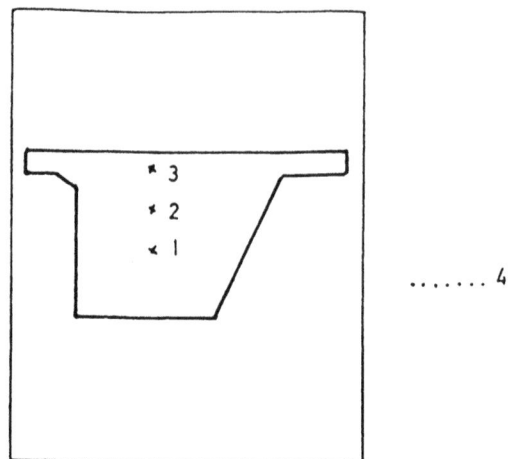

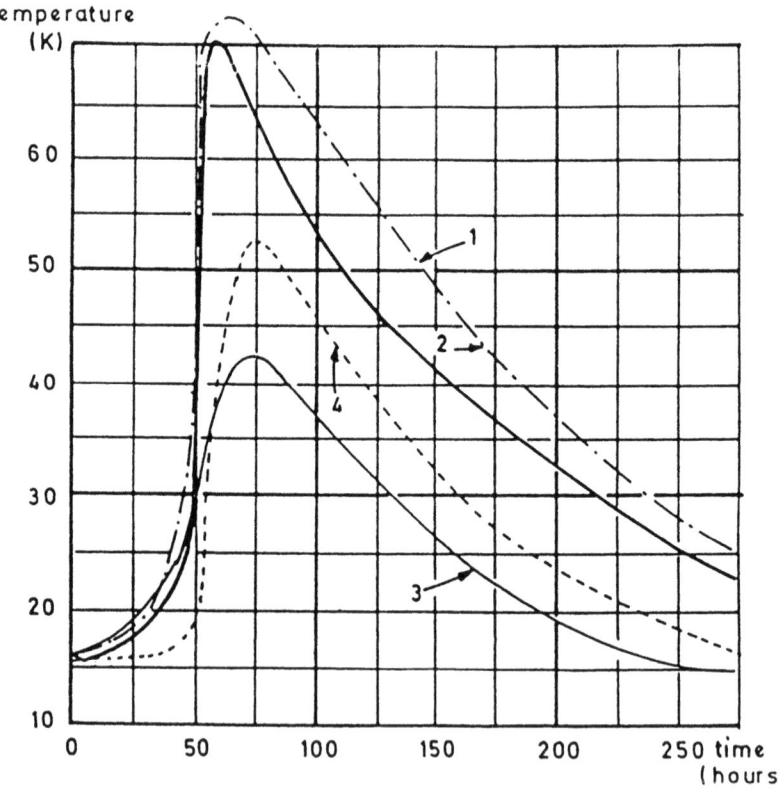

Figure 9 : Evolution of temperature in the deck.

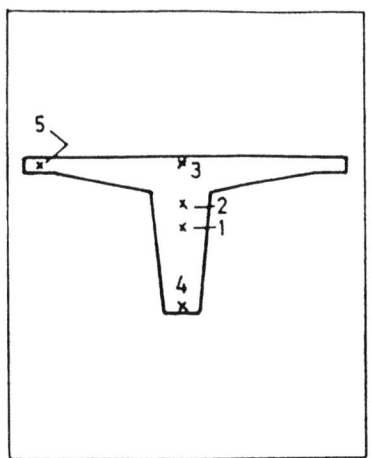

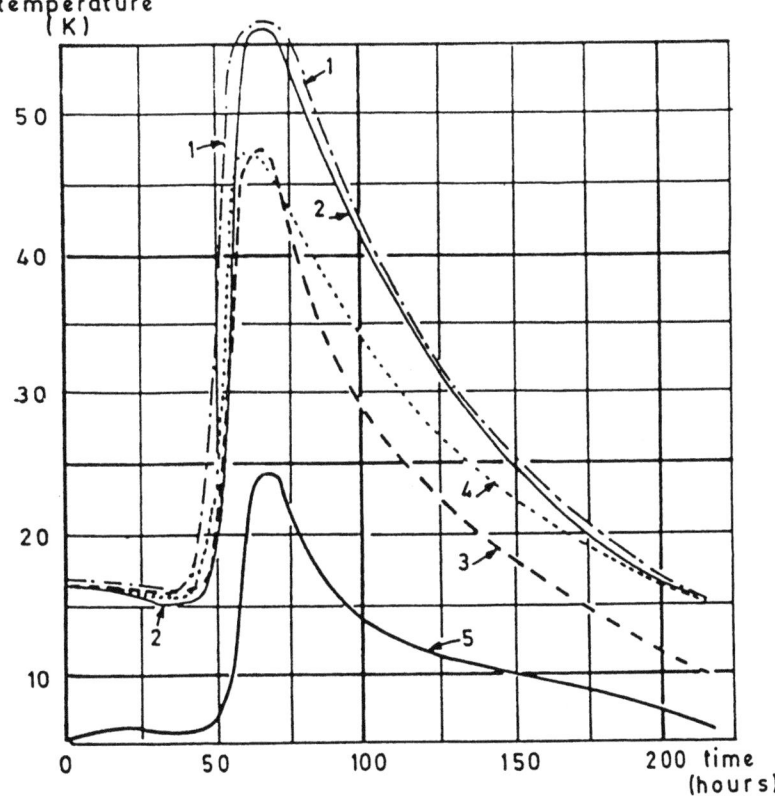

$$\frac{1}{\lambda_{MIN} + 2\,\lambda_c} = \frac{1-\mu}{3\,\lambda_c} + \frac{\mu}{\lambda_s + 2\,\lambda_c}$$

$$\frac{1}{\lambda_{MAX} + 2\,\lambda_c} = \frac{1-\mu}{3\,\lambda_s} + \frac{\mu}{\lambda_c + 2\,\lambda_s}$$

where μ is the voluminal proportion of steel ($\mu = 0.038$), λ_c and λ_s the thermal conductivity of concrete ($\lambda_c = 1.6$ W/m. °C), and steel ($\lambda_s = 54.35$ W/m. °C), respectively λ_{MIN} and λ_{MAX} the minimum and maximum values for the conductivity of reinforced concrete ($\lambda_{MIN} = 1.77$ W/m. °C, $\lambda_{MAX} = 2.97$ W/m. °C. As the concrete conductivity may range in a quite wide interval, a value of 1.77 was adopted, to fit the part of experimental curves corresponding to the late cooling of concrete, when hydration is roughly finished.

  - specific heat of concrete. Similarly, the presence of a high amount of steel reinforcement can be taken into account, with a simple mix law :

$$C_{Rc} = (1 - \mu)\,C_c + \mu\,C_s$$

where C_c is the specific heat of concrete ($C_c = 1.06 \cdot 10^{-3}$ J/kg. °C), C_s that of steel ($C_s = 1.99 \cdot 10^{-3}$ J/kg. °C), and C_{Rc} that of reinforced concrete ($C_{Rc} = 1.10 \cdot 10^{-3}$ J/kg. °C). Two bidimensional calculations have been performed separately, for each instrumented part of the bridge (meshes are shown in Fig. 11).

4.1.3. Comparison between experiments and simulation.
    Temperatures versus time are plotted on Figs. 12 and 13, for both instrumented sections, as predicted by the calculation, and as measured. In the mid-span section, the evolution of temperatures is quite fairly approximated by the FE calculations. Unlike predicted temperatures, the times when actual temperatures reach their maximum values are not the same between the bottom and top of the section. This can be explained by the different ages of concrete (total casting duration was about 24 hours).
    For the end part of the bridge, one can note the shape of the simulated curves near the temperature peak. This is due to the poor accuracy of the end of the adiabatic curve. The value of λ_{Rc} has been fitted, in the previously computed range, to get quite the same speed of temperature decrease in the calculation than in the bridge.

Figure 10 : Semi- adiabatic and adiabaic curves of HSC.

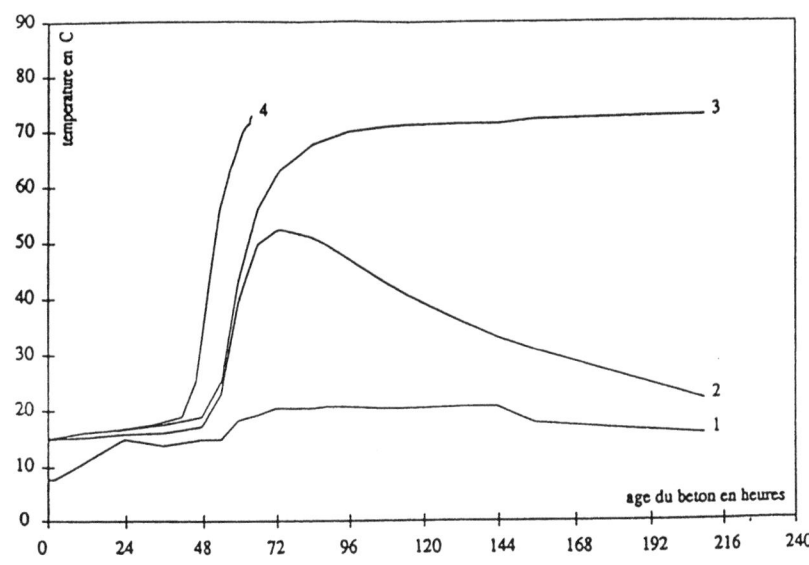

Figure 11 : Meshes used for finite-elements calculations of temperature.

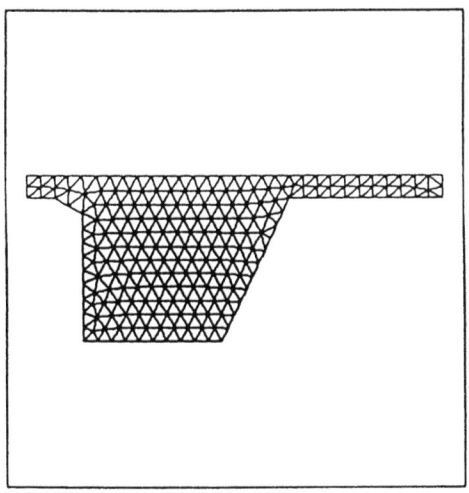

Figure 12 : Calculated temperatures in right-bank end spacer.

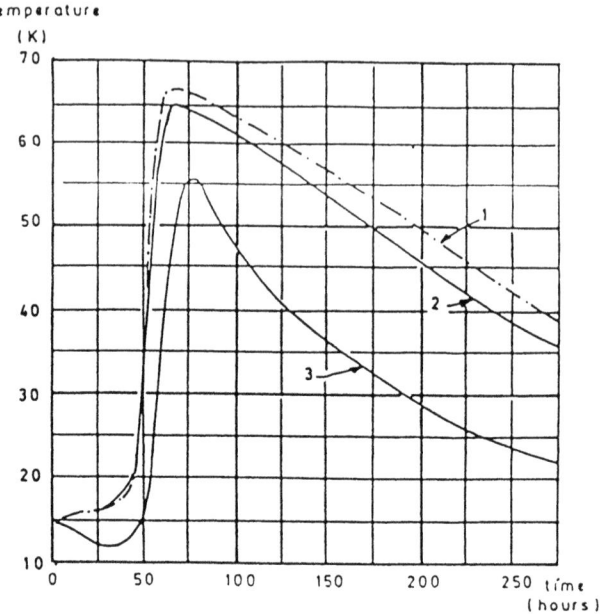

Figure 13 : Calculated temperatures in the deck.

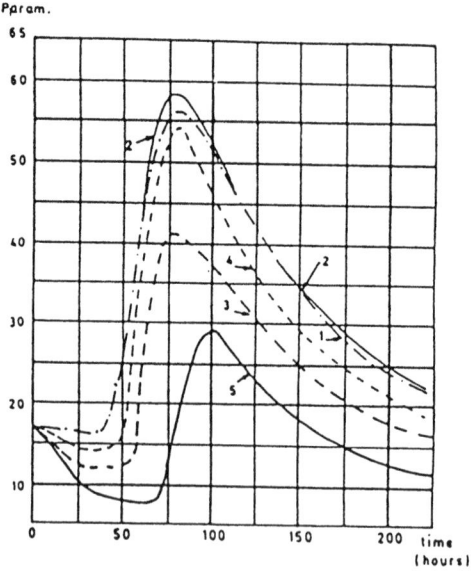

450

Though the minimum value (λ_{MIN}) had been taken, it was impossible to reach the same peak value. It means that the temperature rise in the massive part of the bridge was too high, compared with the one we could expect from the concrete composition and the shape of the structure.

As pointed out in previous paragraph, the consistency of HSC was very liquid (slump > 220 mm). The concrete was first poured near the opposite end, compared with the instrumented one. During the casting of concrete, the free surface of fresh concrete remained practically horizontal. So, a segregation phenomenon could have happened, like in a river where coarse materials remain near the source, and fine materials flows downstream. This mecanism would lead to higher cement content in the instrumented end. A supplementary dosage of about 40 kg/m3 of cement would be sufficient to explain the discrepancy between the peak values. Another consequence of this segregation process is a scattering of the compressive test values, which actually occured in Joigny bridge (the standard deviation was 6.7 MPa at 28 days).

4.2 Measurements of prestressing forces
The cable vibration method of measuring prestressing forces turned out to be operational even on cables that had not yet been grouted.

The tension measurements are rather uniform from one cable to another, but show tensions about 5% greater in the last four cables tensioned (see Fig. 14). A drop of 2% in the tensions of the nine cables already tensioned was also observed when the last cable was tensioned.

The tensions, just after all the cables have been tensioned, are of the order of 510 t per cable near the right-bank anchorage and 520 t near mid-span. The values yielded by the engineering calculations are approx. 510 t near the anchorage and 490 t at mid-span. The difference of about 7% between the measured and calculated values at mid-span can undoubtedly be explained by an overestimate of the coefficients of angular friction at the deviators used in the calculations.

4.3 Deformations of instrumented section

4.3.1 Instantaneous strains due to prestressing
A few points can be made on the basis of the measurements of the deformation of the instrumented section by acoustic strain gauges during the prestressing stages.

In the first two stages (before removal of centering and before removal of formwork, respectively), Figs 15 and 16 show a non-plane deformation of the section and an asymmetry between the two ribs (the upstream rib is more stressed than the downstream rib). These phenomena can be explained by two causes. First one is the presence of the formwork. Second (and main) one is the location of

451

Figure 14 : Cable tensions at mid-span.

Cables	Tension in tons
1	504
2	512
3	501
4	515
5	507
6	517
7	528
8	531
9	531
10	536
11	544
12	553
13	544

Figure 15 : Deformation of instrumented section during tensioning of cables 1 to 5.

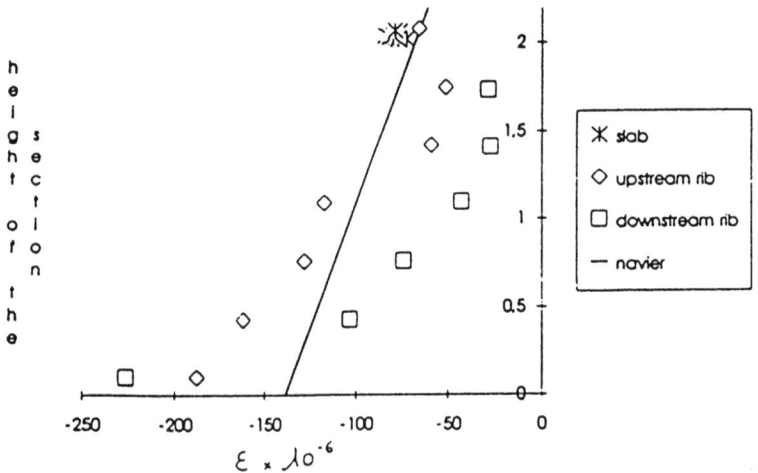

the instrumented section. The distance between this section and the mid-span deviator is about 1.6 m, while its height is 2.2 m. So, a plane deformation hypothesis is no longer valid : force distribution in the ribs does not occur near the deviator especially when it is dealt with the nine first prestressing cables, which are near the symetry axis of the section. Nevertheless, assuming that the Hooke's law is valid, and integrating stresses on the whole section, 88% of the total prestressing force is found.

Unlike the first two stages, the third and fourth stages induced in the section a deformation that was substantially plane and symmetrical (for the third stage, see Figs. 17 and 18). The so-called 'measured Navier' is the diagram of strains which would lead to the same moment and normal effect than the measured strains, while the "calculated Navier" comes from the engineering calculations. The measured Navier, for the two last prestressing stages, are in good accordance with the calculated ones.

On Fig. 19, strains caused by total prestressing and corresponding measured Navier are presented. Mean value of strains is about the same as predicted by the engineering calculations; on the other hand, the distribution is quite different, because of perturbation during the two first prestressing stages.

4.3.2 Long-term deformations
In Fig. 20, strains versus time are plotted, from the end of the construction (March 15th, 1989). After six months, deformations are stabilized, except for little variations due to climatic conditions (temperature and humidity). Additional deformations after one year (see Fig. 21) are about 50% of deformations at the end of the construction (i.e. elastic deformations plus early shrinkage, and early creep caused by the first prestressing cables). Comparisons with calculations performed from creep and shrinkage tests are presently in progress.

5 Conclusion
With regards to the delayed deformation of the Joigny HSC:

- as expected, the shrinkage is not very different from the one of NSC;
- but the creep is higher than expected. It appears very different compared with the one of a previously reported silica-fume HSC.

The conclusions after experimentation on thermal effects are the following :

- temperature rises in HSC structures can be higher than in NSC ones due to the high amount of cement;
- they can be fairly simulated by finite elements calculations ;

453

Figure 16 : Deformation of instrumented section during tensioning of cables 6 and 7.

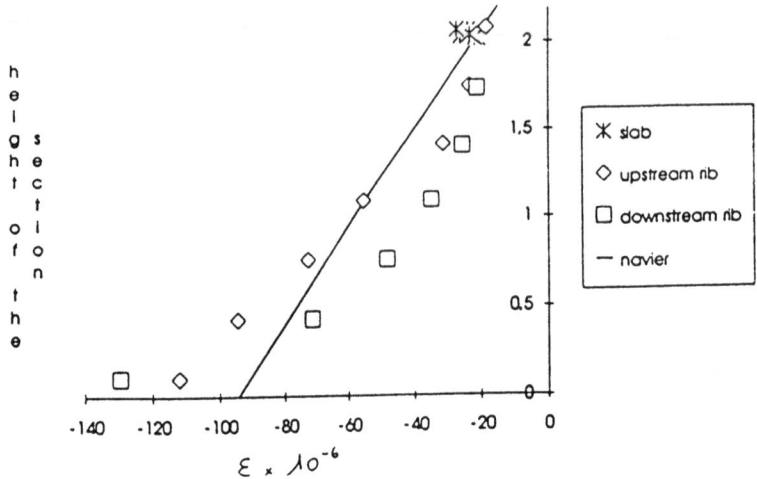

Figure 17 : Comparison of measured and calculated deformation profiles of the sectionfor the cables 8 and 9.

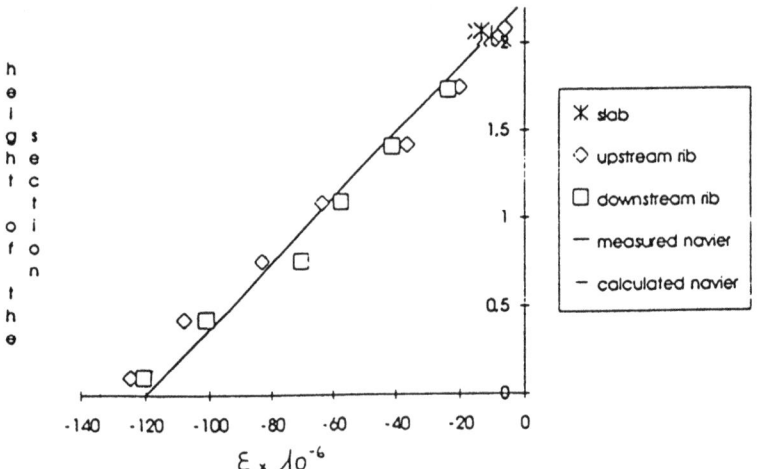

Figure 18 : Comparison of measured and calculated deformation
profiles of the section for the last four cables.

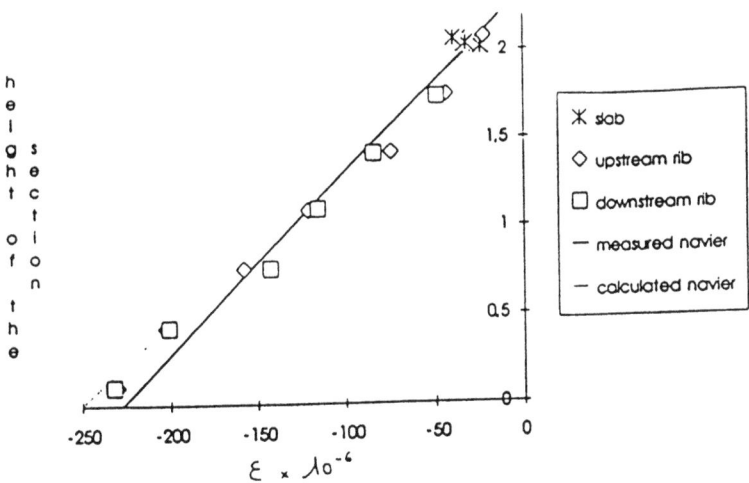

Figure 19 : Comparison of measured and calculated deformation
profiles of the section for the total prestressing force.

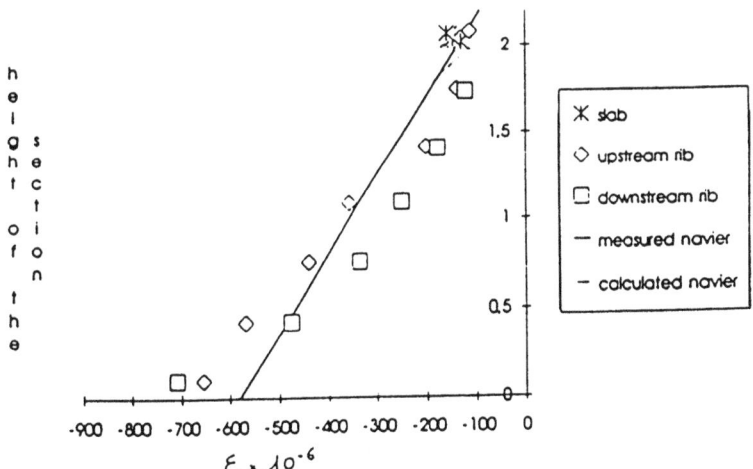

Figure 20 : Evolution of deformations in the instrumented section at different heights.

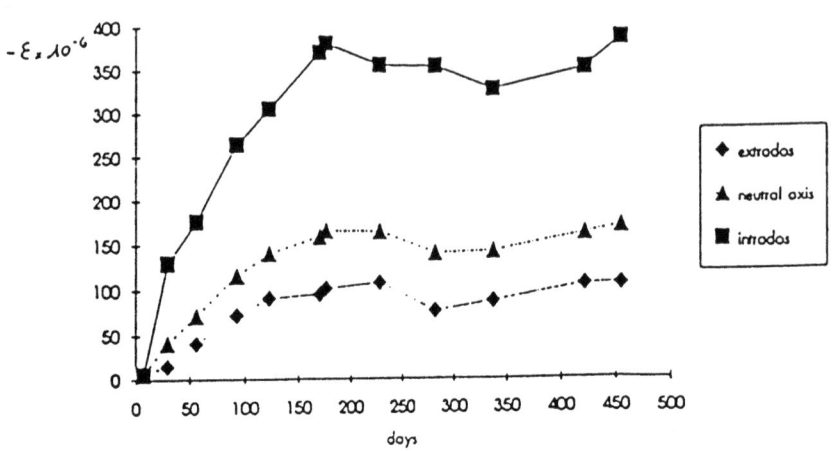

Figure 21 : Deformation profiles at different ages.

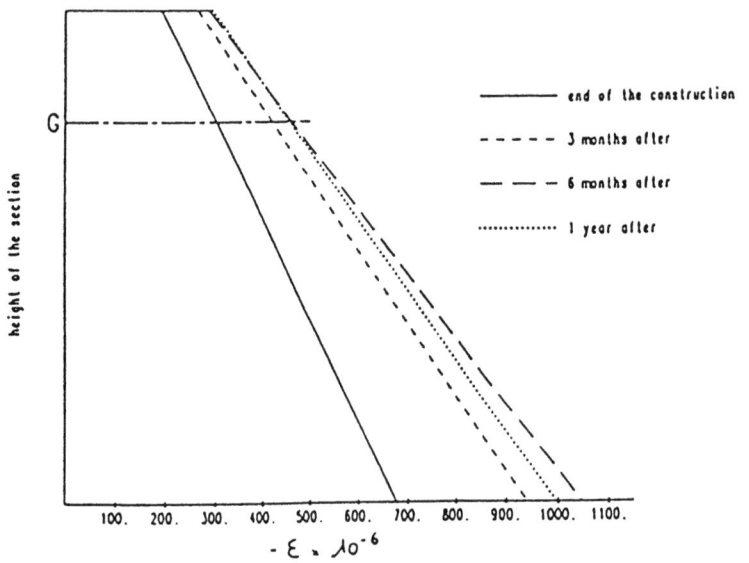

- they generally do not lead to thermal cracking when the newly cast structure is not restrained, and when the thicknesses of the neighbouring parts of the structure are not too different ;
- a certain segregation may happen if the concrete is too liquid, leading to some heterogeneities. In Joigny, this segregation was likely a separation between coarse aggregates and mortar, more than between paste and aggregates, as the concrete did not bleed, even though there was a large superplasticizer dosage.

Dealing with strains at various construction stages:

- as in all concrete structures which have been experimented in the past, measured short-term deformations are less than calculated ones;
- for external prestressed bridges, 'strength of materials' models do not apply in the vicinity of deviators. So, intrumentation should have been put in other deck sections;
- due to the higher specific creep of non silica-fume HSC, the high stress in the bridge service state gives a rather high long-term deformation after six months;
- however, the deformation state of the bridge does not increase anymore after this age, which is consistent with the rapid kinetic of the delayed behaviour of HSC.

References

1. DE LARRARD F., ITHURRALDE G., ACKER P., CHAUVEL D: 'High Performance Concrete of a Nuclear Containment'. Second International Conference on Utilization of High-Strength Concrete, Berkeley, ACI Special Publication, May, 1990.
2. Béton Armé aux Etats-Limites. French Reinforced Concrete Building Code. MELL. 1990.
3. AUPERIN M., DE LARRARD F., RICHARD P., ACKER P.: 'Shrinkage and Creep of High-Strength Concretes - Influence of the Age at Loading' (in French). Annales de l'Institut Techniques du Bâtiment et des Travaux Publics, N° 474, pp. 50-75, May-June 1989.
4. DE LARRARD F. : 'Creep and Shrinkage of High Strength Field Concretes'. Second International Conference on Utilization of High-Strength Concrete, Berkeley, ACI Special Publication, May, 1990.
5. DE LARRARD F., MALIER Y. : 'Engineering Properties of Very-High-Strength Concretes. From the Micro-to the Macro-Structure' (in French). Annales de l'Institut Technique des Travaux Publics, N° 479, December, 1989.

30 HIGH PERFORMANCE CONCRETE UNDERPINNING OF THE GREAT HASSAN II MOSQUE

G. CADORET, C. COURTEL
Groupe Bouygues, Scientific Division , Paris, France

1 Introduction

Throughout the world, most full-scale applications of high peformance concrete capitalize upon the high **mechanical strengths** it offers.

Depending on the situation, high mechanical strength permits either new structural design (e.g. Sylans viaduct, France) or a decrease in quantities or element sections leading to a better utilization of the structure in question (increase in net surface areas, openings in structural elements allowing improved internal circulation areas) (e.g. La Grande Arche, La Défense, Paris).

In other situations, the improved **durability** of concrete can be decisive factor (e.g. offshore platforms).

Furthermore, the **kinetics of binder hydration**. and therefore, strength development functions have been analyzed with a view to enhancing production rates of site precasting or in-situ casting facilities (e.g. Ré Island Bridge, France).

Finally, the **laws describing the behaviour** or high performance concrete in terms of time-dependent deformations (shrink and creep) have been the object of studies and modeling in order to ensure control of structural geometry during the construction phase (e.g., Pertuiset Bridge, France).

On the other hand, to our knowledge, the evolution of time-dependent deformations has never been used as a selection criterion.

We used that criterion in designing the height increase of the Minaret of the Great Hassan II Mosque in Morocco.

Within the framework of the Construction Contract award, the contractor undertook an experimental research program in order to :
- define the concrete mix design
- organize underpinning procedures based on the characteristics of the concrete.

High Performance Concrete: From material to structure. Edited by Yves Malier. © 1992 Taylor & Francis.
Published by Taylor & Francis, 2 Park Square, Milton Park, Abingdon, Oxon, OX14 4RN. ISBN 0 419 17600 4.

459

2 Description of the works
The religious edifice consists of :

- the Mosque itself (200 m x 100 m x 60 m)
- structures in contact with the sea
- the Medersa
- the Hamman
- the Minaret (25 m x 25 m x 175 m)

Although the structural works for the Minaret were completed, the Owner requested that the height of the latter be increased by 25m, for a finished elevation of 200.00 m, including the Jamour. The impact of the height increase (increased vertical loads and higher overturning moments) mainly affected the following structural elements:

- load-bearing walls and columns, which had to be strengthened,
- columns located at the openings, which required strips to thicken column sections in addition to the concrete reinforcement,
- foundations, which had to be underpinned in order to remain within the bearing capacity of the foundation soil.

Strengthening works involved the following :

(a) In the lower section :

- **at 3 corners**, the bearing capacity of the footings had to be increased, leading to the design of ground beams cast through the existing walls, and reinforced at both cantilever ends. They bear on bedrock on either side of the existing footing.
- **the 4th corner (North corner)** involved the most complex operation, i.e., reconstruction of a key load-bearing corner column.

In order to complete this operation, the corner of the Minaret is shored during the construction phase, using two concrete columns (with cross section of 110 x 310 cm) placed on either side of the underpinned column. Loads are transferred to the temporary shoring by means of flatjacks. The load value was estimated taking total concrete deformation into account (elastic and long-term).

(b) Around openings (levels 87 and 107)
The affected structural elements (walls and columns) were demolished after erection of shoring ; loads were transferred to the reconstructed elements by means of flatjacks.
Closure grouting is performed after time-dependent deformations of the concrete have occured.

460

3 Principle of underpinning

3.1 General
Given the age of the concrete elements comprising of the structure, and the loads already applied to them, most of the time-dependent deformations have already occured in the concrete of the Minaret.

Therefore, the characteristics of long-term deformation of the new concrete to be used in underpinning must create as little impact as possible on the existing structure (compatibility of relative deformations).

For this purpose, the underpinning elements, such as ground beams, were only made integral with the existing structure once the estimate of the remaining time-dependent deformations matched the values expected on the existing structure.

This required knowledge of time-dependent deformation functions describing the new concrete, as well as confirmation, on the completed structure, of the accuracy of design assumptions.

3.2 Construction phasing
The timing of operations, such as casting, closure grouting and loading of rebuilt elements is critical.

Dates for these operations are determined based on the following parameters.

- concrete age at closure pours,
- concrete age at load application,
- duration of loading,
- where applicable, age of existing concrete (i.e., not affected by underpinning), in order to minimize the occurence of differential creep between concretes of different ages,
- degree of maturity achieved by the structural concrete at the time of the crucial closure grouting and loading operations,
- results of shrinkage measurements taken on specimens of the concrete used in foundation underpinning of the Minaret.

3.3 Estimated required characteristics of concrete used in underpinning
Due to the deadline requirements of the construction phases and taking into account the data presented above, it was not feasible to use concrete whose creep and shrinkage behavior complies with the French concrete limit state design codes (BPEL).

On the other hand, it was imperative to design a concrete such that:

- shrinkage and creep would take place more rapidly than specified in the french prestressed concrete code, BPEL code (in order to reduce differential deformation between "old" and "new" concretes),

461

- long-term creep would be considerably less than the BPEL requirement, in order to be able to load the concrete at an early and suitable age.

Finally, in order to define construction phases, the time-dependent deformations of concrete needed to be quantified in terms of the concrete age, age at loading and load duration.

4. Time-dependent deformation of concrete

4.1 Comparison of high performance concrete and BPEL limit state codes

Geometrical control of structures built by the balanced cantiliver method (precasting of the Ré Island Bridge) or by in-situ cantilever casting (Pertuiset Bridge) requires knowledge of time-dependent deformation phenomena of high-strength concrete, particularly at an early age.

Recent studies of high performance concrete with silica fume admixtures having average 28-day compressive strength of 80 MPa, initiated and conducted by Bouygues at the Laboratoire Central des Ponts et Chaussées laboratory, revealed shrinkage and creep rates that differ significantly from those specified in the codes.

This research was presented in two articles published in the ITBTP (French Technical Institute of Building and Public Works) Annals (May-June 1989 issue) by the Bouygues Research department and the LCPC.

The following is a comparison between the BPEL laws and the predictive laws derived experimentally, which describe the behavior of high performance concrete containing silica fume (water/cement ratio: 0.38, average strength : 80 MPa, called G1).

Comment :
Like the code, we have expressed creep and shrinkage as the product of a law of evolution through time ranging from 0 to 1, multiplied by a final deformation value.

4.2 Law of evolution of shrinkage (fig. 1)
A considerable percentage of shrinkage in high performance concrete occurs at an early age.

The BPEL code is based on a law describing much more gradual evolution of shrinkage.

Code

$$r\ (t) = \frac{t}{t + 9\ rm} \qquad (rm = \text{average radius})$$

We consider:

the maximum envelope for rm = 10 cm (corresponding to the smallest columns to be underpinned):

$$r(t) = \frac{t}{t + 90}$$

the minimum envelope for rm = 200 cm:

$$r(t) = \frac{t}{t + 1800}$$

4.3 Law derived experimentallly for high perforamnce concrete
(Fig. [1])

It is an expression in the form $e^{-B/(t+2)^{1/P}}$

Shrinkage of wrapped (vapor sealed) specimens (which simulate differential behavior characterizing the core and surface of large elements) complies with the law:

$$r(t) = e^{-4(t+1)}$$

For example, for laboratory specimens (20°C) of G1 concrete, approximately 70% of shrinkage occurs by the time the concrete is 10 days old.

In contrast, applying the BPEL code would require (in laboratory conditions):

. at least 200 days
. at most 10 years

rather than 10 days.

4.4 Law of creep evolution (fig. 2)
Let us recall that creep is equal to:

Efl : Eic Kfl (t1) f (t - t1)

with :   t1 = age of concrete at loading
         t = age of concrete

Figure 1 : Law of shrinkage evolution through time

$$\text{B.E.P.L.} \quad r(t) = \frac{t}{t + 9 \, rm} \qquad t = \text{age of concrete}$$

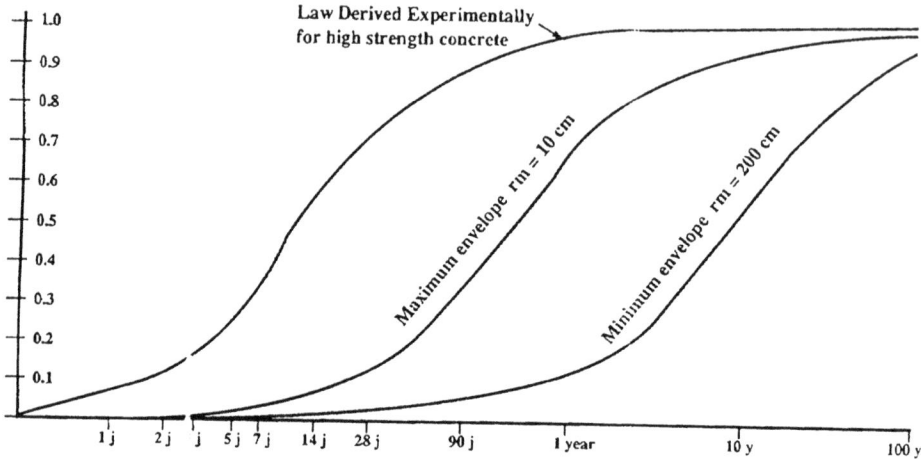

Figure 2 : Law of creep evolution through time

$$\text{LOI B.E.P.L.} \quad f(t) = \frac{(t - t_1)^{1/2}}{(t - t_1)^{1/2} + 5 \, rm^{1/2}} \qquad t - t_1 = \text{load duration}$$

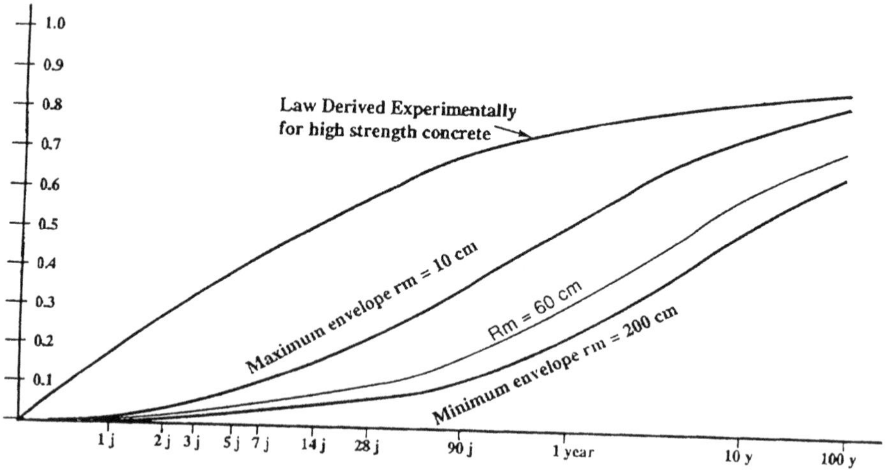

where :

. Eic represents applied stress divided by the instantaneous modulus of concrete at 28 days.
. Kfl represents the creep coefficient, determined in particular by concrete age at loading, t1.
. f (t - t1) is a function of load duration, varying from 0 to 1 as load duration varies from to 0 to infinity.

The code specifies :

$$f (t - t1) = \frac{(t - t1)^{1/2}}{(t - t1)^{1/2} + 5 \, rm^{1/2}}$$

in which (t - t1) is expressed in days and rm in centimeters.
We take:

maximum envelope : rm = 10 cm
minimum envelope : rm = 200 cm

4.5 Law derived experimentally for high performance concrete
(Fig. 2):

$$f (t - t1) = \frac{e^{- (1,7/ (t - t1) + 0.027)^{1/3}} - e^{-4}}{1 - e^{-4}}$$

For example, in concrete containing a silica fume admixture and having average compressive strength fo 80 MPa, approximately 60% of creep occurs after 14 days of loading (in a 20°C controlled atmosphere).

In contrast, applying the BPEL code would require, rather than 14 days (in laboratory):

. at least : 1 year
. at most : 40 years

4.6 Coefficient of creep, Kfl (Fig.3)
The coefficient of creep of a high performance (80 MPa) concrete containing a silica fume admixture varies rapidly with the date of loading up to the equivalent of 8 days of age under standardized controlled conditions (20°C).

The Code, on the other hand, shows that the coefficient of creep decreases much more slowly as a function of concrete age at loading than the decrease observed on the high peformance concretes studied.

Figure 3: Concrete EG1 - Coefficient of creep Kfl(tl)

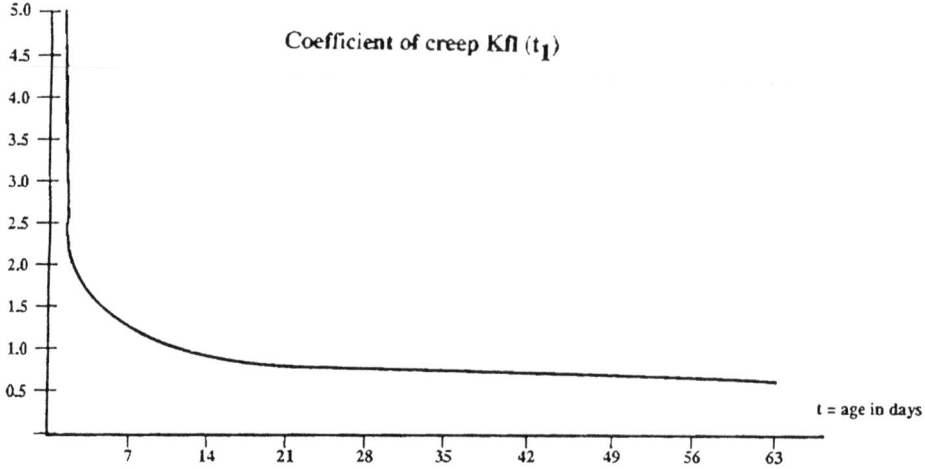

5 Definition of concrete mix design

5.1 Desired characteristics
In view of the desired deformation characteristics, we decided to use high peformance concrete with a silica fume admixture; the water and silica fume contents as well as the type of cement used are similar to the G1 concrete, object of our analysis of time-dependent deformations in 1988.

In other words, we were using a class of concrete with which we were already familiar, and whose behavior in time seemed to be compatible with the strengthening procedures defined.

5.2 Concrete design
In view of the average mechanical strength (40 to 43 MPa at 28 days) of the cement delivered to the site, we decided to import rapid-hardening artificial portland cement whose characteristics qualify it as suitable for marine use (French CPA HP PM NF-VP).

As for aggregates, those commonly used on site proved suitable. Gravel from the Yquem Wadi is crushed limestone, with "Bouskoura" sand obtained by grinding and screening calcarenite.

The resulting concrete mix design is as follows:

. Cement	425 kg/m^3
. Silica fume	40 kg/m^3
. Aggregates (5/15) mm	1045 kg/m^3
. Sand (0/5 mm)	755 kg/m^3
. Water	165 liters/m^3
. Plasticizer	13.8 liters/m^3
	(4% of cement weight)
. Retarder	2 liters/m^3

The average 28-day strength obtained on the trial mix is 92.5 MPa. Slump was 24 cm at T0 and 22 cm at T0 + 60 min. In view of the high reinforcement density (400 kg/m^3) and the planned concreting rates, the concrete had to have, and retain, high workability.

5.3 Inspection procedures
In addition to the written instructions relevant to manufacture and quality control of concrete, an inspection procedure was defined to check the concrete placed in each section of the structure.

This procedure specifies and defines systematic follow-up of the maturity and strength characteristics of the concrete. In addition, for some sections of the structure (ground beam V3, V4, temporary column, corner half-column) shrinkage measurements were taken by the site laboratory).

6 Inspection of the structure

6.1 General
In order to verify the design assumptions used in calculating the pressure of flatjacks and determine exactly how old the concrete should be at time of closure grouting and loading operations (based on the maturity of the concrete), we installed the following instruments:

- Acoustic strain gauges, namely in a temporary prop and the two reconstructed corner half-columns;
- Thermistors and maturity meters in all underpinned parts of the structure.

In addition, a "Perrier retractometer" is used to take shrinkage measurements in an element having constant temperature, on specimens made of site concrete, measuring 7 cm x 7 cm x 28 cm, sealed to prevent water exchanges simulates the conditions of the actual structure (differential shrinkage occurring in the core and surface of large elements).

As many authors have observed, the phenomenon of cement hydration is affected by temperature.

The relationship between the degree of binder hydration, and therefore, the development of the physical and mechanical properties of the concrete on the one hand, and temperature on the other, is represented by the Arrhenius function.

Thus, calibration and measurement of concrete characteristics of shrinkage and strength were performed in the laboratory. The follow-up and decisions regarding progress on site where organized essentially from maturity measurements taken on in situ concrete.

6.2 Definition and inspection of construction

For each underpinned part of the structure, we wrote procedures concerning :

- measurements (maturity measurement procedure, strain gauge procedure, shrinkage measurement procedure) with, in particular, definitions of instrument layout and/or frequency of measurement.
- methods of interpretation of measurements
- methods of construction for underpinning, defining in particular the dates of closure grouting and/or loading of flatjacks.

The time lapse required between casting and closure grouting or loading of flatjacks where applicable is directly related to the interpretation of measurements taken on the structure.

6.3 Example of application

(a) Pier Cap

- Casting of ground beam
- Shrinkage of ground beam
- Thermal expansion of concrete
- Closure grouting.

Based on the load history and shrinkage characteristics of the pier cap concrete, virtually identical to those of the G1 concrete (cf. shrinkage curve measured on ground beam concrete), calculations indicate that closure grouting must take place when concrete is at least 22 days old (under laboratory conditions, i.e., 20°C).

Temperature follow-up and interpretation of temperature as an indication of maturity indicated that structural concrete should be at least 11 days old at time of closure grouting.

(b) Corner V1/V2

The reconstruction of corner V1/V2 is undoubtedly the most complex of all the strengthening works.

Indeed the major operations such as demolition of a corner column and construction of the extra height could not be undertaken until measurements (temperature, shrinkage, relative deformation) taken on the relevant parts of the structure are processed and analyzed.

In this manner, we realized, through calculations and follow-up of measurements, that optimum timing consists of pre-loading the 10-day old concrete, for eight days, using water. Permanent loading of flatjacks is done with grout.

Indeed, in accordance with the law of creep behavior of high peformance concrete, more than 50% of creep is eliminated without requiring additional jack loading.

7 Conclusion

Works are in progress.

Phasing of operations, especially as regards the lapse of time necessary between two tasks (to allow for shrinkage, creep and thermal expansion) are precisely defined through measurements taken on the structure and in the site laboratory.

Thanks to additional funding provided by the Projet National VNB, the conclusions drawn on this jobsite will significantly exceed the framework strictly required for construction of the works.

Additional instrumentation on site as well as tests on concrete will enhance our knowledge of high peformance concrete.

31 PREFABRICATION OF HIGH PERFORMANCE PRESTRESSED CONCRETE BUILDING COMPONENTS

G. CHARDIN
Saret-PPB, France

The industrial manufacture of concrete building components prestressed by bonded wires is moving fast towards a wider use of high performance concretes. This paper presents the industrial experience of Saret-PPB within its factories in France.

The PPB group manufactures floorbeams and slabs for flooring, beams for buildings and bridges, as well as industrial cladding.

1 Prestressing by bonded wires

The principle of prestressing by bonded wires consists in the tensioning of high strength wires (indented wires or strands), lying on elements external to the item being manufactured. After the concrete hardens, the tensile force is transferred from the wires to the concrete, by cutting the wires from the external elements. The bonded wires, unable to go back to their original length (to a non existent tension), realize by their bonding to the concrete, the prestressing of the beam.

The technology related to this principle is well known. It consists of tension the bonded wires lying on the anchorage heads, which are fixed when starting the tensioning operation. This operation is realized by one-wire hydraulic jacks as far as the beams are concerned, and thanks by a traction head for all the bonded wires as far as floorbeams and plates are concerned. The anchorage heads are separated from the formwork and anchored in the extremities of the concrete bed on which the formwork lies.

After casting and drying the concrete in the moulds, the anchorage heads are turned towards the beams, and the bonded wires are therefore slowly and progressively detensioned.

2 High performance concrete in precasting

The use of HPC changes the manufacturing means and methods, as well as the dimensioning of components and associated building

High Performance Concrete: From material to structure. Edited by Yves Malier. © 1992 Taylor & Francis.
Published by Taylor & Francis, 2 Park Square, Milton Park, Abingdon, Oxon, OX14 4RN. ISBN 0 419 17600 4.

systems. It would not have been practical for us to change at the same time in our Licensees' plants the type of concrete for all PPB components: floorbeams, slabs and beams. So, we decided to start our study with our smallest component, that is to say the floorbeam.

3 The floorbeam component

The 'Floorbeam' component, in association with concrete or polystyrene infill blocks and with the cast in situ concrete, allows the construction of floors extensively used in housing and improvement works. This market being a highly competitive one, it was imperative that the use of HPC should not be limited only to the exploitation of its high compressive strength, but be extended thus allowing us to renew our range of products, and in so doing help us be even more competitive on the floorbeam market.

We took advantage of this renewal to ask our bonded wire suppliers to design indented bonded wires ϕ4mm and ϕ5mm, of a higher class than the ones we are using: ϕ 4 class 1860 instead of 1770, and ϕ 5 class 1770 instead of 1670, see Table 1.

Now, the HPC can tolerate bonded wires of a higher prestressing tension, without any increase in their bedding length.

Table 1 : Characteristics of the bonded wires

	Am/m²	F_{prg} kN	F_{peg} kN	ρ % 1000 h
ϕ 4-1860-VGR	12.6	23.4	21	2.5
ϕ 5-1770-VGR	19.6	34.7	31	2.5

3.1 First floorbeam study

Our first study has been concenrtrated on the PPB 110 Floorbeam, heel width 10 cm, web width 5 cm, height 11 cm, prestressed with ϕ5/1770/TBR. The use of HPC allows us to reduce the width of the floorbeam down to 4 cm for the heel, and 4 cm for the web, while maintaining a 11 cm height, and to use ϕ4/1860-TBR.

With this new floorbeam, the deflection characteristics are nearly identical to the ones obtained with the existing floorbeam during the casting of the compression table of the floor, with or without props. The mechanical performance regarding the flexibility of the floor, and as far as the shear stress is concerned, are also identical to the present floors. There is also an decrease of the floorbeam weight, and a large increase of the precasting bed productivity.

3.1.1 Flexibility performances

The use of HPC, despite the reduction of the floorbeam width, allowed bending moments identical to the old floorbeams. These moments are taken into account for calculation of the maximum propping distances or of the maximum span without propping, see Table 2.

Table 2 : Bending moments in daN.m

Floorbeam height and nr. of ϕ	Floorbeam with 10 cm base, ϕ4-1770-TBR normal concrete	Floorbeam with 8,5 cm base, ϕ4-1860-TBR HP concrete
112	290	270
113	330	325
114	360	350

These moment values must be guaranteed at 80%, with a 95% probability, by the manufacture control operations.

We did not use the high compressive strength of the HPC in prestressing floorbeams at high level and so improve the performances of the floors. There was no use in doing so, for in floorbeamed floors, the Serviceability Limit States (SLS) verification is always of first importance. The new floorbeams have prestressing values of a similar range to the previous ones.

A small improvement in floor performance is obtained thanks to the new bonded wiring having higher ultimate strength, but this is merely a consequence which is not directly linked with the HPC of the floorbeam (the cast in situ concrete is plain concrete).

3.1.2 Shear stress performances

For the higher flexibility floorbeam with four bonded wires, despite a reduction of the web, the HPC allows shear stress performances of a range similar to the previous one. The allowable shear resistance of normal concrete (for a 10 cm floorbeam) is usually 9 bars, whereas the one obtained with the HPC designed by Saret (for a 8.5 cm floorbeam) is 13 bars.

3.1.3 Weight performances
The 10 cm basis floorbeam weighs 18.3 daN/ml, and the 8.5 cm base floorbeam weighs 15;3 daN/ml, so the weight saving is 16%,thus improving its handling by two labourers only, and providing a big advantage in alteration works and craftsmanship.

3.1.4 Productivity
It was up to now possible on our 2.5m wide precasting bed to produce 22 floorbeams of 10 cm width, it is now possible to manufacture 26 of the new floorbeams, which increases the productivity of the bed by 18%.

3.2 Second floorbeam study
For our second study, we concentrated on the PPB 140 floorbeam, with a 10 cm flange width, and a 5 cm web width, height 14 cm, prestressed with ϕ5-1670-TBR   This floorbeam meets the requirements of floors whose beams must be laid without propping (i.e. suspended ground floor and floors laid at a level higher than the ground).

The shape of the floorbeam remains unchanged. The HPC, thanks to its resistance to compression and tension, associated with the use of ϕ5-1770-TBR, improves the performance of the floorbeam without propping, as well as the floor's resistance to shear stress.

3.2.1 Flexibility performance
The use of HPC allows an increase in bending moments of the floorbeams only, which has been taken into account for the calculation of the maximum span without propping, Table 3.

Table 3 : Bending moments in daN.m

Floorbeam height and nr. of ϕ	Floorbeam with 14 cm base, ϕ5-1670-TBR normal concrete	Floorbeam with 14 cm base, ϕ5-1770-TBR HP concrete
145	790	935

This improvement in bending moments allows for a given floor an increase in the maximum span without propping of about 9%.

3.2.2 Shear stress performance

The increase in allowable shear stress from 9 bars (NC) to 13 bars (HPC) faises the minimum span for which transverse wiring inside the floorbeams is necessary, hence an overall improvement of the product quality.

3.2.3 Distortion performance

In both our studies, the result of using HPC is to reduce equally the flexural camber of the floorbeams, again improving the overall product quality.

3.3 Material

All the benefits described above have needed a few modifications of the wiring material used for floorbeams. Some improvements have also been brought about by the use of silica fume in order to prevent the concrete from sticking to the mould parts.

3.4 The concrete and its curing

The HPCs used are made of limestone or silica agregates rolled and/or crushed (depending on the factories), gravel 6/12mm maximum, sand, cement (HR or HPR), densified silica fume and highly plasticizing admixtures.

Silica fume HPCs investigated by other laboratories do not entirely apply to the small industrial components manufactured by extrusion, whose production rate is high. A specific 'know how' due to the extrusion process, as well as a particular study are therefore necessary.

This is why Saret has proceeded in its own laboratory to a full study of HPC submitted to heat curing, and designed a basic concrete using silica fume and admixtures which meets all the criteria specific to its extrusion manufacturing process, allows heat curing and enables the strength at 7 and 28 days required for the prestressing to be obtained for the above described floorbeams.

As for normal concretes, heat cured HPCs have at 28 days a slightly lower strength than the same non-heat cured concretes.

For each individual plant, and according to the aggregates used by them, Saret has devised from a basic mixture, a specific composition.

HPC allows us to reach high strength shortly after prestressing, that is to say 10 hours after the manufacture starts up, and provides a good assurance of achieving the required strength. Since the prestressing values of floorbeams have not been increased, a certain 'comfort' in production is therefore provided.

Table 4 : Resistance table in MPa on a 10x10 cm cube (heat cured concrete)

	At prestressing (arerage 3 cubes) age = 10 hours	At 7 days (strength guaranteed on a	At 28 days large number of test tubes)
Floorbeam 110 4 φ4. 1860	31	50	55
Floorbeam 140 5 φ5.1770	34	60	65

You will doubtlessly understand why I cannot communicate our concrete composition, or describe our 'know how', commercially sensitive, in setting up, extrusion and heat curing operations.

3.5 Industrial production
All our factories in France manufacture this new type of floorbeam, which was awarded in July 1990 with the Technical Advice of the Specialized group nr 3 from the Ministerial Commission whose head office is at the CSTB (Centre Scientifique et Technique du Bâtiment). The production using HPC will represent for 1990 about 8 to 9 million linear meter out of a total production of 12 million meters of floorbeams.

4 Plates and beam components

The following study concentrates on HPC used in the manufacture of plates and beams. For these products, various orientations are considered :

4.1 Plates
For these components we want to use the high strength in short term compression to :

- either reduce the duration of manufacturing cycles for an equal strength.
- or at a higher strength level, and for a given cycle, increase the number of bonded wires to be cast in the plate, giving therefore higher performances to the finished floor.

4.2 Beams

For these components, two cases are to be considered :

4.2.1 Self-resisting beams, purlins, girders.
In this case, the characteristics of HPC can be used efficiently for strength calculation, either to reduce the section of beams of similar performance, or for a given beam insert in the section a greater number of prestressing bonded wires. Should the case of a smaller section be considered, savings in concrete could be realized, bringing lower weights and prices. In the case of extra bonded wires, we could obtain higher performances of the beam.

4.2.2 Beams used in composite sections, with a cast in situ
     compression table
In these sections, the SLS verification of the complete section is of most importance, and is limited by the final shrinkage of the cast in situ concrete. So, the performance of this section will only be very slightly changed with an over-reinforced beam, using to the full all the HPC characteristics, but using plain cast in situ concrete.

In order to benefit fully from all the advantages of HPC used in the manufacture of beams, the compression table must also be made of HPC. To start with, this will only be possible for civil works and building works undertaken by very large companies, this possibility will then be extended to all building works when the supply of HPC by the ready mixed concrete manufacturers is generalised in France.

5 Conclusion

As far as the PPB floorbeams are concerned, the use of HPC has enabled us to conceive a new, lighter floorbeam with the same strength as the previous floorbeam, to improve the productivity of our production facilities and to increase the performance, without propping, of another floorbeam from our range.

Work still has to be done in order to achieve perfection in the use of HPC, both in production and performance research concerning all the other PPB products.

To my mind, among all the possible uses of HPC, precast components certainly represent an important potential for this concrete.

32 OFFSHORE APPLICATION OF HIGH PERFORMANCE CONCRETE SPACE FRAME

C. VALENCHON
Bouygues Offshore, Paris, France

After a brief historical review of triangulated structures, the use of concrete trusses is illustrated with the presentation of the main features of three civil work realizations, showing the evolution of both design and construction methods.

Three structural concepts developed for offshore application are then presented to illustrate potential applications of concrete trusses, concrete space frames and concrete tubular trusses.

Finally, the construction and mechanical testing of concrete space frame and tubular truss structures, performed as part of an internal R & D programs carried out by Bouygues Offshore, are briefly discussed.

Emphasis is placed on the correlation of the space frame technology with development of high performance concrete.

1 Introduction

Development of high performance concrete opens the way for new concepts in concrete offshore structures, able to compete with conventional structures in a wide range of applications.

The use of triangulation to reduce the weight of large span structures is not new. The first lattice structures were made of wood, then in the 19th century, iron was used (the Eiffel Tower) and shortly thereafter steel.

The advantages of lattice structures are their lightness and their constructibility by successive assembly of the constituent elements. There are however only few applications of concrete lattice structures, mainly because of practical difficulties encountered during construction.

Today, post-tensioning techniques combined with the use of highly workable, high strength concrete, and industrial precasting methods, have made concrete trusses and space frames not only feasible but also economical.

High Performance Concrete: From material to structure. Edited by Yves Malier. © 1992 Taylor & Francis.
Published by Taylor & Francis, 2 Park Square, Milton Park, Abingdon, Oxon, OX14 4RN. ISBN 0 419 17600 4.

The successful use of triangulated concrete trusses in civil works construction led us to investigate potential applications to offshore structures, concrete having proven to be an excellent material for marine use.

The following new concepts of concrete lattice structures proposed to offshore operators are presented hereafter :

- Concrete truss Module Support Frame (MSF) for North Sea GBS or TLP.
- Gravity Based Structure (GBS) for Arctic applications using a unique system of internal concrete space frame.
- Concrete tubular truss structures.

2 Concrete trusses and space frames in building and civil works

Although concrete trusses have been used in bridge construction since the beginning of the century, this technique has not been widely used because practical difficulties outweighed the theoretical advantages. Such difficulties included the high density of reinforcement in tension members due to the use of smooth bars and low strength steel, and the lack of efficient admixtures and poor compaction techniques available, so concreting operations required great skill.

Solutions to the practical difficulties encountered earlier stemmed from the development of concrete technology in the past decade and particularly:

- high strength deformed bars and couplers,
- post-tensioning techniques,
- steel formwork,
- vibrators,
- precasting techniques,
- highly workable, high peformance concrete made possible by the use of new admixtures and silica fume.

The evolution of the design of concrete truss structures is parallel with the evolution of the concrete technologies.

The space frame concept exemplifies the most appropriate structure drawing the maximum advantages of the excellent performances of high performance concrete (1).

This evolution is illustrated hereafter by three projects carried out by Bouygues on which different types of concrete trusses have been used: the roof of the Tehran stadium in Iran (1974), the Bubiyan bridge in Kuwait (1983) and the Sylans viaduct in France (1988).

2.1 The Tehran Stadium, Iran

In this project, the economic situation of IRAN, the need for completion in record time and the seismic history of the country (which made lightweight roofing a necessity) all contributed to the decision of building concrete triangular truss girders of over 80 m span (2) (see Fig. 1).

These truss girders constituted the secondary frame of the roof over a swimming pool and an enclosed stadium. Placed parallel to one another, they were fixed between two main box girders, resulting in a square roof of 108.4 m side, including cantilever sections of 13 m.

Diagonal members of the trusses, which were tensioned under dead loads, were prestressed using longitudinal tendons harped at the lower nodes.

Precasting was used whenever possible in construction. The nodes, diagonal members, vertical members and flanges were precast. A closure was cast at each end on a casting bed. After tensioning, the girder was lifted in position and made integral with the main girders.

2.2 The Bubiyan Bridge, Koweit

The superstructure of the Bubiyan bridge which provides a 2500 m long highway link between mainland Kuwait and Bubiyan island is composed of a three-dimensional truss made of reinforced concrete with external post-tensioning (3,4,5).

The longitudinal solid webs of conventional bridges are replaced by an open lattice system of trusses (see Fig. 2).

The truss webs are oriented longitudinally in inclined planes which create common node points in adjacent trusses. This spatial geometry then forms, in the transverse direction, another system of trusses. As a result, the flanges are connected by a system of inclined orthogonal trusses and, therefore, the structural behaviour of the bridge with respect to distribution of loads resembles that of a two-way slab in building construction.

The diagonal and vertical members of the truss were precast together horizontally as triangles. The reinforcement dowels of the triangles were embedded into the slabs when the deck was cast. The longitudinal post-tensioning cables, all external, were deviated at the lower nodes, thereby exerting vertical forces to counterbalance the dead loads. They were located within a water-tight duct and grouted.

The bridge was erected span by span using a cable-stayed launching girder. Match cast segments were assembled without epoxy.

The combined use of concrete truss, precast segmental construction with match-cast joints and external post-tensioning, led to savings of 20% in concrete and 30% in steel when compared

Figure 1 : Tehran Stadium. General view of a truss beam

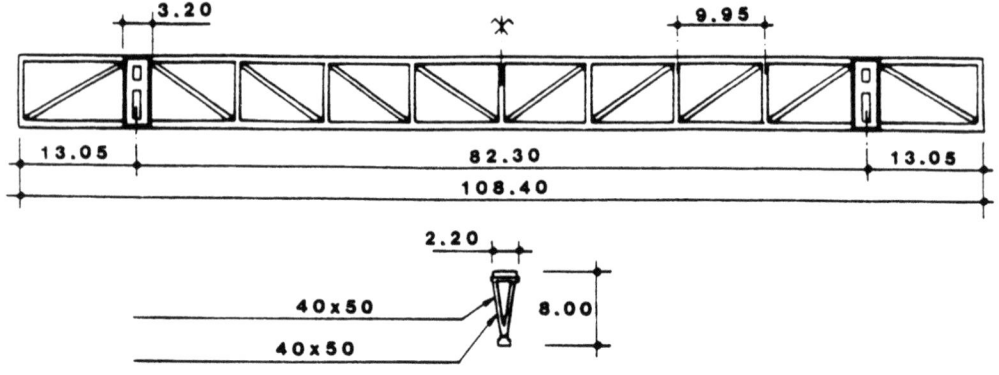

Figure 2 : Bubiyan Bridge. General view of a span and section

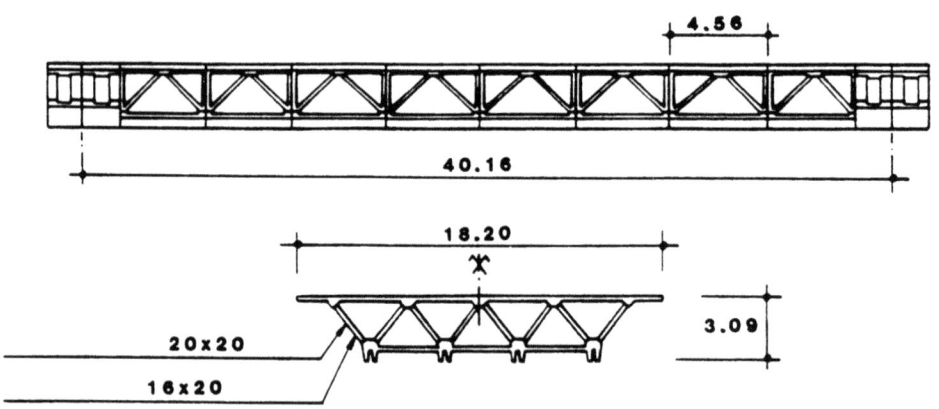

to a conventional box girder solution (cost savings are documented in (3)).

In conjunction with this project, a full-scale test model was built and a series of test loading applied. Analysis of the results have proved that the performance of the structure under service loads was completely satisfactory and that the linear and reversible nature of the deformations throughout the various loading phases was remarkable (6).

2.3 The Sylans Viaduct, France

With a total length of 3000 meters, the Sylans and Glacieres viaducts on the highway linking Macon (France) to Geneva (Switzerland) marked a new step in the use of triangulated concrete trusses (7,8).

The viaducts are located in a mountainous region where the sheer wall of the gorge led to the construction of a separate deck for each traffic direction, which resulted in two relatively narrow decks with only four slanted webs in the cross section (see Fig. 3).

The slanted webs are all composed of "X" shaped elements forming a lightweight and graceful structure. All diagonals were post-tensioned. Longitudinal post-tensioning was partially external.

The construction method consisted first of precasting the "X" shaped elements vertically in battery of 8, followed by their positioning in the formwork for casting of the entire segment. Fully match-cast sections of 423 m were cast, requiring very strict organization for control of geometry.

Erection was carried out by sequential cantilevering using a launcher.

3 Concrete trusses in offshore

3.1 Concrete truss MSF

A design study of a concrete Module Support Frame (MSF) for a North-Sea concrete Gravity Based Structure(GBS) with Sleipner A characteristics was carried out in 1987 by Bouygues offshore on behalf of Statoil as an alternalte to a steel MSF made of a combination of trusses and bulkheads (9).

The design of the concrete MSF was derived from the Sylans project, using the same 'x' bracing principle, but with larger member sizes.

The concrete MSF, as shown in fig. 4 is comprised for:

- two main longitudinal truss girders carrying the load of the modules and other deck equipment
- seven secondary truss girders stiffening the MSF transversally, with two of these girders transferring part of the topsides load to the risers and utility shafts.

481

Figure 3 : Sylans Viaduct. General view of a span and perspective view of a segment.

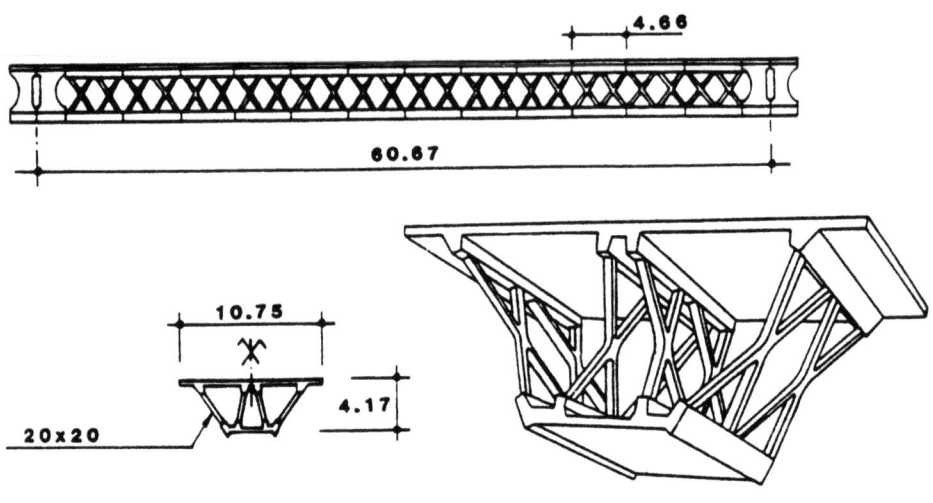

Figure 4: Concrete truss MSF

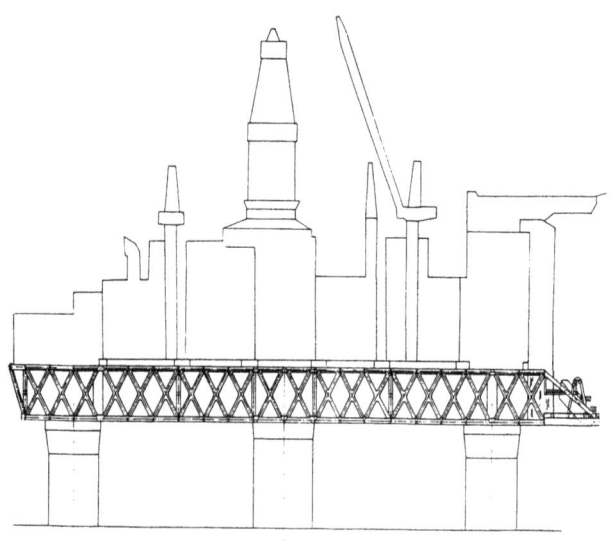

All the trusses are made of an upper flange and a lower flange, connected together by "x" bracings.

The overall dimensions of each truss are approximately 14 m high by 1.2 m wide. Size of members - bracings or flanges - ranges between 0.9 m x 0.9 m for the smallest bracings to 1.2 m x 1.6 m for the largest vertical bracings (above supports).

High strength, normal weight concrete with a characteristic compressive strength on cylinders of 64 MPa at 28 days has been assumed for the calculations.

Post-tensioning would be applied to the concrete MSF by 27 K 15 cables located in the upper and lower flanges as well as in some of the "x" bracings.

The concrete MSF has been designed to bear 31,800 tonnes of modules load (in 7 main modules) together with 6,900 tonnes of mezzanine load and 7,100 tonnes of cellar deck load.

The main advantages of using concrete versus steel for a MSF are that its construction does not require a heavy industrial infrastructure and that its cost of construction is lower.

However, the weight of a concrete MSF is slightly higher and temporary post tensioning could be required during certain construction or installation phases.

3.2 Concrete space frame arctic structures

A new production structure design for Arctic operations that makes use of a concrete space frame has been developed by Bouygues offshore. The space frame endows the structure with high resistance to ice pressure, using minimum concrete volume.

The concrete substructure consists of a number of elements (fig.5) :

- a peripheral sandwich-constructed, ice-resistant wall composed of steel plate at inner and outer surfaces with concrete in-filling a bottom slab, stiffened locally by monolothic ribs, arranged in a triangular pattern,
- an internal concrete space frame, providing the overall stiffness to the structure,
- transition walls connecting the ice wall to the concrete space frame,
- ballast walls dividing the structure into independant compartments for damage stability purposes during towing of the structure.

In the proposed concept, the concrete space frame replaces the internal solid walls which would be necessary in a conventional design. The frame provides lateral support to the peripheral ice wall and transfers the ice forces to the supporting subgrade.

The concrete space frame technology is combined with the use of high performance concrete of 70 MPa compressive strength on cylinders, which allows further reduction of concrete volume in the structure.

The peripheral ice resisting wall is made of composite steel/concrete panels in order to improve the structure ductility and to provide better resistance to ice abrasion.

3.3 Construction methods

Production structure for the Beaufort Sea would be built in temperate zones located south of the 50th parallel, and towed to the point of installation.

Construction of the concrete sub-structure would be performed in a dry-dock using a combination of conventional in-situ casting for the slabs and walls and a newly developed method for construction of the concrete space frame.

The concrete space frame is designed as an assembly of identical precast concrete elements called crystals which consist of a node and twelve half-bars. Casting and assembly of the crystals utilize new mechanized construction methods developed and tested as part of a research and development project carried out by Bouygues Offshore.

Crystals are precast in enclosed steel molds by injecting fluidizied concrete, after placement of the preassembled reinforcement for the bars and node (fig. 6).

Crystals are assembled to form the space frame, first by inserting reinforcement bars in the sleeves located at the ends of the bars to be connected, and then by injecting a special grout in the sleeves and in the space between the bars.

Construction of the space frame of an Arctic production structure would require 2,000 to 3,000 crystals, each having a unit weight of 15-30 tonnes according to its location within the structure.

Crystals would be fabricated in a dedicated precasting plant installed close to the dry-dock. The rate of fabrication would be one crystal per day per mold .

The high performance concrete space frame appears well adapted for the construction of Arctic production platforms, because this design concept criteria :

- in service, they must be strong enough to withstand the high pressure of the pack ice,
- the structures must be as light as possible in order to have a towing draft in the range of 10 m. The latter is important because of the navigation conditions at the entry to the Beaufort Sea.

In addition to weight savings, concrete space frames offer other advantages over conventional concrete or steel designs such as :

Figure 5 : Zee Star 120 - Elevation

Figure 6 : Zee Star 120 - Partial section

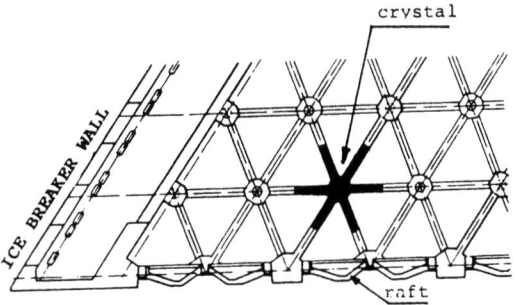

- industrialized precasting,
- high-quality construction and long-service life,
- easy inspection of all parts of the structure and a lower level of maintenance than conventional steel structures,
- high level of redundancy of the space system,
- overall cost savings on the project.

3.4 Concrete tubular truss structure

After presentation of several constructions or projects in which the members of the trusses are of solid sections, a new type of offshore structure consisting of a truss assembly of hollow cylindrical members will be presented.

This concrete tubular truss structure concept has been developed as an alternate to conventional steel jackets or minimal concrete gravity platforms. The new concept is illustrated hereafter through a paticular stucture intended for use as a well-head platform in the southern part of the Norwegian section of the North Sea in 70 m water depth (10).

The concrete tubular structure is composed of an assembly of three types of precast components including tubes for legs, tubes for bracings and leg nodes. The elements are joined together by means of post-tensioning cables.

3.5 Design, construction and installation principles

The preliminary development work performed on this new stuctural concept has led to the following design and construction principles :

- The structure would be assembled horizontally on a quay, loaded on a conventional launching barge by skidding, launched, up-righted and installed with the assistance of a 500 T crane vessel.
- The use of concrete leads to relatively large size elements ; therefore, in order to minimize weights, the legs have been designed to provide support of the stucture during fabrication, loading and launching. As a result, they are parallel.
- The structure shall be able to float on its own structural elements (legs, bracings, pile sleeves), without any additional buoyancy tanks.
- Foundation of the structure is ensured by means of skirt piles post-tensioning forces in the members are such that they cancel any tension in the concrete tubes under all the loading cases steel is used for piles, pile-sleeves and conductor guides fabrication.

3.6 Description of the structure

The structure has a simple shape of a vertical legged tripod fixed to the sea bed by means of steel piles (see fig. 7).

Figure 7 : Crystal fabrication

The legs are spaced 25 m, they are interconnected by four horizontal levels at elevation 0 m, 50 m and 75 m. The structure is stiffened by diagonal bracings between the horizontal levels. The legs are approximately 4 m in diameter and their thickness varies between 0,2 and 0,3 m.

Two pile-sleeves are fixed symetrically about each leg to the nodes of the two lower levels. The concrete volume of the tubular structure amounts to 1,400 m^3.

3.7 Precasting and assembly

Precasting of the tubular elements, legs and bracings would be performed using either a conventional prefabrication technique by filling a formwork with concrete after positioning of reinforcing steel, or a centrifuge method as for concrete piles fabrication (this type of piles are often used in the construction of harbours and wharves).

Once the precast elements have gained sufficient strength, assembly can be performed, starting with the two legs on the skid-ways. The leg elements and nodes are positioned, the joint between them is cast, cables are threaded and tensioned. The upper leg is assembled in the same way on top of scaffoldings.

Bracings are placed between the nodes and connected using the same method.

3.8 Post tensioning

Assembly of precast elements to form the concrete tubular structure is ensured essentially by post-tensioning.

Post-tensioning forces are applied by external cables : in the tubular elements, legs or bracings, the cables run in ducts which are located inside the tube but outside the concrete.

The cables and their ducts are always protected from the outside environment by a concrete wall.

All the cables are anchored at the nodes, either inside the legs for the leg cables, or in a recess provided in the node and located at the opposite end of the corresponding bracing.

The cables run within a high density polyethylene duct and are grouted after tensioning. The anchor heads are sealed with a steel cap which in-turn is embedded in concrete after grouting.

Use of external cables is very well adapted in the present case. This technique is being used more and more often even in conventional bridge construction because it presents several advantages with respect to ease of installation and quality of construction, such as:

- improved concreting operations, thanks to the suppression of duct in the formwork,
- easier cable threading as the cables are usually straight, lower losses by friction, thanks to straightness,

- improved grouting operation, as the duct is watertight over the whole cable length,
- possibility of replacement of a damaged cable.

3.9 Topsides load

The structure presented hereabove has been designed to provide support for 500 T of topsides. However, this structure could support up-to 2,000 tonnes of topsides without structural weight increase. Increase of topsides weight would lead to reduction of pile length as the pile penetration is governed in the present case by tension loads.

Concrete tubular structures provide an interesting alternative to conventional steel jackets. The advantages of this new concept are:

- reduced construction cost,
- increased topsides load capacity,
- reduced maintenance costs resulting from the use of high performance concrete,
- high quality of construction, thanks to precasting which allows simple and efficient quality control.

The structure presented in this paper shows only one example of concrete tubular structures. Different water depths, environmental conditions, operational criteria can be considered, providing other shapes and dimensions of structures.

4 Prototype construction and mechanical testing

4.1 Concrete space frame

An R & D project undertaken by Bouygues Offshore and completed in 1988 has allowed production of crystals on an industrial scale, using concrete with compressive strength on cylinders ranging from 70 to 100 MPa. The crystals produced during this program have bar length of 2.5 m for horizontal bars and 3.5 m for inclined bars (see Figs. 6 and 8). A program of mechanical testing on the crystals and the connections has been carried out at the Structural Testing branch of the CEBTP near Paris, first for the purpose of examining whether the placing technique adopted affects the mechanical properties of the concrete, then to study the mechanical behaviour of the three-dimensional truss under various static and cyclic loadings.

The CEBTP first conducted a series of compression and split tensile tests on bar sections taken from both reinforced and non-reinforced crystals. The same tests were carried out for comparison on standard test specimens made from the same batch of concrete.

Figure 8 : Crystals assembly

These tests showed that the bar concrete had a homogeneous consistency even in the case of reinforced elements, and that injecting the concrete did not modify its mechanical properties.

Compression tests performed on pairs of bars cut from a crystal and connected using the same procedure as for crystal assembly demonstrated that the connection has a higher compressive strength than the standard sections of the bars.

A series of tensile tests was then conducted on crystals, consisting of the node and six coplanar bars. Static and cyclic tensile loads were applied to the crystal through an insert embedded in grout at the end of the bars, simulating a connection between two crystals (see fig. 9). The tensile tests performed on the connection-bar-node-bar-connection assembly demonstrated that, as intended, the connection showed higher tensile strength than the bar itself.

In the final phasis of the test program, compression tests were conducted on crystals of the same type as those tested previously. In these tests, carried out under either static or cyclic loads, failure always occured by buckling of a longitudinal reinforcement element in one of the bars.

Tensile and compression tests were carried out in turn on adjacent bars of a single crystal, in order to verify that failure of the crystal in one direction did not affect its strength in the other directions.

The fatigue testing in tension and in compression confirmed that current design practices with respect to fatigue strength of concrete structures are applicable to the present case of concrete space frames. Therefore, fatigue is not a determining criterion for the design of this type of structure.

All of the tests demonstrated the quality of the completed structure from crystal precasting to connection, thus proving that applied forces were properly transferred through the assembly consisting of the node of a crystal, its bars and connections with adjoining crystals.

4.2 Concrete tubular truss structure

In order to demonstrate the feasibility of the proposed concept and to finalize all details of precasting and assembly of components, Bouygues Offshore has undertaken a R & D project aimed at the construction of a prototype concrete tubular structure.

This project includes prefabrication of concrete tubes for legs and bracings, prefabrication of nodes, followed by assembly of precast components to obtain a prototype structure including a complete leg element with two nodes and one tube in-between, two other nodes coplanar with the lower node of the leg, three horizontal bracings and two diagonal bracings. The prototype structure is shown on figure 10.

Concrete used for the prototype has a characteristic strength on cylinders of 70 MPa.

The elements have been assembled using 13-strand cables able to provide a 280 T prestressing load.

491

Figure 9 : Mechanical tests

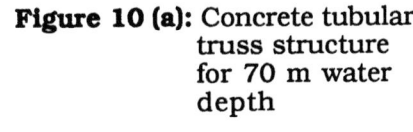

Figure 10 (a): Concrete tubular truss structure for 70 m water depth

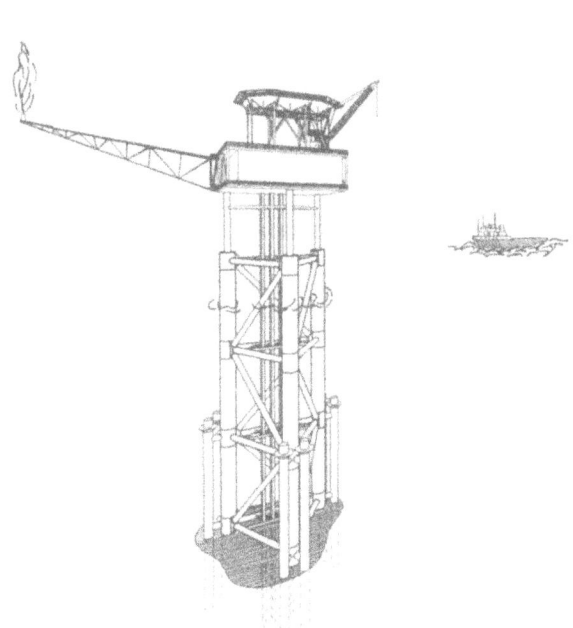

Figure 10 (b): Concrete tubular truss structure prototype

5 Conclusion

The successful construction of the concrete space frame and of the concrete tubular truss structure prototypes, the positive results of the mechanical testing program and the encouraging results of the various feasibility studies performed pave the way for new concepts of high performance concrete structures for the offshore industry, able to complete with steel.

Concrete trusses and space frames present major advantages with respect to conventional concrete designs or steel structures, especially when combined with high performance concrete.

The use of high performance concrete, thanks to its excellent workability, allows economical fabrication of precast elements of highly complex and accurate geometry. In addition, it allows the reduction of the size of the factor.

Use of high performance concrete also improve durability of the structures.

References

1 RICHARD P., HUARD G., VALENCHON C., 'Best Use of High Strength Concrete thanks to Suitable Experiment, Design and Fabrication' - Utilization of high strength concrete, Stavanger, Norway, June 1987.

2 RICHARD P. 'Triangular Concrete Trusses' IABSE Proceedings 1983.

3 PODOLNY W. Jr - MIRELES A.A. 'Kuwait Bubiyan Bridge. A3-D Precast Segmental Space Frame' - PCI Journal January/February 1983.

4 LACOMBE H. 'Presstressed Concrete Triangular trusses' IABSE Proceedings 1984.

5 'Concrete Truss Cuts Weight' from Engineering News Record, November 18, 1982.

6 BRUNEAU J. - CAUSSZ G. - RASPAUD B. - RADIGUET B. 'Test Loading of a Concrete Truss' IABSE Proceedings 1984.

7 'The Sylans and Glacières Viaducts' - J. BOUDOT, PHAM XUAN THAO, B. RADIGUET - International Association for Bridge and Structural Engineering IABSE Symposium Paris/Versailles 1987.

8 Civil Engineering, ASCE Publication, March 1988 - 'Swiss Cheese Box Girders' (Sylans Viaduct) by P. RICHARD

9 VALENCHON C. - WARLAND T.A. - GUDMESTAD O.T. 'Offshore Applications of High Strength Concrete Space Frame' OMAE 1990, HOUSTON, Texas

10 VALENCHON C. - Gudmestad O.T. - SKJAEVELAND H. 'Concrete Tubular Truss Structure' OMAE 1991, STAVANGER, Norway.

33 HIGH PERFORMANCE LIGHT WEIGHT AGGREGATE CONCRETE IN BLAST RESISTING STRUCTURES

F. HANUS, J.C. FAURE, G. PEIFFER
SOGEA, Paris, France

1 Introduction

Over the last few years concrete has been used in special applications in which its main characteristic, mechanical resistance to compression, is no longer the main criteria for its choice.

In the abbreviation HPC, performance naturally applies to compression resistance but also to other physical-chemical and mechanical properties: tensile strength, fatigue resistance, permeability, low or high density, chemical resistance, etc...

The definition that Mr John Chapon gives to HPC in the beginning of this reference work: "obtain the characteristics which make the material the best adapted to the use for which it is made" generalizes this notion of HPC and effectively translates the current trend in this domain.

In fact, it is a question of adaptation and therefore the development of that irreplaceable material, hydraulic concrete, to suit new requirements.

The major progress made over the last few years with regards to workability, thanks to fluidizing type additives (superplasticizers) and in mix proportioning procedures : Baron-Lesage method, grout method., B. Bache method, V. Milewski, etc... now make it possible to extend the use of hydraulic binder base concretes to applications which, up to to-day, were reserved for other materials, such as steel.

The purpose of this article is to demonstrate, with a concrete example, how the application of these new methods has enabled SOGEA to achieve of a concrete solution with a technical and economical performance level greater than that obtained, up to to-day, with metal structures.

High Performance Concrete: From material to structure. Edited by Yves Malier. © 1992 Taylor & Francis.
Published by Taylor & Francis, 2 Park Square, Milton Park, Abingdon, Oxon, OX14 4RN. ISBN 0 419 17600 4.

2 The Problem

In certain military or civil buildings, structures are designed to resist, in particular, the two following types of effect:

blast effect caused by a powerful explosion at a certain distance, piercing effect caused by an explosive charge placed against the structure.

In order to fulfill their purpose, these structures must, in addition, be airtight to the overpressure waves caused by these explosions.

Moving parts (for example, wide doors with a 4 x 5 m opening) were traditionally made of steel which enabled them to resist the blast effect while being sufficiently light to remain operable. In fact, the types of missions of these buildings make it necessary for opening and closing manœuvres to be carried out in very short periods of time. On the other hand, these metal panels while being of a reasonable weight, provided average performances with regards to the piercing effect of contact charges.

3 Loads to be taken into account

These are mainly the contact charge and blast effects.

3.1 Blast effect

An explosion is reflected by the incident overpressure wave hitting the structure, either directly or after reflection or focussing as a MACH wave shape. Reflected wave pressure on structure is much higher than the incident pressure. The amplification ratio varies from 2 to 14 in certain cases.

The explosion is characterized by its impulse (see figures 1 and 2).

$$I = \int P^* \, dT \qquad \textbf{I} \text{ is generally expressed in Pa x seconds}$$

The impulse time may vary from around 15 milliseconds for a traditional explosive located 20 to 30 meters to more than 200 ms for a nuclear explosion.

Therefore, the impulse value depends on the explosive charge, its type and the distance of the explosion.

The structure is deformed under the reflected wave effect and the period in the fundamental mode is of the following type:

Figure 1 : Mechanisms of blast and impulse damage to structures

DYNAMIC LOADING INVOLVES BENDING MOMENTS
AND SHEAR FORCES IN THE STRUCTURES.

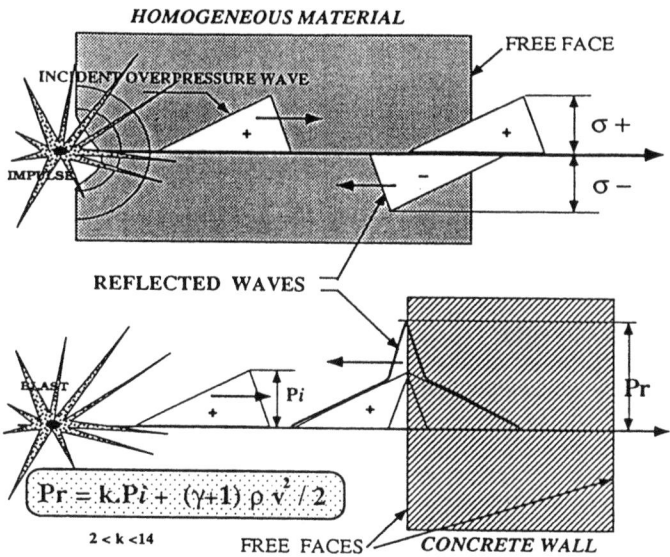

$$Pr = k.Pi + (\gamma+1)\,\rho\,v^2 / 2$$

$2 < k < 14$ FREE FACES   CONCRETE WALL

Figure 2 : Reflected overpressure wave

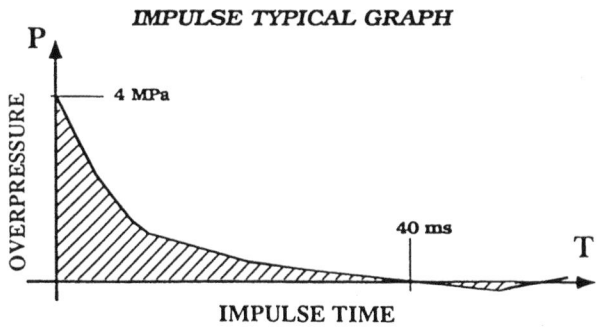

$$T = K2\pi \sqrt{\rho.h/EI} \qquad h = \text{thickness}$$

$$E = \text{modulus of elasticity}$$

$$\rho = \text{density}$$

A low modulus of elasticity and low inertia are therefore very favorable with regards to induced stresses.

The panel may be supported on three or four sides with special antirebound devices.

During the tests, the panel is subject to reflected dynamic pressures varying from 1 to 5 MPa. Corresponding alternating stresses are very high. Acceleration may reach + 900 g thus subjecting the concrete to high dynamic traction, especially during the rebound.

An example of these stresses to which a flat panel, supported on four sides is submitted during an explosion, is given in figures 3 and 4.

3.2 Contact charges

When an explosive charge is placed against the structure, overpressure waves are always formed and therefore very high local pressures attempt to pierce the material. However, in this case, alternating induced stresses are also formed by the wave trains generated in the structure. These waves propagate from the explosion area to the free opposite edge with a speed of approximately $C = (E/\rho)0.5$. When the compression wave reaches the free edge it is reflected as a traction wave of an equal intensity. When these waves are returned they are combined with the incident wave train and the resulting wave is the algebraic sum of these waves. In the beginning, compression decreases and then traction becomes preponderant.

If the amplitude of traction exceeds the rupture break point level of the material, the material ruptures.

The type of material has a considerable effect on the induced stress intensity. To this end, it has appeared that the presence of bubbles (discontinuity) or metal strips is a determining factor. The tensile strength of the material is very important in relation to its compression resistance.

In this type of explosion, the structure also suffers from a large thermal effect which material components must be designed to resist.

3.3 Stress calculation

The dynamic calculation of structures is complex and only accessible to a small number of software programs (in France, Hercule). However, this type of program is slow and costly and difficult to use.

Figure 3 : Dynabeton - Bending moments

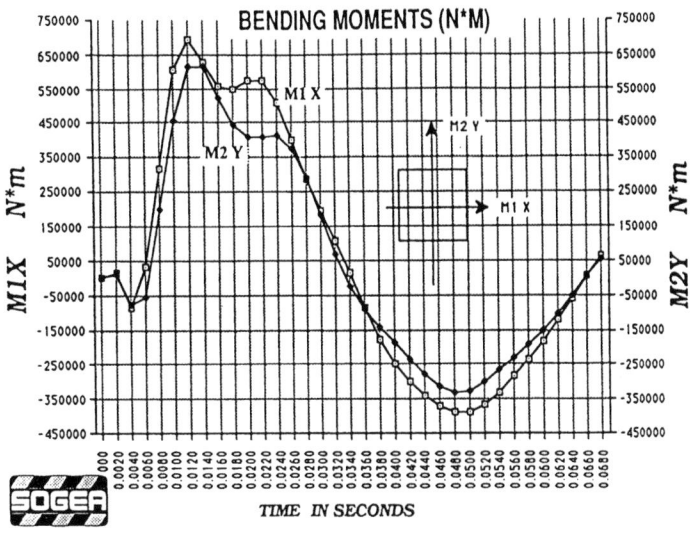

Figure 4 : Dynabeton - Acceleration

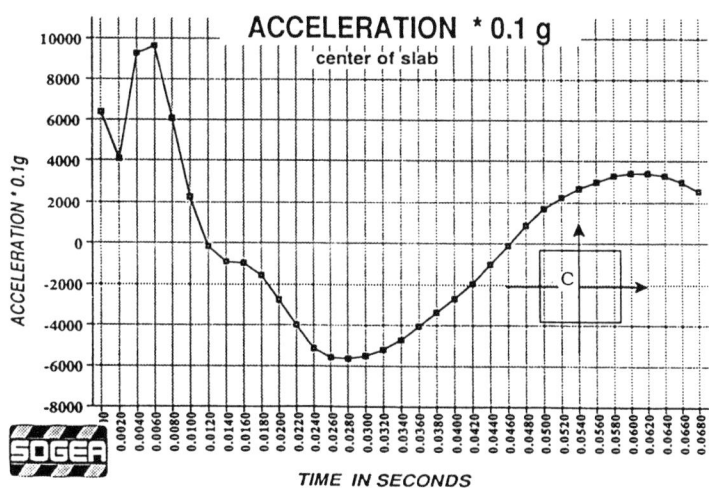

In order to enable fast and more economical research, the technical department of SOGEA has developed dynamic calculation software, specifically for panels supported around all or part of their contour.

This program may be used on a microcomputer and enables the stresses, accelerations and deformations in the structure, during the explosion, to be determined.

4 Purpose of the research

For several years, SOGEA has been working on a research program in the fibre reinforced concrete domain, especially with glass fibres (second generation PGRMC) and amorphous metal glass fibres (invented and produced by Saint-Gobains Pont-a-Mousson). SOGEA had the opportunity to develop a hydraulic concrete reinforced with these fibres that provided satisfactory resistance to blast effects and piercing effects, with a density less than 1.6 T/m^3.

The structures to be built with the new material must be systematically tested at life size by the client. The explosive charges were of around 850 kg of TNT placed at 8 to 30 meters from the wall in order to test its blast resistance and of 20 to 30 kilos to test its contact charge resistance (see photos).

Complementary tests on shell penetration and the degree of fire retarding protection, were also carried out.

At the end of the test, the panels and related control devices must remain operational, remain tight to the incident overpressure wave and open and close within the scheduled time limits.

Naturally, after reading the previous section, it is obvious that compression resistance is not the key feature of this concrete. On the other hand, criteria such as those listed below are preponderant:

rupture ductility
cracking
tensile strength
lightness

Finally, research was oriented towards a concrete with the following nominal composition:

OPC cement
microsilica (silica fume)
river sands
light weight aggregates and sand
amorphous metal-glass fibres of appropriate sizes
specific rheological additives and admixtures
water

Obviously, the mixing of this concrete must be easy while still providing the required mechanical performance. Consequently, special attention was paid to workability (see photo of a Dynabeton sample).

5 Formulation of a high performance light weight concrete - Dynabeton*

The concrete should satisfy the technical requirements concerning fresh concrete workability, ease of casting, ened hard concrete density and required mechanical characteristics. The cost of the material satisfying these requirements conditions its effective application.

The type of stresses to which this concrete will be submitted in service oriented the study towards light weight aggregate based formulas reinforced with stainless steel fibres.

* trade mark

5.1 Light weight aggregates and high performance
Useful data on light weight concretes and aggregates are to be found in reference (1). Figure 5 taken from this reference work, compares the performances of the concretes, formulated using various aggregates, by means of a diagram (dry density-compression resistance).

The use of appropriate techniques to make high or very high performance concretes (2), (3) using normal density aggregates, especially silica fume or fluidizing additives, should make it possible to increase the level of resistance reached at a given density.

The use of fibres entails a modification to the concrete granular structure. The studies of Serna Ros (4) and more recently Rossi (5) concern the formulation of normal aggregate concretes with metal fibres. The authors have applied the Baron-Lesage experimental method to the fibre concrete. Naturally, this method consists in optimizing the consistency of the concrete, for a given binder volume and composition, by simply varying the proportion of sand in relation to gravel. The principle is illustrated in figure 6.

The consistency of the concrete is measured, as per (5) in relation to the flow time, using the ICL workability meter.

The optimum proportion of sand and gravel depends on the type of aggregates and, for fibre concretes, the type, length and percentage of fibres used.

In Figure 6, we have added the data obtained from (5) and the data obtained in our laboratory. In the latter case, the aggregates were crushed at a density around 2.65.

Figure 5 : Relationship between the compressive strength and the density of light weight aggregate concretes

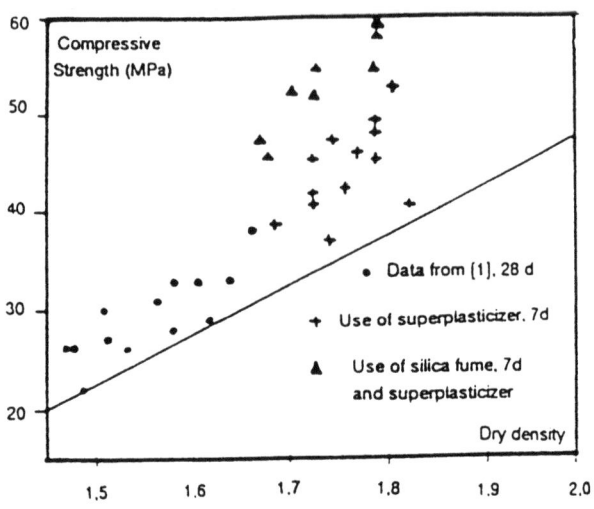

Figure 6 : Mix design curves for different aggregate and fibre characteristics

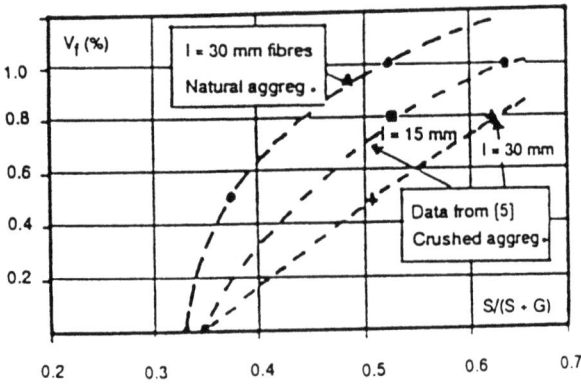

These curves link the percentage of fibres used to the optimal proportion of the granular skeleton. They were established for amorphous steel fibres in ribbon format, commercialized by Pont-a-Mousson S.A. The main findings, likely to guide the choice of concrete constituents, are as follows:

1. For a given aggregate size and type, the granular correction (percentage of sand) increases with the percentage and length of fibres.
2. The granular correction is less when crushed aggregates are used. The deduction is not rigorous as the size of large rolled or crushed aggregates differ. However, it corresponds to other practical observations.

Also it seems that the correction is less if sand fineness increases.

The workability effectively obtained with some aggregates during this operation must later be dropped to the value required for concrete casting.

This adjustment may be obtained by varying the volume of the binder. This method offers the advantage of preserving a reference formulated binder, thus avoiding downgrading resistance by adding water.

Lastly, the volume of binder required to obtain a given workability increases with the percentage and length of fibres.

5.2 Practical formulation of a light weight concrete reinforced with fibres

The Baron-Lesage method, validated for normal aggregate concretes, cannot be directly applied to light weight concrete formulation. According to Table 1, grouping the volumes of the main constituents envisages, the deviation in densities is very high between sand and expanded clay aggregate may be remarked.

Therefore we preferred to use an approximative mehtod by analogy with the results available for normal aggregate concretes. In fact, the granular corrections required resulted mainly from a geometric type interreaction between the constituent components.

The approach seemed valid and was revealed to be efficient in the case of fibres for which we have graphs, such as those in figure 6. Then all that is required is to use the same proportions by volume. In other cases, the need to work quickly led us to adjust the formulas with several test mixes.

This approach imay be used with little experience. In all cases, it is facilitated by the use of the general tendencies of fibre concretes mentioned above.

In practice, control of concrete density is a delicate operation. The density increase due to the weight of the fibres is accentuated by the necessary increase in the proportion of sand. In one of the studied variants, and, finally the one selected, we modified the sand by using a hollow glass microball additive. This fine constituent limited the granular correction and reduced the density while maintaining a sufficiently high compression resistance.

5.3 Selection of an explosion resistant material

Compression resistance at 7 days of light weight concretes formulated with or without silica fumes are added to the diagram in Figure 5. Observe that the silica fume base variants are under the cloud of dots representing concretes without additives. This increase in performance is made possible by selecting a high resistance expanded clay aggregate.

The acknowledgement in sizing of the various mechanical data obtained in the laboratory (bending behavior, modulus of elasticity) led us to select a possible formula for the dynamic tests so as to be able to characterize the material under the most representative conditions.

The compression resistance at 28 days on a 16 x 32 cylinder was approximately 50 MPa for a 1.6 dry density (approx.).

The main composition data (average) is given in Table 2.

5.4 Mechanical behavior during dynamic, static and direct traction

In order to be representative of the tested material, the direct traction behavior tests were carried out on core borings taken from cubin blocks.

The results in this document concern cylindrical test specimens with an axis perpendicular to the block casting direction. This is in order to stress the material in the same relative direction as the structure.

The static behavior is revealed by the stress - displacement curve in Figure 7. The direct traction test is carried out at a set displacement speed on a prenotched test specimen.

The resistance plateau recorded after the first peak remains relatively high up to the maximum displacement selected, being 0.2 mm. This post peak ductility may mainly be allocated to the separation of the fibres.

The dynamic behavior at high stress speeds were studied using the Hopkinson bar from the University of Delft (7).

This equipment is schematically described in Figure 8. The cylindrical test specimen is bonded between aluminium bars, overall length approximately 10 m. The traction stress is generated at the base of the lower bar by dropping a weight: landing reception conditions make it possible to modify the type of impulse, if necessary.

The graph in Figure 9 groups the displacement-constraint curves obtained at a nominal stress speed of 48 MPa/ms.

The dynamic behavior of a light weight concrete without fibres has been obtained using the same formula as the reference

Figure 7 : Stress - deformation curves under static loading (7)

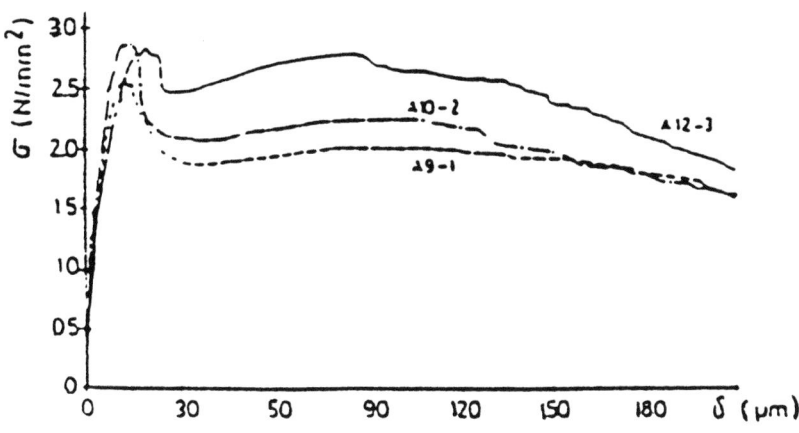

Figure 8 : Split Hopkinson Bar (7)

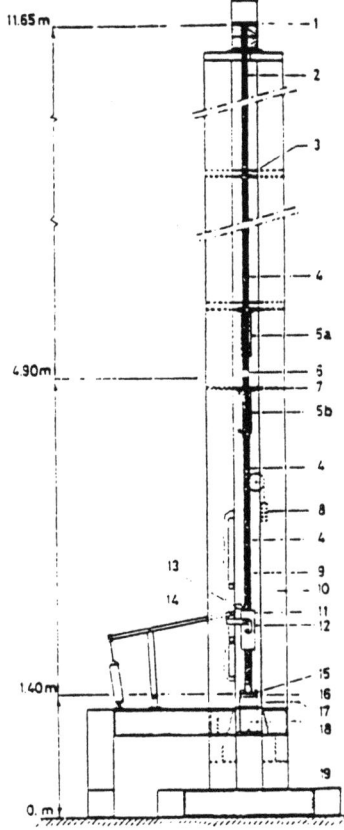

1   Buffer
2   Upper bar
3   Guide
4   Strain gauge
5a Upper cooling jacket
5b Lower cooling jacket
6   Concrete test specimen
        0 76x100 mm
7   Working platform
8   Counter weight
9   Lower bar
10 Frame
11 Drop weight
12 Coupling
13 Uncoupling
14 Lifting device
15 Demping material
16 Anvil
17 Guide tube
18 Pneumatic jack
19 Frame base

505

Figure 9 : Stress - deformation response of light weight

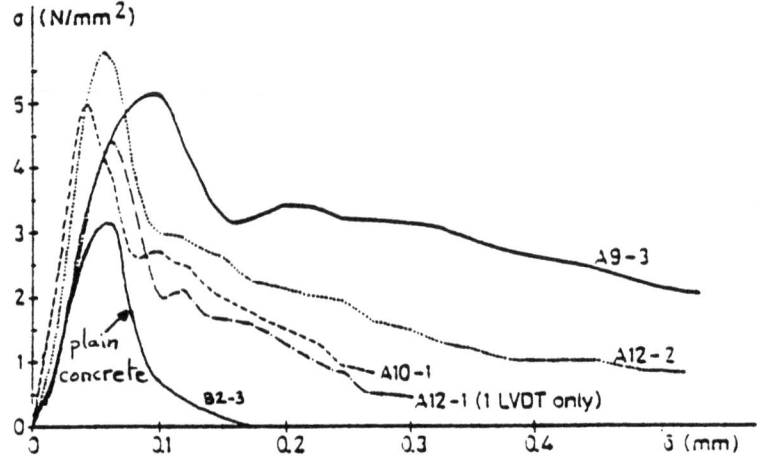

concrete, excluding the fibres, has been added. The increase in resistance due to the fibres clearly appears at high stress speeds.

The static tensile strength of the fibre concrete has been multiplied by 1.9 at a 48 MPa/ms load speed.

This experience has also made it possible to measure the light weight concrete dynamic moduli. With values between 8000 and 12000 MPa the moduli are very low in relation to the usual static modulus for this type of concrete.

Table 1 : Density of main constituents

Constituent	Density
0-5 m river sand	2.65
4-12 mm expanded clay	1.05
Microballs	0.70
POC 55 cement	3.15
Amorphous metal-glass	7.20

Table 2 : Composition of high performance light weight concrete

Constituent	%
Volume (sand+ microballs)/Gravel	0.60
Water/cement	0.40
Silica fume/Cement	0.10
Fluidizer/Cement	0.02
Amorphous metal-glass ribbions	0.7

6 Behavior of Dynabeton

Traditional concretes have a brittle behavior with regards to this type of stress. They are completely disorganized and non reinforced parts are ejected. After a blast test there are a limited number of cracks, very open and directly transverse. The panel often is broken by shearing along a support.

Dynabeton, consisting of amorphous metal glass ribbons reinforcement and light weight aggregates has, on the other hand, a ductile behavior. There are numerous cracks but they do not directly cross the panel. Crack routing is similar to consecutive Zs. The result is a considerable increase in their path distance which consequently absorbs the energy. It is found that the fibres satisfactorily react as crack arresters during the initial stages and then as bridges.

The deformation in the center of a Dynabeton panel is similar to that of a traditional concrete panel, in spite a much lower modulus of elasticity.

The Dynabeton panels successfully passed through live contact explosion and blast effect tests which is a considerable improvement in relation to conventional panels.

Therefore, these panels are truly a progress as their performance levels are higher and they cost less.

Following the tests, it was impossible to demolish the Dynabeton panels with a pneumatic hammer, in a reasonable period of time. Nevertheless, even though it is as easy to push the point of the hammer in a Dynabeton panel, as it is in a traditional concrete panel, it is not, on the other hand, possible to split it or break it down. This interesting and surprising performance may be explained by the presence of fibres, the exceptional adhesiveness of this concrete to the reinforcements and the light weight aggregates which give it high rupture ductility.

Also, torch cutting was tested and also proved to be very difficult to use. Slag forms which blocks the heat of the flame and stops cutting. Dynabeton has excellent fire resistance and this is one of the important properties in its field of applications.

7 Advantages of Dynabeton

Besides the good ductility and cracking resistance performance, this type of concrete cast with metal profiles and reinforcements must have good durability (sufficient impermeability and alkaline reserves). The ageing tests carried out were always satisfactory. It is therefore possible to imagine that the use of OPC cement, microsilica and stainless steel fibres are determining elements from this standpoint.

Photo 1: Concrete with amorphus metal glass fibers

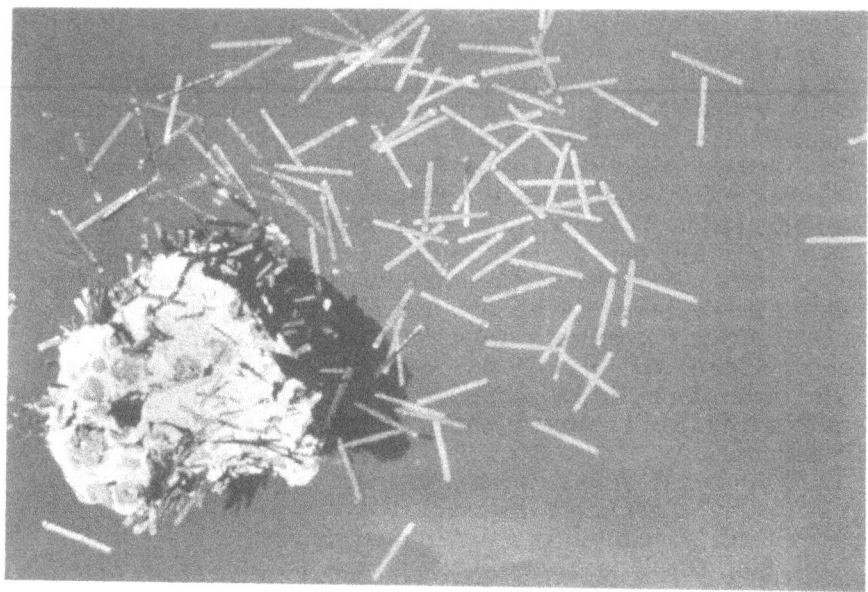

Photo 2: The experimental structure and the explosive charges (850 kg of T.N.T.)

8 Conclusion

The composition methods and the new products available (ultrafine reagents, superplasticizers, fibres, additives) make it possible today to adapt the properties of concrete and make its use possible in domains which, up to now, were inacessible.

SOGEA has thus been able to develop a high performance light weight concrete that is exceptionally efficient in moving parts of structures designed to resist explosions, in compliance with very precise specifications.

This example confirms importance of the field of applications available to HPC in the future.

References

1. Arnould, Virlogeux. Light concretes and aggregates, Presses de l'Ecole Nationale des Ponts et Chaussées, Paris, 1986.
2. F. de Larrard, C. Puch. Formulation of high performance concretes. The casting method. Bull. liaison labo. P. et Ch., may-june 1989, pp. 7583.
3. J.M. Pedeches. Study of mechanical properties and shrinkage of high performance concrete. Thesis, 1988. Université Paul Sabatier, Toulouse.
4. P. Serna Ros. Study of the contribution of steel fibres to the improvement of concrete shearing behavior. Thesis de doctorate de l'ENPC, Paris, 1984.
5. P. Rossi, N. Harrouche, A. Belloc. Metal fibre concrete composition method, Ann. ITBTP, 475. june-july 1989, pp. 37-44.
6. A.M. Paillere, B. Godart. Contribution of metal fibres to the improvement of the behavior of concrete structures. Proc. 1st Int Rilem Cong., Chapman and Hall, Vol. 2, pp. 668-693.
7. J.G.M. Van Mier. Contract Report, Delft University of Technology, Nov. 1988.

34 HIGH PERFORMANCE CONCRETE IN THE ARCH OF THE BRIDGE OVER THE RANCE

J-F. DE CHAMPS and P. MONACHON
Campenon Bernard, Paris, France

1 Introduction

The expressway from Pontorson to Lamballe, intended to link Normandy to the Northern route across Brittany, will cross the Rance south of Saint-Malo, just upstream of the well-known tidal power station, between the villages of Port Saint-Jean and Port Saint-Hubert. The gap is 300 meters and the tidal range can reach 10 meters between low and high water. Sound granite outcrops at a shallow depth under the banks and under a few meters of silt, sand and gravel in the river bed.

Several approaches to crossing the river were investigated beginning in 1974 by SETRA at the request of the District Public Works Agency of the Ile-et-Vilaine department, working with the Sites Commission and the architect Charles Lavigne.

The approach finally chosen in 1986 was a reinforced concrete arch bridge having a span of approximately 260 meters, to be built in successive cantilever cast-in-situ segments, with an independent top deck having a steel-and-concrete structure, the metal framework of which, consisting of two cross-braced I girders under the pavment, would be pushed into place and the concrete deck then cast in situ.

The bridge carries two 3.50˙m traffic lanes with a 2.00˙m shoulder.

The construction of the bridge was awarded to the contractor Campenon Bernard, which proposed replacing the B.40 concrete of the arch called for in the specifications by a high-strength B. 60 concrete, which would substantially increase the bearing capacity and durability of the arch while adding relatively little to the cost of the bridge.

High Performance Concrete: From material to structure. Edited by Yves Malier. © 1992 Taylor & Francis.
Published by Taylor & Francis, 2 Park Square, Milton Park, Abingdon, Oxon, OX14 4RN. ISBN 0 419 17600 4.

2 Presentation of the project and methods of execution

2.1 Characteristics (Figures 1 and 2)

- Total length of bridge :   421.00 m
- Width of deck :   12.00 m
- Span of arch at springers :   261.00 m
- Rise of arch : approx.   35 m
- Running section of box of arch
  . Width :   7.50 m (increased to
                                               12.00 m at springers)
  . Height :   4.20 m
  . Thickness of webs :   0.40 m
- Volume of high-strength concrete in arch (HSC) : 2,800 m³.

Figure 1 : Cross sections of deck and arch

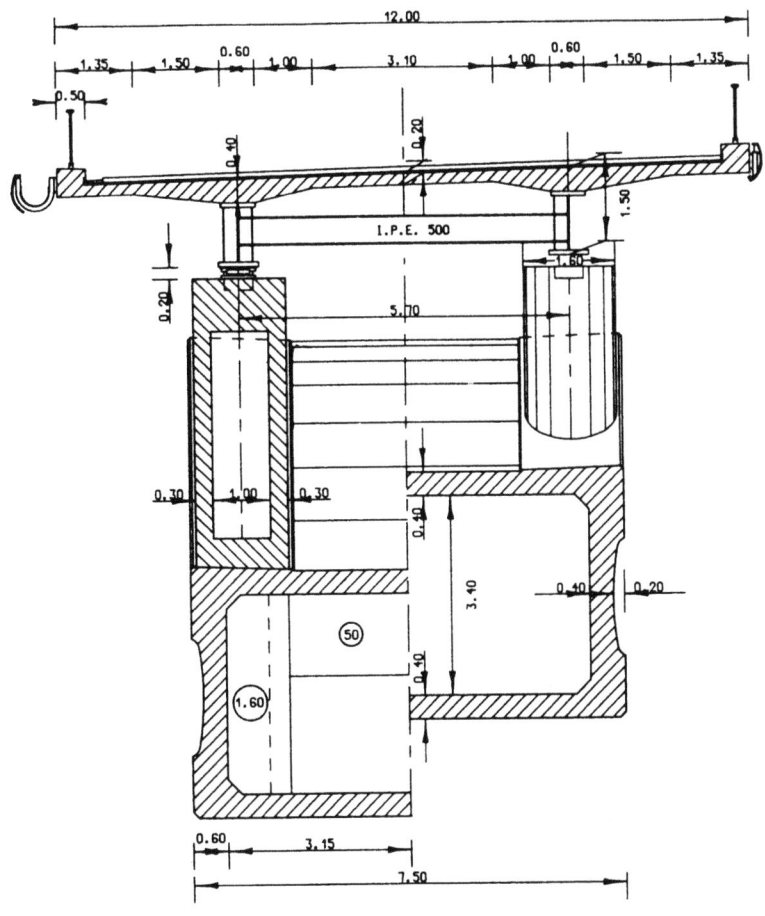

Figure 2 : Elevation of structure and plan view of arch

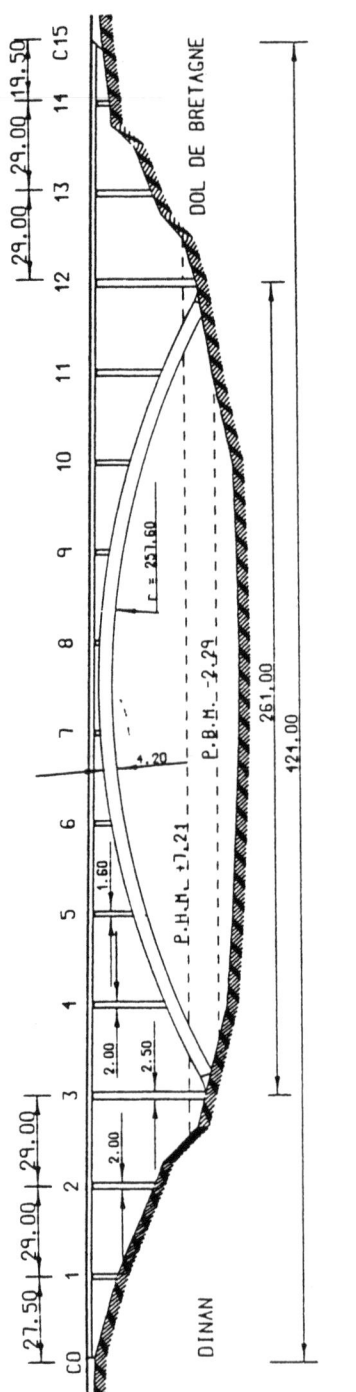

The mean stress on the concrete of the arch at the service ability limit state does not exceed 11 MPa, but the regulation check of the buckling stability of the arch at the ultimate limit state would just have been passed with a concrete having a characteristic strength of 40 MPa.

2.2 Method of construction of arch (Fig. 3)
Very succintly, it is broken down into 6 stages :

Stage 1
Construction of cofferdam supporting block.
Casting of segments 1 to 6 on centering.
Castering of segment 7 in form traveller on small temporary bent.

Stage 2
Casting in successive cantilevers of segments 8 to 14, with prestressing in extrados.
Placement and adjustment of jacks on temporary bent under segment 14 (force given).
Casting of segments 15 and 16.

Stage 3
Erection of staying mast.
Casting of stayed segments 17 to 26.
Removal of small temporary bent.
Casting of stayed segments 27 to 31.

Stage 4
1st jacking at crown.
Removal of odd-numbered stays.
2nd jacking.
Removal of even-numbered stays.

Stage 5
Dismantling of masts and temporary bents.
Construction of small piers.
Pushing of metal girders of deck.
3rd jacking.

Stage 6
Casting in sections of reinforced-concrete slab.
4th jacking.
Placement of superstructures.

Figure 3 : Construction process

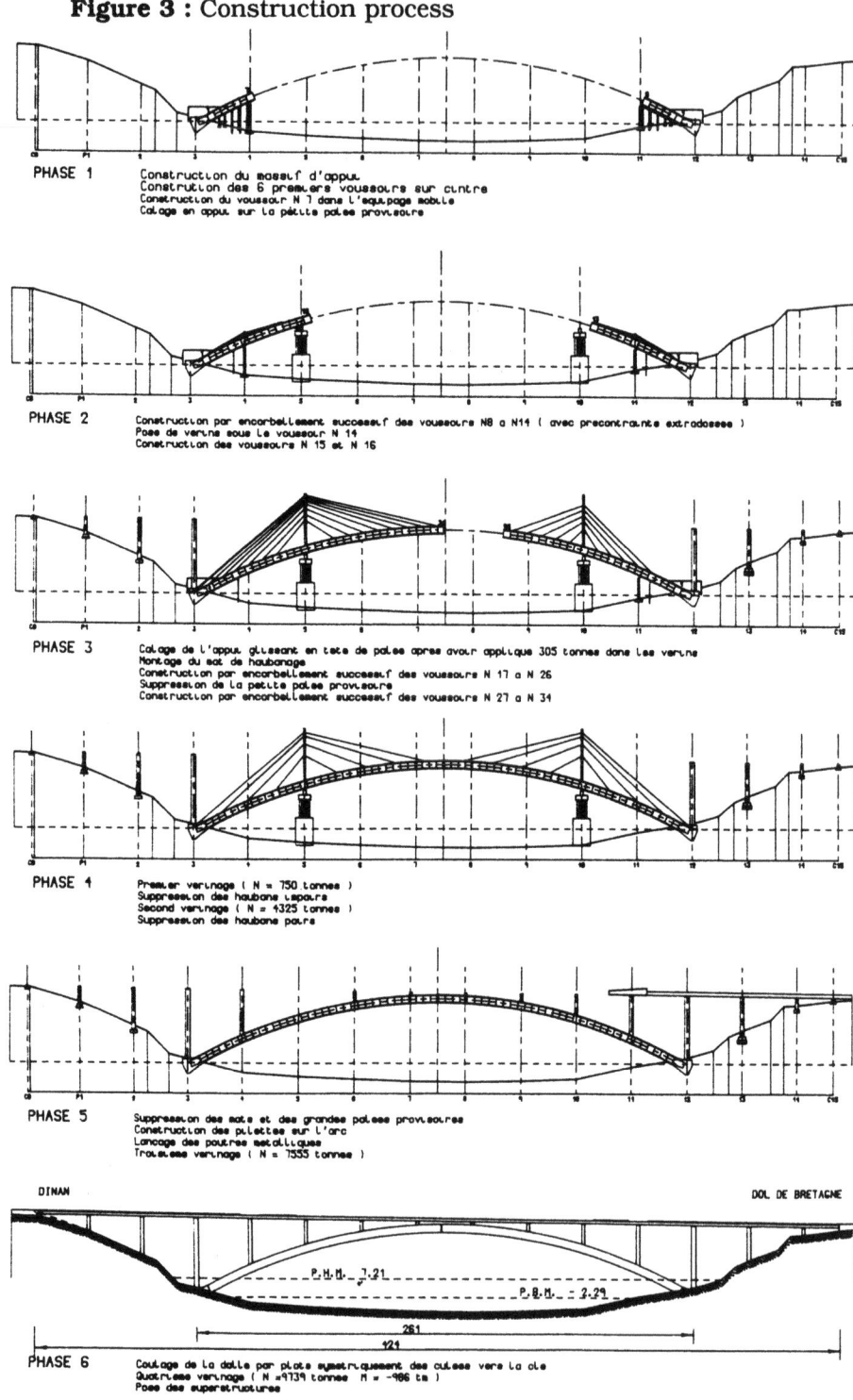

PHASE 1
Construction du massif d'appui
Construction des 6 premiers voussoirs sur cintre
Construction du voussoir N 7 dans l'équipage mobile
Calage en appui sur la petite palée provisoire

PHASE 2
Construction par encorbellement successif des voussoirs N8 a N14 (avec précontrainte extradossée)
Pose de vérins sous le voussoir N 14
Construction des voussoirs N 15 et N 16

PHASE 3
Calage de l'appui glissant en tête de palée après avoir appliqué 305 tonnes dans les vérins
Montage du mât de haubanage
Construction par encorbellement successif des voussoirs N 17 a N 26
Suppression de la petite palée provisoire
Construction par encorbellement successif des voussoirs N 27 a N 34

PHASE 4
Premier vérinage (N = 750 tonnes)
Suppression des haubans impairs
Second vérinage (N = 4325 tonnes)
Suppression des haubans pairs

PHASE 5
Suppression des mâts et des grandes palées provisoires
Construction des pilettes sur l'arc
Lancage des poutres métalliques
Troisième vérinage (N = 7555 tonnes)

DINAN   DOL DE BRETAGNE

P.H.M._7.21

P.B.M._3.29

261

124

PHASE 6
Coulage de la dalle par plots symétriquement des culées vers la clé
Quatrième vérinage (N =9739 tonnes M = -986 tm)
Pose des superstructures

514

The total duration of the work, including setting up, will be thirty months, including about ten months for the casting of the high strength concrete arch.

3 The specific advantages of HPC

The specification called for a concrete having a characteristic strength of   40 MPa (B.40) containing 400 kg of seawater-setting CPA 55 cement per cubic meter, with a slump of 5 to 9 cm at placing and a 0/25‾mm grading.

The contractor had planned to bring in ready-mix concrete from Saint-Malo (about 30 km away), with the drawback of a transport time that would be rather long, and even uncertain in the summer holiday season.

Moreover, the concrete was to be distributed by a cableway carrying a one-cubic meter skip, leading to rather slow concreting (5 to 7 m³/h), which in turn would have forced the trucks to wait a long time to unload. All this meant that it would be difficult to control the slump. In addition, the mediocre quality of locally available aggregates risked making it difficult to meet the specifications.

Finally, and most important, all parties concerned attached some importance to improving the durability of the bridge through an excellent density and closeness of the concrete. This is why Campenon Bernard proposed to the Project Supervisor, which accepted, that the arch be made of high-strength B. 60 concrete with a 0/12‾mm grading and the addition of silica fume, which would have the following advantages:

- **greater security of the arch**, especially with respect to buckling, through the increase of the characteristic strength from 40 to 60 MPa,
- **greater precision of the adjustment of the arch**, because the high strength concrete exhibits less creep, and it occurs earlier and so can be compensated for as the work proceeds,
- **better duraibility of the concrete**, thanks to the high density attained through the low water/cement ration and the presence of silica fume:
- permeability 10 times smaller, according to Maage (Réf.1), ensuring excellent protection of the reinforcements very low penetration of chlorides and low carbonation,
- corrosion resistance from 2 to 5 times greater, according to Berke (Réf. 2),
- increased resistance to potential damage by the alkali reaction in the presence of seawater (15 times less at 6 months according to Asgeirsson, Olafsson and Gudmundsson, Réf. 3 and 4),

- **substantial increase in early strength**, so obviating thermal curing, which always impairs long-term strength and sometimes causes microcracking,
- **reduction of thermal cracking**, since tensile strength increases rapidly with no increase in the heat of hydration (see table in appendix 1, Record of temperature on 2 segments),
- **complete freedom from drying shrinkage cracking**, thanks to curing of all sides of the concrete because of the full formwork, including the tops of the slabs, making the skin highly impermeable after form removal,
- **ease of placing and freedom from segregation**, in spite of the highly angular crushed gravel and localised high reinforcement density, thanks to the excellent cohesion, fluidity and homogeneity of the fresh concrete; but external vibration had to be substituted for full-depth vibration.

Finally, on-site preparation, in a well-equipped on-site plant used only for this purpose, ensured output of **excellent regularity**, as attested by the statistical results given below.

4 High performance concrete

4.1 Site specifications
Characteristic strength :   60 MPa
Strength at form removal :   18 MPa at 20 hours
Strength at tensioning of stays :   30 MPa at 3 days
Slump 18 ± 2 cm at placement, maintained for 1 hour
Placing rate :   5 to 7 m³/h.

4.2 Choice of materials
Cement: PMCP HS CPA
(see analysis in appendix 2)

Aggregates:
* 0/4 mm sedimentary sand
* 3/8 and 8/12 mm crushed magmatic rock gravel
* Densified Silica fume (Pechiney)
(see analysis in appendix 2)
Admixtures:
Superplasticizer + Retarder

4.3 : Formula of the high performance concrete

Table 1 : Mix design

	Segments 1 to 10 kg/m³	Segments 11 to 31 kg/m³	
Cement	400	430	
Silica fume	32	34	
0/4 mm sand	724	785	
3/8 mm gravel	-	463	
8/12 mm gravel	1,159	550	
Superplasticizer	12 kg (1)	(12 (1) then	(21.5 (2)
Retarder		(0.9 (3)	(1.29 (4)
Total water SSD	148	150	

(1)   Superplasticizer A
(2)   Superplasticizer B
(3)   Retarder C
(4)   Retarder D

The formula of the high strength concrete given in Table 1 was developed in accordance with the following principles :

1. Optimization of the continuity of the granular skeleton by the Baron Lesage method.
2. Determination of the optimum superplasticizer content by the grout method.
3. Combination of the superplasticizer and the retarder, the quantity of which must be adjusted according to the temperature and the length of time workability to be maintained.

4.4 Concrete plant and mixing sequences

For the prepartion of the high performance concrete alone, the site set up an Arbau Euro 55 batching plant equipped with a 1 m³ vertical shaft mixer and level 3 accessories (FASC. 65 - CCTG), with special arrangements for the storage and precise proportioning of the silica fume, admixtures and water. (The rheology and other properties of highstrength concrete are highly sensitive to small variations in the proportions of the components).

The total duration of the mixing cycle was 5 minutes, with the following sequences:

- introduction of aggregates + cement + silica fume
- introduction of water + retarder
- introduction of superplasticizer.

4.5 Statistical results

4.5.1 Strength values. Strengths are given in Table 2.

Table 2 : Strength

	Number of tests	Mean (MPa)	Standard deviation	Coef. of variatio
Compressive st.				
20 H	35	24.3	8.36	34.4
22 H	36	31.2	9.45	30.3
24 H	35	36	8.37	23.1
3 D	26	58	5.44	9.4
7 D	51	67.5	4.7	7.0
28 D	49	81.3	2.65	3.3
90 D	3	88	-	-
Tensile st.				
28 D	1	6.9	-	-

Note : Each test is represented by the mean of the results obtained on three 160 x 320 mm cylinders, capped with sulphur in the usual way. This means of preparation gives a strength 10 % lower than normal way (with grinding wheel).

Conformity to fascicule 65 - CCTG

1. Mean = fc 28 $\geq$ fc 28 + 1.3 S
$$60 + 1.3 \times 2.65 = 63.4$$

2. Minimum = fc i $\geq$ fc 28 - K2
$$60 - 3 = 57$$

These two conditions are amply satisfied.

4.5.2 Workability
Observed slump value = 18 $\pm$ 2 cm.

5 Comments on difficulties encountered

At the start of the work, the contractor encountered some deviations from the findings and results of the concrete studies. There is reason to believe that the origin of these discrepancies lies mainly in variations in the basic component materials, on which the concrete depends.

Cement

Strength variations are shown in Table 3.

Table 3 : Variation in cement strength

Age	Standard deviation announced by maker	Observed variations
1 D	1.3 MPa	6.3 MPa
2 D	1.6	2
7 D	1.9	7.3
28 D	2	6.7

- Fineness variation: 8.8 %
- Variation of workability: 100 %
- Variation of C_3A content : from 2.5 to 5 %

Silica fume
- Variation of ignition loss: 85 %
- Variation of fineness: 18.4 %
- Variation of total carbon: factor of 58

Aggregates
- Sand: variation of fineness modulus =   27 %
- Gravel: variation of flatness coefficient = 113 %

It was in an attempt to alleviate these variations, or at least mitigate their consequences, that Campenon Bernard increased the cement content slightly (30 kg) and decided to separate the admixtures according to function.

However, it is well known that the use of set retarder to maintain the slump for the time required for placing has the inevitable indirect consequence of increasing the setting time and reducing the early strength values: the concrete started setting only after 16 hours, or even 20 or 22 hours in cold weather.

Finally, we had some fears fortunately baseless, of the risk of an alkali-aggregate reaction, a doubt arising because of the divergent results of preliminary tests of the potential reactivity of the aggregates by three different laboratories. This doubt was finally eliminated only after the six months of the standard test, which established that the aggregates were perfectly innocuous. This was very fortunate, because in the meantime more than half the arch had been poured. And what could have been used in their place, since the aggregates from the area are all geologically identical ?

This points to the need for an accelerated test.

6 Conclusions

The decision to build the Rance arch bridge of high strength concrete was a successful one in a context made difficult by the geometry of the structure and the special conditons of concreting by cableway.

In spite of the known mediocre quality of aggregates in the region, the HSC produced a concrete of far better quality than the ready-mix initially planned. Could a ready-mix even have achieved the required strength values ?

The advantages found may be summed up as follows :

- great regularity of workability and strength properties,
- placing made easier by a highly fluid consistency,
- large margin of safety on strength values, once setting has begun,
- substantially improved durability.

In such circumstances, the presence of an on-site plant provides great flexibility and excellent regularity.

The addition of silica fume and, with it, a superplasticizer, gives the concrete better performance, excellent cohesion, and good workability, but requires the use of suitable vibrating equipment.

References

1. M. MAAGE : Efficiency factors for condensed silica fume in concrete - P. 783 - 3rd International Conférence - CANMET/ACI. Trondheim 1989.
2. N.S. BERKE. Resistance of microsilica concrete to steel corrosion, erosion, and chemical attack. P. 861 - 3rd International Conférence - CANMET/ACI. Trondheim 1989.
3. Asgeirsson, H., and Gudmundsson, G. 'Pozzolanic activity of silica dust'; Cement and Concrete Research 9:249-252 ; 1979.
4. Olafsson, H. 'Effect of silica fume on the alkali-silica reactivity of cement'; Report BML 82.610 ; The Norwegian Institute of Technology, Trondheim, Normay ; pp. 141-149 ; Feb. 1982.

1. Casting of segments in successive cantilevers with mast and stayed cables.

2. Construction of small piers and pushing of metal deck girders.

3. Superstructures.

4. The bridge is completed.

Appendix 1 : Temperatures record

	Segment 9	Segment 10
Ambient temperature at time of pouring	20° C	14° C
Maximum temperature of concrete (1)	53° C at 21 hours	48° C at 21 hours
Minimum ambient temperature	12° C at 11 hours	6° C at 11 hours

(1) The temperature was measured at the core of the thickest part (60 cm) of a web of a segment. This temperature is held for about 6 hours and decreases slowly thereafter.

Appendix 2 : Characteristics of the cement and of the silica fume

	HS CPA cement	Silica fume
Ignition loss	1.21 %	3.15 %
SiO_2	22.51	93.86
Al_2O_3	3.77	0.13
Fe_2O_3	3.92	1.19
TiO_2	0.15	0.01
MnO	0.07	0.19
CaO	64.74	0.22
MgO	0.93	0.66
SO_3	2.33	0.04
K_2O	0.22	0.75
Na_2O	n.d.	0.20
P_2O_5	0.34	0.03
CO_2	-	0.34
Specific surface area	3620 g/cm³	-
% $\geq$ 80 μm	0.1 %	-
% $\geq$ 40 μm	5.1 %	-
Initial setting time	3.20 H	-
Final setting time	4.25 H	-
False setting	no	-
LCL workability	3 sec.	-
Comp. strength 1 day	11.4 MPa	-
2 days	25.0 MPa	-
7 days	46.2 MPa	-
28 days	65.3 MPa	-
Shrinkage, 7 days	290 μm/m	-
28 days	570 μm/m	-
Heat of hydration (12 h)	147 j/g	-

35 THE ROIZE BRIDGE

G. CAUSSE, S. MONTENS
Scetauroute, France

1 Introduction

The bridge over the Roize river is situated on the interchange
between the A49 motorway (Valence-Grenoble) and the A48
motorway. It carries one of the A49 access roads. The Roize river
runs between high banks and is slow flowing most of the time, but
can be subject to heavy flooding. It runs into the Isère river not far
from the interchange. The bridge also crosses underground gas
pipelines which dictate the layout of its piers. This site has been
marked down since 1987 as a possible place to build an
experimental bridge.

2 Natural and functional data

2.1 Soil conditions
The bedrock is thickly covered with alluvions from the Isère river
and its tributaries. In successive order can be found sandy-clay
backfills with scattered characteristics, silty aggregate of medium
characteristics, then very consolidated sandy aggregates with
ultimate pressure of about 5 MPa. The bridge is founded on this
layer at about 10 m below ground level.

2.2 Geometry
The road has a spiral layout and a convex parabolic vertical profile.
The transverse slope varies between 0 and 3.5%. In order to
simplify the bridge's design, the deck was given the following
geometry: the lower flange axis is situated on a circle in a plane
slightly inclined from horizontal. The variable slope is achieved
by a progressive rotation of each deck segment around this axis.
Thus all the segments have the same geometry. An over-width of the
top slab (700 mm) has of course been provided, in order to contain
the spiral of the actual roadway layout.

High Performance Concrete: From material to structure. Edited by Yves Malier. © 1992 Taylor & Francis.
Published by Taylor & Francis, 2 Park Square, Milton Park, Abingdon, Oxon, OX14 4RN. ISBN 0 419 17600 4.

2.3 Functional transverse profile

The deck carries a mono-directional, 7-m wide, 2-lane roadway bordered on the left by a 1-m wide band and on the right by a 2.5-m wide hardshoulder.

The safety barriers are normal BN4-type galvanised steel barriers.

3 Choice of deck structure

3.1 General ideas

The final choice resulted from about 10 years' research by numerous engineers concerning the lightening of medium-span bridge decks. Their aim was to replace prestressed concrete webs used in traditional decks by lighter elements such as steel webs, either plane or corrugated, or concrete or steel trusses. This latter idea was chosen, but material saving has progressed even further by reducing the dead load by two methods: using a steel lower flange, and reducing the top slab thickness by using high performance concrete (HPC), prestressed with bonded strands. Other ideas have led to a construction process which can be extrapolated to larger bridges.

3.2 Brief description of the deck

The deck is a prestressed composite truss made of prestressed concrete and steel. It comprises:

- a single lower flange made of a hexagonal steel tube;
- two triangulation planes (Warren type) inclined and secant on the lower flange axis, made of reconstituted, welded, rectangular, steel sections;
- transverse floor beams to which diagonals bearing the top slab are connected. These are made of reconstituted, welded, I-shaped, steel sections;
- a high performance concrete slab constituting the upper member and roadway, comprising precast panels assembled with cast-in-place joints above the floor beams.

The originality of the structure lies in its modular conception. The steel frame is composed of factory-built tetrahedrons which are brought to the site and then assembled. The concrete elements are also precast segments assembled on site.

Each tetrahedron comprises a floor beam, four diagonals and a 4-m section of lower flange. Thus the assembly of steel elements on site is simply a question of welding the lower flanges together end to end.

The assembly of steel and concrete elements and slabs is achieved by concreting joints situated above the floor beams.

4 Test Model

Although their shape is unusual, the design of the steel parts uses no innovation either in their calculation or their construction.

The structure's originality results both from:

- the type of connection between the truss and the slab obtained by building a relatively rigid steel element into the concrete;
- the use of relatively slim (considering their span), high-performance concrete slabs, prestressed with bonded strands.

A test model was built, therefore, on which to perform a certain number of tests in order to increase knowledge of these two aspects. This test model, consisting of elements whose design and dimensions are identical to those of the actual bridge, will be load tested in order to study the fatigue behaviour of the slab under local loads. It will then be tested with much greater loads in order to test its behaviour beyond elasticity so as to confirm the good behaviour of the whole.

5 Description of the bridge

5.1 Layout and general characteristics
The bridge comprises a continuous truss girder with three spans of 36 m, 40 m and 36 m, resting on two piers and two abutments. This girder is circular with a 535-m radius. The deck has a constant width of 12.20 m and is 2.30 m high. It is protected by a waterproof layer, and coated with a bituminous concrete wearing course.

5.2 Piers and abutments
The abutments are buried abutments, founded on three 1-m diameter piles. They are hollow inside to enable work to be done on external prestressing tendons. They are equipped with a transition slab.

The piers are each composed of 2 reinforced concrete columns resting on a connecting footing and founded on two 1.30-m diameter piles.

The shape and layout of the piers were researched by M. Berlottier, the architect responsible for the majority of the bridges on the A49 motorway.

5.3 Steel frame
The floorbeams are simple I-shaped steel sections with connectors comprising an angle with a bent bar which enable the transmission of shear due to transverse bending while improving the fastening of

527

Figure 1 : Bridge characteristics

ELEVATION

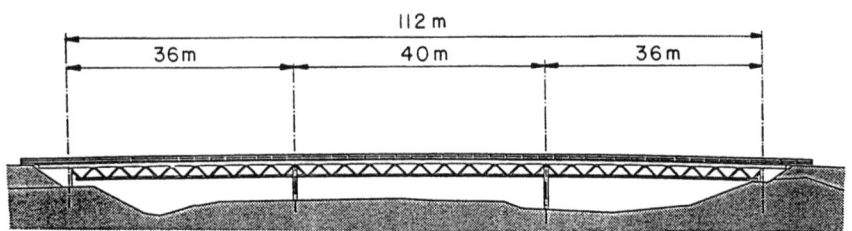

CROSS SECTION

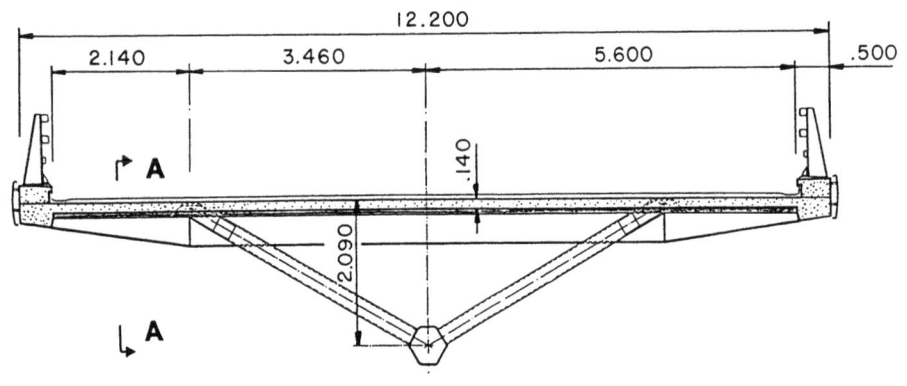

SECTION A-A

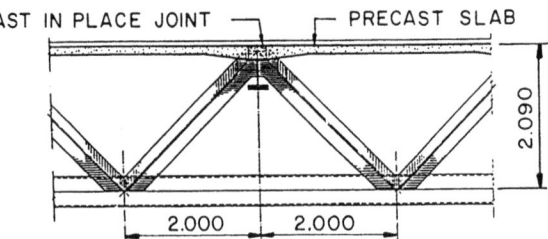

slabs against lifting. These sections have triangular notches cut into them to enable the connecting node to be positioned. Their upper flange will bear the edges of precast slabs, and be used as formwork for the concrete joint.

The diagonals comprise four metal sheets 16 to 30 mm thick, assembled in a rectangular section.

The upper node comprises two thick rectangular metal sheets which prolong the diagonals' webs and are deeply embedded in the concrete of the joint. Various sheets welded onto these two thick sheets enable the transmission of loads towards them. The node will be filled with concrete when the joint is concreted, which will improve its rigidity.

The hexagonal lower flange comprises two folded metal sheets assembled by continuous longitudinal welding whose thickness varies between 20 and 30 mm. It is stiffened by four diaphragms located under the impact points of the diagonals. The truss elements are assembled end to end on site by deep welding of the lower flange on a lath.

5.4 Slabs

These are made with concrete whose characteristic strength at 28 days is 80 MPa. The strength taken into account for calculations is only 60 MPa. The slabs are 12.20 m long (total deck width) and their width varies between 3.72 m (inside curve) and 3.82 m (outside curve). The slope variation means that the slabs are warped, the fourth angle being located 14mm from the plane of the other three angles.

Slab thickness varies, increasing from 140mm to 220mm around the floorbeams because of a 600 mm-long triangular gusset.

The are prestressed by 54 T13 bonded strands, parallel to the longitudinal deck axis. They are also transversely prestressed after concreting the joints, by two 4T15 flat-duct tendons located on either side of the floor beam.

5.5 Longitudinal prestressing

Once the deck is entirely assembled, it is prestressed using five external longitudinal 12T15 tendons. These are Freyssinet tendons comprising 12 sheathed, greased monostrands in a polyethylene tube filled with cement before stressing.

This technology was chosen as it enables tensioning of the tendons to be phased very easily and limits the size and weight of the tensioning equipment.

The tendons are deviated on the lower flange by means of bent pipes fastened thereon with diaphragms. They are deviated at the top by the floorbeam resting on a pier which is strengthened for this purpose.

6 Construction

6.1 Steel frame

Steel frame are built in the Alès workshop of J. Richard Ducros, holder of the "frame" works contract. The assembly principle is the following :

1. construction of the basic elements : top node, floorbeam, diagonals using E355R steel sheets, and lower flanges with A52FP steel sheets. This latter type of steel was chosen to enable it to be folded after slight pre-heating. An E355R steel would make folding much more delicate and would probably require considerable pre-heating;
2. assembly on an assembly gauge of the V comprising the top node and two diagonals, then welding and testing,
3. assembly on another gauge of the full tetrahedron, then welding and testing,
4. test assembly of several elements in the workshop.

The tetrahedrons are then sent to the site and laid on scaffolding. Each tetrahedron rests on the scaffolding by means of jacks located under the floor beams around the top nodes. The jacks enable very precise adjustments to be made to the geometry, taking into account in particular the necessary camber.

The slabs can then be positioned, the lower members welded and the joints concreted.

6.2 Precasting HPC slabs

Slabs are concreted in pairs on a precasting bench equipped with two moulds situated between two anchoring abutments for bonded strands. A precasting cycle comprises the following operations :

- placing reinforcement cages in moulds,
- tensioning bonded strands,
- concreting,
- adding a curing compound,
- leaving the concrete to mature for around 30 hours,
- transferring the bonded prestress to the slabs,
- stripping and stocking.

6.3 Assembly - finishing touches

The deck is laid on scaffolding in three phases. The first and second phases are symmetrical and consist in assembling the side spans: laying the steel frame and slabs using a crane, welding and concreting joints, then decentring.

At this stage, no longitudinal prestressing is used but the structure remains stable. The HPC top slab in compression and the lower steel flange in tension are amply sufficient in size.

Figure 2 : Edge of precast slab and steel module

Figure 3 : Upper connection

The third phase can then begin: assembly of the central span, moving the side spans in order to bring them into contact with the middle part and finally, concreting the whole.

The longitudinal prestressing tendons are tensioned enabling decentring of the central span.

The final assembly of lateral equipments can be carried out, and the waterproof layer, wearing course and expansion joints laid.

7 High performance concrete

7.1 Study tests

Preliminary tests have been performed on 10 concretes with different mixes. Their variable parameters were: sand (660 to 736 kg/m^3), cement (450 to 500 kg/m^3), water (130 to 145 l/m^3), silica fume (45 to 60 kg/m^3), admixture (19 to 25 kg/m^3), and possibly a small quantity of fibres.

Mean compression strengths at 28 days have always been higher than 95 MPa. The choice between various mixes has been made according to ease of processing.

The chosen mix is as follows:

- Sand 0/5 mm	736 kg
- Aggregate 4/10 mm	451 kg
- Aggregate 10/14 mm	667 kg
- CPA/HPR cement from Vicat-St Egrève factory	450 kg
- Condensil (silica fume)	45 kg
- Water	130 l
- Plasticizer retarder	
- RH 2000 PF	19.5 kg
- Ratio W/C	0.289

The study tests actually consisted in making concrete from the basic mix with the following variations:

- sand variation	+/- 10 %
- water variation	+/- 10 l /m^3
- silica fume variation	+/- 5 kg /m^3
- admixture variation	+/- 2 kg /m^3
- cement variation	+/- 25 kg/m^3

Mean compression strengths obtained for the basic mix design were:

36 MPa at 24 hours
72 MPa at 7 days
111 MPa at 28 days

These strengths are relatively insensitive to variations of the various parameters. Workability, however, is notably reduced when water or admixture are under-dosed.

7.2 Suitability tests
Concrete was made in a ready-mix concrete plant located near the site (Union Béton).

Suitability tests were performed on the basic mix defined by the study tests, with variations of water content +/- 5 1/m³.

Compression strength was measured at 7 days and 28 days. Workability was measured at 0h, 30mn, 1h, 1h30mn.

It was observed that a low dosage of water does not affect strength, but has a great influence on workability. Mean compression strength at 28 days was 120 MPa.

7.3 Creep tests
Slabs prestressed at an early age (30 hours), and previous tests [1] performed on HPC loaded at an early age having shown that their creep coefficient was high, it was found necessary to perform creep tests on test specimens. These tests were performed by INSA in Lyon with test specimens of dia. 11 cm, L 22 cm loaded at 24h, 36h, 48h, 72h with a stress of 10 MPa.

If the creep speed law mentioned in [1] is adopted as a kinetic reference, the following creep coefficients Kfl (refering to BPEL French Code) can be deduced :

loading age	Kfl
24h	5.2
36h	3.4
48h	3.1
72h	2.8

7.4 Slab manufacture
Slabs were manufactured between 15 June and 5 September 1990. Workability time was about one hour. Slump was 22 +/- 2cm. Concrete set was late (about 20h) but setting time was quick (about 3h).

Despite the curing compound used, considerable crazing was observed.

Starting with the 5th slab, compression strength fell considerably (around 40%). After checking, it was found that the ready-mix concrete plant had been provided with HPR cement from Montalieu instead of from St Egrève. HPR cements, contrary to other cements, have no maximum specified strength, and any two HPR cements can have widely differing strengths.

Compression tests results were very scattered. For some slabs, strengths measured on test specimens were quite insufficient. Therefore, coring was performed on the relevant slabs. Testing these cores gave better results which enabled the slabs to be accepted. Examination of the low stength test specimens showed them to have defects (segregation, cracking) due to faulty fabrication.

The longitudinal elastic modulus at 28 days was 43370 MPa, which is less than the BPEL French Code formula value with fc_{28} = 80 MPa. On the other hand, the value of 0.28 found for the Poisson ration is high. Density is 2.46.

References

1. M. Auperin, F. de Larrard, P. Richard, P. Acker. Shrink and Creep in High Performance Concretes. The Influence of the Loading Age. ITBTP Annals N° 474, pp. 50-75, May-June 1989.

Figure 4 : General view

Table 1 : Quantities of material

Foundations, piers, abutments	concrete	300	m³
	reinforcement	27/0	t
Deck	B80 concrete	250	m³
	reinforcement	44 .0	t
	bonded strands	4.2	t
	transverse prestressing	3.6	t
	longitudinal prestressing	8.0	t
	structural steel	153.0	t
Gross area		1366	m²
Net area		1176	m²

Table 2 : Participants

Owner	Area
Project Management Consultant	Scetauroute, Grenoble Branch Office, Tullins Works Division
Design and Shop Drawings	Scetauroute, Bridge Engineering Division
Control of Shop Drawings	CETE Lyon
Architect	Mr Berlottier
Civil Engineering Contractor	Campenon Bernard
Steel Framework Contractor	J. Richard Ducros

Table 3: Study tests of concrete

| | THEORETICAL MIXES | | | | | | | | WORKABILITY | | COMPRESSION STRENGTH | | |
| | Aggregate | | | Cement | Silica Fume (Condensil) | Water | Admixture | W/C | Slump | LCPC | 24h | 7 d | 28 d |
	0-5	4/10	10/14	CPA HPR			RH 2000 PF						
BASIC 1	736	451	667	450	45	130	19.5 kg	0.289	23.0 cm	3s2	34.7	67.5	111.1
BASIC 2									22.5 cm	3s8	36.8	75.0	108.7
BASIC 3									22.5 cm	3s6	36.0	75.0	114.0
+10% SAND	810	421	623	450	45	130	19.5 kg	0.289	22.5 cm	3s2	32.8	75.0	108.2
-10% SAND	662	481	711	450	45	130	19.5 kg	0.289	21.5 cm	5s1	43.8	74.0	115.0
-10 liters WATER	736	451	667	450	45	120	19.5 kg	0.267	17.0 cm	8s4	45.2	80.0	113.5
+10 liters WATER	736	451	667	450	45	140	19.5 kg	0.311	>23 cm	2s6	38.0	79.8	109.8
+5 kg Condensil	736	451	667	450	45	130	19.5 kg	0.289	22.0 cm	4s2	30.2	70.0	105.8
-5 kg Condensil	736	451	667	450	45	130	19.5 kg	0.289	21.0 cm	4s6	33.3	68.2	107.5
+2 kg RH 2000 PF	736	451	667	450	45	130	21.5 kg	0.289	22.5 cm	3s8	30.8	67.5	109.3
-2 kg RH 2000 PF	736	451	667	450	45	130	17.5 kg	0.289	19.0 cm	5s6	36.0	72.8	110.3
-25 kg CEMENT	736	451	667	450	45	130	19.5 kg	0.306	23.0 cm	4s9	30.2	72.7	107.2

AUTHOR INDEX

SUBJECT INDEX

This index uses keywords assigned to the individual chapters as its basis. The numbers are the page numbers of the first page of the relevant chapter.

538

Milton Keynes UK
Ingram Content Group UK Ltd.
UKHW021928071024
449327UK00022B/1730